STUDENT SOLUTIONS MANUAL

to accompany

STATISTICS

From Data To Decision

Second Edition

Ann E. Watkins
California State University, Northridge

Richard L. Scheaffer
University of Florida

George W. Cobb
Mount Holyoke College

WILEY

John Wiley & Sons, Inc.

COVER PHOTO: Bruno Morandi/Getty Images

ISBN-13   978-0-470-53060-3

10 9 8 7 6 5 4 3 2 1

Printed and bound by Bind-Rite, Inc.

Table of Contents

Chapter 1

Practice Problem Solutions

P1. This plot is rather striking; the older hourly workers were far more likely to be laid off in Rounds 1 through 3 than were the younger workers.

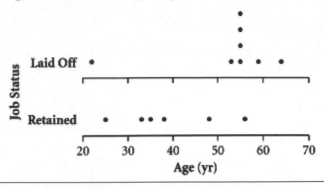

P2. B.

P3. Overall, the patterns in the two displays are strikingly similar. Specifically, both graphs show that older workers were especially likely to be targeted for layoff in the earlier rounds, and that younger workers were more likely to be targeted in Rounds 4 and 5. Both displays also show that most of the layoffs were planned in the first two rounds. This sequence of layoffs for hourly workers provides stronger evidence for Martin's case. In the first three rounds, seven of the eight people laid off were over 50, while only one of the six retained in these three rounds was over 50.

P4. **a.** The average age is now 52.67. On the dot plot, this is the column that is just a bit more than half-way between 50 and 55. That is, it is the tallest column in the group of seven columns of dots *between* 50 and 55. The number of repetitions out of 200 that gave an average age of 52.67 or larger is 37. Thus, the estimated probability of getting an average age of 52.67 or larger is 37 out of 200 or about 0.185.

b. There is no evidence of age discrimination because an average age of 52.67 or larger is relatively easy to get just by chance.

P5. **a.** The ten workers laid off were ages 22, 33, 35, 53, 55, 55, 55, 55, 59, and 64. Their average age was 48.6 years.

b. The chance model is that 10 workers were selected at random for layoff from 14 workers with ages 22, 25, 33, 35, 38, 48, 53, 55, 55, 55, 55, 56, 59, and 64. Write these ages on 14 cards of equal size. Place the cards in a bag or box, mix the cards up by shaking the bag, and then draw 10 of the cards at random. Calculate and record the average age of the ten picked cards. Then, replace the cards, mix, and repeat many times, recording the average age each time. Make a dot plot showing the distribution of these average ages and compute the proportion of times you get an average age of at least 48.6.

c. 45 of the 200 dots are at or above 48.6, for a proportion of 22.5%.

d. No, this evidence does not help Martin's case. Martin wants to show that the company systematically targeted older workers for layoff. These results show that we would expect an average age of 48.6 or more, purely by chance, about 22.5% of the time, which is not a rare occurrence.

Exercise Solutions

E1. a. There were 6 hourly workers under age 50, and 3 were laid off. So 50% of the hourly workers under age 50 were laid off. 50% were not laid off.

b. 10 of the hourly workers were laid off. 3 were under age 50, so $3/10 = 0.3$ of the hourly workers who were laid off were under age 50. 70% were age 50 or over.

c. You should compute the proportion of hourly workers under age 50 who were laid off and the proportion of those age 50 or over who were laid off. 50% (3 out of 6) of those under age 50 and 87.5% (7 out of 8) of those ages 50 and over were laid off. Thus, a disproportionately high proportion of workers age 50 and over were laid off.

d. For the salaried workers, 37.5% of those under age 50 were laid off and 60% of those age 50 and older were laid off. For both hourly and salaried workers, the proportion of workers age 50 and over who were laid off is disproportionately high, but the difference is more extreme in the case of the hourly workers and so appears to provide more evidence for Martin.

However, note that the number of hourly workers in each age category is very small. Thus, a small change in the number laid off in each category would make a large change in the difference in the proportions laid off. For example, suppose Westvaco had laid off one more worker under age 50 instead of one of the workers age 50 or over. Then the difference in the proportions would be 4 out of 6 or 67% of those under age 50 were laid off and 6 out of 8 or 75% of the workers age 50 or over were laid off, now, there is not a very big difference.

E3. a.

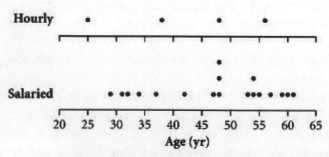

Only four hourly workers kept their jobs, making comparison difficult. However, the distribution of salaried workers who kept their jobs generally falls to the right of the distribution of hourly workers who kept their jobs. Indeed, the average age of the hourly workers who kept their jobs was 41.75, whereas it was 47.17 for the salaried workers.

2

b. We cannot conclude this from the dot plots in part a alone. We would have to take into consideration the age distributions for hourly and salaried workers *before* the layoffs began, as well as after. All we can say is that the salaried workers who kept their jobs tended to be older than the hourly workers who kept their jobs.

E5. a.

Round	Number Laid Off	40 or Older	Percentage 40 or Older
1	11	9	82%
2	9	8	89%
3	3	2	67%
4	4	2	50%
5	1	0	0%

b. Most of the layoffs came early. Of the 28 who were chosen for layoff, 20 were identified in the first two rounds. Only 8 were chosen for layoff in the last three rounds altogether.

Early rounds hit the older workers harder; later rounds tended to have higher percentages of younger workers. In the first two rounds, the percentage was very high among those over 40 who were laid off, at 85% (17 of 20). In the later rounds, the percentage for those over 40 dropped to 67% in Round 3 (2 of 3), 50% in Round 4 (2 of 4), and 0% (0 of 1) in Round 5.

The pattern is consistent with what you would expect to see if the department head who planned the layoffs was trying to cut costs by laying off the older, more experienced, and thus possibly more expensive workers first.

E7. a. The cases are the five individual players and the variables are games, at bats, hits, home runs, stolen bases, and batting average. Brock's batting average is 0.293. Hamilton's number at bats is 6276 or 6277. Raines' number of hits is about 2608.

b. The cases are the seven individual stocks and the variables are closing price on 10/28, closing price on 10/30, change (in price), percentage change (in price), and volume (of trading). Coca-Cola change is –8 5/8, Coca-Cola percentage change is –6.30. Eastman Kodak closing price on 10/30 is 170. General Motors closing price on 10/28 is 47 1/2. General Motors percentage change is –15.79. US Steel change is –12. US Steel percentage change is –6.45.

E9. a. Simulations will vary. For example, a student might draw the following 10 ages: 22, 25, 33, 48, 53, 55, 55, 55, 56, and 59. The summary statistic would then be 7 because 7 of the 10 are age 40 or older.

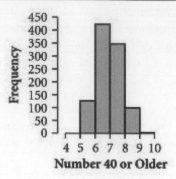

b. 24 of the 50 samples (48%) contained 7 or more people of age 40 or older.

c. It is quite likely to randomly draw ten ages from the fourteen and get 7, 8, or 9 who are aged 40 or older. In fact, nearly half of the samples in the dot plot contained 7 or more people of age 40 or older. Such a result is thus quite likely and provides no evidence for Martin.

NOTE: *Choosing an Estimator*

A general rule in statistics is that you should not throw away information that might be relevant. Only using whether the person is 40 or older throws away information—the person's exact age. For example, suppose Westvaco had laid off everyone over 60 and kept everyone under 60. That would be age discrimination, but might not show up in an under/over age 40 analysis since all those aged 40–59 were kept.

Here is a specific example of that principle: In the Martin case, there were 10 hourly workers involved in the second round of layoffs; 4 of these workers were under 40. Three were chosen to be laid off; all 3 were 40 or older. To see whether this result could reasonably be due just to chance, consider drawing 3 tickets at random from a box with 10 tickets, 4 of which say "Under 40," and the 6 of which say "40 or older." A natural summary is the number of tickets out of 3 that say "40 or older." Simulating this process a large number of times shows that the chance that all 3 tickets say "40 or older" is about 1 / 6. (In fact, it is possible to count the possible outcomes. There are $\binom{10}{3} = 120$ possible outcomes, of which $\binom{6}{3} = 20$ have all 3 tickets saying "40 or older," so 1 / 6 is the exact probability.)

This probability is large enough that we cannot reasonably rule out the possibility that such data could arise by chance alone. Thus, this conclusion differs from the one conclusion on the actual ages, which was analyzed in an earlier activity. Because the analysis based on the actual ages uses more of the available information, it is more informative and trustworthy. The analysis based on the actual ages (the mean) uses more of the available information. On the other hand, it might be easier to convince a judge or jury by using 40 or over because that is the federal protected class.

E11. C. 45 of the 200 samples, or 22.5%, had an average weight of 15 ounces or less. Since this could happen a little more than once out of every five times if the loaves are selected randomly, she should probably give her baker the benefit of the doubt.

E13. **a.** Only four pair give an average age of 59.5 or older:

25 33 35 38 48 **55** 55 55 55 56 **64**
25 33 35 38 48 55 **55** 55 55 56 **64**
25 33 35 38 48 55 55 **55** 55 56 **64**
25 33 35 38 48 55 55 55 **56 64**

b. Thus, the probability of getting an average of 59.5 or more is 4/45 ≈ 0.09.

c. The evidence of age bias is somewhat weak.

E15. **a.**

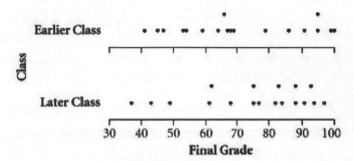

b. There does not appear to be a great difference between the sections. The mean score of students in the earlier class (71) is only slightly lower than that of students in the later class (74.8), and the range and variability of the scores are very similar, so it appears that the difference between class averages could reasonably be attributed to chance. The teacher need not look for an explanation.

E17. B is the best choice of those given, but the question remains whether such a difference is a reasonably likely outcome due to chance alone.

E19. **a.** ii.. Martin's case would benefit from showing that it's pretty unusual to get 12 or more older workers if workers had been selected completely at random for layoff. That would mean that Westvaco could be asked for an explanation of these unusual results.
b-c. Results will vary.

d. This is 13% of the trials. A random selection process would generate a result at least this extreme about once every 7 or 8 trials. This is not a small enough percentage to cast serious doubt on a claim that those laid off were chosen by chance.

E21. **a.** $_{14}C_{10} = \begin{pmatrix} 14 \\ 10 \end{pmatrix} = 1001$ ways.

b. They could have laid off 5, 6, 7, 8, or 9 older workers.

c. $\begin{pmatrix} 9 \\ 7 \end{pmatrix} \cdot \begin{pmatrix} 5 \\ 3 \end{pmatrix} = 36 \cdot 10 = 360$, $\begin{pmatrix} 9 \\ 8 \end{pmatrix} \cdot \begin{pmatrix} 5 \\ 2 \end{pmatrix} = 9 \cdot 10 = 90$, $\begin{pmatrix} 9 \\ 9 \end{pmatrix} \cdot \begin{pmatrix} 5 \\ 1 \end{pmatrix} = 1 \cdot 5 = 5$

d. $\dfrac{360 + 90 + 5}{1001} \approx .4545$. There is a 45.45% chance of getting seven or more workers age 40 and over if 10 workers are selected randomly from the 14 hourly workers.

E23. **a.** 7/1000 = 0.007

b. Dear Society Members,

As you know, I've been suspicious ever since the two winners of last week's drawing were Filbert's cousins. To see how unlikely such an event would be, I made my own 50 raffle tickets, just like those in last week's drawing, and marked four of them to represent the four that belonged to Filbert's cousins. I then mixed the tickets up well and drew two of them. I did this 1,000 times. I got two tickets representing those belonging to Filbert's cousins only 7 times out of the 1,000 drawings. This gives me an estimated probability that both tickets would belong to Filbert's cousins of 0.007, which is a really small chance. This leaves us with the following possible explanations:

- The tickets weren't well mixed. Does anyone know if, for example, Filbert's cousins bought the last four tickets and these ended up on top? Or maybe they bought the first four tickets, which were on the bottom?
- A really unlikely event occurred.
- I don't like to use the word "cheating," but that is another possible explanation.

c. Even the most unlikely event is *possible*, so, yes, it's possible. In fact, the event that both winning tickets were held by Newman's cousins happened seven times in the simulation.

d. There are $\begin{pmatrix} 4 \\ 2 \end{pmatrix} = 6$ ways to draw two winning tickets both held by Newman's cousins.

There are $\begin{pmatrix} 50 \\ 2 \end{pmatrix}$ or 1,225 ways to draw two tickets from the 50. So the probability is

$\dfrac{6}{1,225} \approx 0.005$.

Chapter 2

Practice Problem Solutions

P1. a. The data were collected by a statistics class; a case is a penny; the variable is the age of the penny.

b. The shape is strongly skewed right, with a wall at 0. The median is 8 years, and the spread is quite large, with the middle 50% of ages falling between 3 and 15 years; however, it is not unusual to see a penny that is more than 30 years old.

c. If the same number of pennies is produced each year and if a penny has the same chance of going out of circulation each year, then the height of each column would be a fixed percentage of the previous height. The distribution is skewed right since pennies most recently manufactured would naturally be more readily available in circulation. The older the penny, the less likely it would be for it to arise in a sample simply because it is likely that fewer of them remain in circulation.

P2. a. w: skewed to the left;
x: mound-shaped or approximately normal;
y: two mounds, possibly representing two groups of objects

b. x because it tapers symmetrically on either side of the maximum of the mound, while the other two are either skewed or have two mounds.

c. The mean is around 52 because most values are equally spaced around 50, with two extra values to the right of 60 that will serve to increase the mean a bit. The standard deviation is around 5.

P3. Student estimates will differ somewhat from the actual means and standard deviations given here.

a. A typical SAT math score is roughly 500, give or take about 100 or so.
b. A typical ACT score is about 20, give or take 5 or so.
c. A typical college-aged woman is about 65 inches tall, give or take 2.5–3 inches or so.
d. A typical professional baseball player had a single-season batting average of about .260 or .270, give or take about .040 or so.

P4. There are 180 dots, so assuming the data are ordered from lowest value to highest value, the lower quartile coincides with the 45th dot, the median is the average of the 90th and 91st dots, and the upper quartile is the 135th dot. As such, we have median = 19; lower quartile = 17; upper quartile = 25, Also, note that the distribution is skewed toward the higher ages.

P5. a.

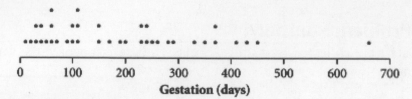

b. The distribution is skewed right with no obvious gaps or clusters. There is a wall at 0 days because no mammal can have a smaller gestation period. The elephant is the only possible outlier.

c. About half of the mammals have a gestation period of more than 160 days, and half have a shorter period. Those in the middle half have gestation periods between 63 and 284. Large mammals have the longer gestation periods.

P6. a. The distribution is approximately normal with mean around 46 and standard deviation around 6.

b.

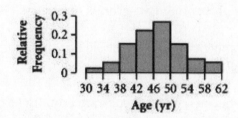

c. The proportion of people who are at least 50 years old corresponds to the sum of the heights of the three rightmost bars in the histogram in (b), namely about 0.24, or about 24%.

P7. a. The proportion of countries with a life expectancy of less than 50 years is about 0.008 + 0.01 + 0.05 + 0.04 = 0.108, or about 10.8%.

b. The number of countries with a life expectancy of less than 50 years would then be (0.108)(223) = 24.08, or about 24 countries.

c. The shape of this distribution is skewed left toward the smaller values. The median is between 70 and 75 and the middle 50% of the life expectancies are between 60 and 75 years.

P8. a.

```
           Back-to-back stem-and-leaf plot
              Ave Long |    Max Long
    ─────────────────────┼──────────────────
                     431 | 0 | 4
                88776555 | 0 | 588
          222222222000 | 1 | 3344
                65555555 | 1 | 8
                    0000 | 2 | 0000334
                       5 | 2 | 67778
                         | 3 | 0004
                       5 | 3 | 7
                       1 | 4 | 0
                         | 4 | 557
                         | 5 | 00000344
                         | 5 |
                         | 6 |
                         | 6 |
                         | 7 | 0
```

5|2|6 represents a mammal
with an average life span of
25 years and a mammal (not
necessarily the same mammal)
with a maximum life span of
26 years.

b. The average longevity distribution is skewed right, with two possible outliers, while the distribution of maximum longevity is more uniform but has a peak at 20–30 years and a possible outlier. The center and spread of the distribution of maximum longevity are larger.

c. Maxima are larger than averages by definition; maxima tend to spread out more because of the possibility of extremely large values, not constrained by averaging.

P9. Three pairs of histograms for average longevity (1st row) and maximum longevity (2nd row) using bar widths of 4, 8, and 16 years are as follows:

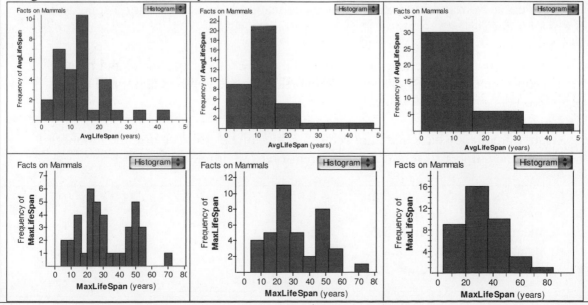

9

The shape of the maximum longevity distribution is quite different from that for average longevity. The average longevity distribution is skewed right with two possible outliers at 35–40 and 40–45. The distribution of maximum longevity is more uniform but with a peak at 20–30 years and a possible outlier at 70–80 years. As must be the case, the center of the distribution of maximum longevity is much higher than the center of the distribution of average longevity—about 30 years compared to about 15 years. The spread of the distribution of maximum longevity is also much larger.

The bar width that seems most appropriate in this scenario is 8 years since it preserves the general shape of the graphs using a bar width of 4 years (as compared to distorted graphs obtained using a bar width of 16 years), while not focusing on too much specific detail that is not important to understand the general trend.

P10. The number of deaths per month is fairly uniform, with about 190,000–220,000 per month. Summer months have the smallest numbers of deaths, and winter months the largest.

P11. A case is a student in your class. The quantitative variables are age, number of siblings, and number of miles he or she lives from school. The categorical variables are hair color and gender.

P12. The last digits of social security numbers are essentially random digits, so they should be fairly uniform over the range 0 to 9, as these are. Here we would say that the distribution is approximately uniform with about five–six students with each digit from 0 to 9.

P13. **a.** The cases are the individual male members of the labor force aged 25 and older, and the variable is their educational attainment. The distribution shows an increasing proportion of males through the first three levels of education with a huge jump at the high school graduation level. After high school, the proportions in each education category decrease regularly with increasing education levels, except for a spike at bachelor's degree.

  b. The distributions for males and females have much the same shape and much the same proportions. Females have lower proportions in categories 8 and 9 and higher proportions in categories 4 and 5.

  Relative frequency bar charts are better for this comparison because the number of males and the number of females in the labor force are different.

P14. **a.**　　mean: 2.5　　　median: 2.5
b.　　mean: 3　　　median: 3
c.　　mean: 3.5　　　median: 3.5
d.　　mean: 49.5　　　median: 49.5
e.　　mean: 50　　　median: 50

P15. The new mean height will be about 4 feet 4 inches. The median should not change much because it will still be one of the 3rd graders, who all are about 4 feet tall.

P16. a. The median life expectancies are 55 for Africa and 80 for Europe.
b. For Africa, the median is smaller than the mean (56.5) because of the skewness toward the larger values. For Europe, the mean (79.6) is slightly lower than the median because of the left skew.

P17. a. quartiles: 2 and 5   IQR: 3
b. quartiles: 2 and 6   IQR: 4
c. quartiles: 2.5 and 6.5   IQR: 4
d. quartiles: 2.5 and 7.5   IQR: 5

P18. a. The median, lower quartile, and upper quartile (in that order) for predators and nonpredators are given by:

predators: 12, 7 and 15;
nonpredators: 12, 8 and 15

b. The distribution of the average longevity of predators is mound-shaped, centered at about 12 years, with 50% of the values falling between 7 and 15 years. The distribution of the average longevity of nonpredators is centered at exactly the same place and has about the same spread, but it has two outliers on the high side. Its shape is essentially mound-shaped although the two outliers make it appear skewed towards the larger values.

P19. The distribution from the European countries shows a little skewness toward the smaller values with a median around 68, indicating that most countries have predominantly urban populations. Liechtenstein is a small, independent, principality in the Alps between Switzerland and Austria with a population of only 33,400. The distribution from the African countries shows a strong skewness toward the larger values, with a median near 40, indicating that most countries have predominantly more rural populations. The spread in the African distribution is larger than the spread in the European distribution, i.e., there is more variation among the percentages for African countries.

P20. The fact that the median line is not present suggests that it coincides with one of the quartiles. There is much more variability in the average longevity for wild mammals than for domesticated mammals. The side-by-side boxplots for the speeds of wild and domesticated mammals are below:

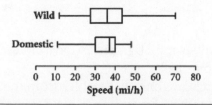

11

P21. a. The five-number summary for the average longevity of this set of mammals is

       minimum: 1

       lower quartile (Q_1): 8

       median: 12

       upper quartile (Q_3): 15

       maximum   41

Once again, software packages may give quartiles slightly different from those done by hand. Here are the results from Minitab:

```
Variable N    N*    Mean   Median  TrMean StDev SEMean
Ave long 38   1     13.13  12.00   12.32  8.00  1.30

Variable Min  Max   Q1     Q3
Ave long 1.00 41.00 7.75   15.00
```

b. IQR = 15 − 8 = 7

c. $Q_1 − 1.5 \cdot$ IQR = 8 − 1.5(7) = −2.5. There are no outliers on the lower end.

d. $Q_3 + 1.5 \cdot$ IQR = 15 + 1.5(7) = 25.5. The life spans of 35 years for the elephant and 41 years for the hippopotamus are outliers. The largest value that isn't an outlier is 25. This is where the upper whisker will end.

e.

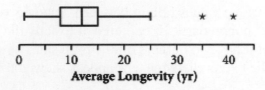

P22. The mean is 4.4, and the deviations from the mean are shown in this table. The sum of these deviations is 0. To get the standard deviation, find the (n − 1) average of the squared deviations, $\frac{41.2}{4} = 10.3$, and take the square root to get about 3.21.

Value x	Deviation from Mean: $x − \bar{x}$	Squared Deviations: $(x − \bar{x})^2$
1	1 − 4.4 = −3.4	$(−3.4)^2 = 11.56$
2	2 − 4.4 = −2.4	$(−2.4)^2 = 5.76$
4	4 − 4.4 = −0.4	$(−0.4)^2 = 0.16$
6	6 − 4.4 = 1.6	$(1.6)^2 = 2.56$
9	9 − 4.4 = 4.6	$(4.6)^2 = 21.16$
Sum	0	41.20

P23. a. i. 0, because all of the deviations from the mean are 0

b. iii. 0.577

c. iv. 1.581

d. vii. 5.774. Note that the values are 10 times as far from the mean as those in part b, so the standard deviation is 10 times as large.

e. ii. 0.058. Note that the values are one-tenth as far from the mean as those in part b, so the standard deviation is one-tenth as large.

f. v. 3.162. Note that the values are twice as far apart as those in part c, so the standard deviation is twice as large.

g. vi. 3.606. It may be hard for students to distinguish part f from part g. If so, they should compute the standard deviation to check their answer.

P24. a.

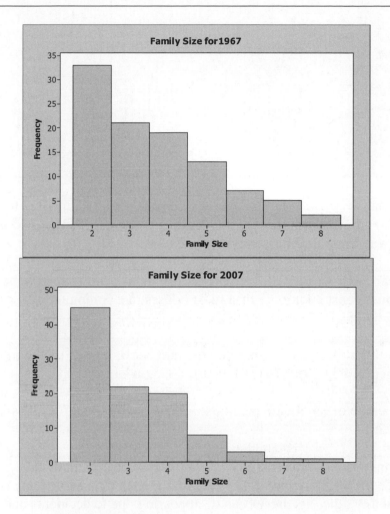

b. 1967: mean = 3.63, SD = 1.61
2007: mean = 3.09, SD = 1.29;

Observe that family sizes got smaller and less variable from 1967 to 2007.

P25. a. The median is the average of the 50th and 51st data value (assuming the data is ordered from lowest to highest); this value is 3 for each year.
b. Note that the lower quartile is the 25th data value, the upper quartile is the 75th data value, and the IQR is the difference of these two. Computing therefore yields

$$1967: Q_1=2, Q_2=3, Q_3=5, IQR = 3$$
$$2007: : Q_1=2, Q_2=3, Q_3=4, IQR = 2$$

c.

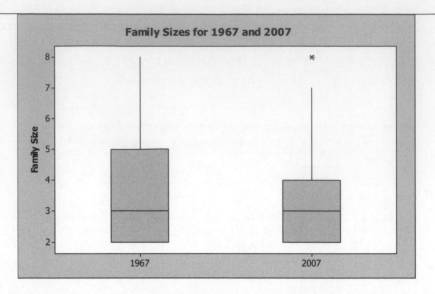

Family Sizes for 1967 and 2007

d. Family size has decreased slightly over the 40 year period from 1967 to 2007. The median size is 3 for both years, but the mean size has decreased from 3.63 to 3.09. In addition, there was less variability among family sizes in 2007 than there was in 1967.

P26. a. The mean is larger than the median because house values tend to be skewed right—some houses cost a lot more than most houses in a community. Very few houses cost a lot less.

b. To get the total property taxes, multiply the number of houses by the mean value by the tax rate to get ($392,059)(9,751) (0.0115) = $43,964,124.

c. This is an average of $4,508.68 per house. (This assumes that the assessed value is equal to the price.)

P27. a. Medians were used in this story because the distribution of car ages is strongly skewed right. There are more brand-new cars on the road than cars of any other age (because cars of any other age have been disappearing due to accidents and mechanical problems). A few people drive very old cars.

b. Vehicles are proving more durable, so cars can be driven for a longer time. Also, people are possibly choosing to spend their income on other things or are forced to spend their income on other things.

P28. a. Divide each of the summary statistics by 12 so then the mean is 4 feet, the median is 3.75 feet, the standard deviation is 0.2 feet, and the interquartile range is 0.25 feet.

b. Add 2 to the measures of center so the mean is 50 inches, and the median is 47 inches. The standard deviation and interquartile range do not change.

c. For the measures of center, add 4 then divide by 12 so the mean is $4\frac{1}{3}$ feet, the median is $4\frac{1}{12}$ feet. The measures of spread are divided by 12 (adding 4 to each height does not change the spread) so the standard deviation is 0.2 feet, and the interquartile range is 0.25 feet.

P29.
a. mean: 12; SD: 1
(Add 10 to the mean from the given example, SD is the same as in the example)
b. mean: 20; SD: 10
(Both the mean and SD are 10 times those in the example)
c. mean: 110; SD: 5
(Each value in the example has been multiplied by 5 and then had 100 added to it. Thus, the mean is 5 times that in the example, plus 100, and the SD is 5 times that in the example.)
d. mean: 900; SD: 100
(Each value in the example has been multiplied by 100 and then had 700 added to it. Thus, the mean is 100 times that in the example, plus 700, and the SD is 100 times that in the example.)

P30. a. Outliers occur above $-30 + 1.5(21) = 1.5$. Hawaii, at 12, is an outlier.
b. The count will decrease by 1 to become 49.

```
Summary of Lowest Temperature without Hawaii
              No selector
       Percentile    25
            Count    49
             Mean   -41.5
           Median   -40
           StdDev    16.2
              Min   -80
              Max    -2
            Range    78
  Lower ith %tile   -51
  Upper ith %tile   -32
```

The minimum, median, quartiles, and interquartile range should remain the same or about the same. With a sample of size 50, removing one data value will have little effect on the median but could have greater effect on the quartiles as they are, in essence, the medians of samples of size 25. The mean should go down by a bit more than one degree because the difference between Hawaii's temperature and the mean is about -52 degrees and there are 49 states remaining. The standard deviation should go down only slightly, but this is difficult to predict. The range will decrease from 92 degrees to about 80 degrees. The maximum will decrease from 12 to a little less than zero.

P31. a. 18th
b. The middle 90% is between about 325 and 730, while the middle 95% between about 275 and 750.
c. about 510

P32. 90% of cases lie between the 5th and 95th percentiles (since 5% of cases lies to the left of the 5th percentile and 5% lie to the right of the 95th percentile, by definition). The middle 95% of data is enclosed between the 2.5th percentile and the 97.5th percentile.

P33. The quartiles are about 450 and 650; the median is about 550; the IQR is about 650 − 450 = 200. The minimum of 200 and the maximum of 800 complete the five-number summary. A boxplot is shown here:

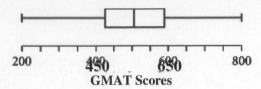

GMAT Scores

P34. a. 1.29% **b.** 4.75% **c.** 34.46% **d.** 78.81%

P35. a. −0.47 **b.** −0.23 **c.** 1.13 **d.** 1.55

P36. a. $0.9279 - 0.0721 = 0.8558$ or 85.58%.
b. $0.9987 - 0.0013 = 0.9974$ or 99.74%.

P37. a. The z-score that has 5% of the values below it is −1.645, and the z-score that has 5% of the values above it is 1.645. So the interval is −1.645 to 1.645.
b. −1.96 to 1.96

P38. a. The death rate for cancer is more standard deviations below the mean, so it is more extreme.

$$\text{heart disease: } z = \frac{180-219}{46} = -0.8478$$

$$\text{cancer: } z = \frac{151-194}{30} = -1.4333$$

b. The death rate for cancer is more standard deviations above the mean, so it is more extreme.

$$z_{\text{heart}} = \frac{260-219}{46} = 0.8913$$

$$z_{\text{cancer}} = \frac{228-194}{30} = 1.1333$$

c. The death rate for heart disease in Colorado is more extreme than the death rate for cancer in Georgia.

$$z_{\text{CO heart}} = \frac{135-219}{46} = -1.8261$$

$$z_{\text{HI cancer}} = \frac{158-194}{30} = -1.2000$$

P39. **a.** The z-score for the height 6 feet (or 72 inches) is $\frac{(72-70.4)}{3.0} = 0.53$. The area under the normal curve to the right of this point is $1 - 0.7019 = 0.2981$. Thus, about 29.8% of U.S. males between 20 and 29 are taller than 72 inches.

b. The z-score for the 35th percentile is –0.385. The height that corresponds to that z-score is

$$x = mean + z \cdot SD = 65.1 + (-0.385)2.6 \approx 64.1 \text{ inches}.$$

So, a woman would need to be 64.1 inches tall to be at the 35th percentile.

P40. **a.** $219 \pm 1.645(46)$ or about 143 to 295

b. $219 \pm 0.99(46)$ or about 173 to 265

P41. Recall from P39 that the z-scores are computed as follows:

$$\text{Male: } z = \frac{x-70.4}{3.0} \qquad \text{Female: } z = \frac{x-65.1}{2.6}$$

a. The z-score is 2.87, which is outside both intervals.
b. The z-score is 1.12, which is not outside either interval.
c. The z-score is -1.80, which is not outside either interval.
d. The z-score is 1.88, which is not outside either interval.

Exercise Solutions

E1. I. Graph D, because on an easy test most people get high scores.
II. Graph A, because the distribution of heights has two modes (mothers and daughters).
III. Graph C, because most countries in the Olympics get no medals at all and only a very small number of countries get multiple medals.
IV. Graph B, because the weights should be mound-shaped. Most chickens will be clustered near a central weight with decreasing numbers having lower or much higher weights.

E3. **a.** A case is one of the approximately 92 officers who attained the rank of colonel in the Royal Netherlands Air Force. There is only one variable: the age at which the officer became a colonel.

b. This distribution is skewed left with no outliers, gaps, or clusters. The median is 52 years, and the quartiles are 50 and 53. So we would say that the middle half of the ages are between 50 and 53, with half above 52 and half below.

c. In questions like this, students aren't expected to "guess" the "right" answer. Instead, they are expected to generate many possibilities that could then be investigated. For example, some military services have an "up or out" rule. It may be the case in the Royal Netherlands Air Force that if you haven't been promoted to colonel by your 55th birthday, you must retire. The armed services have very little use for 55-year-old privates. Alternatively, there might be a mandatory retirement at age 55. A third possibility is age discrimination against older people in the service.

E5. a. This distribution is strongly skewed left. The distribution for an actual year follows, and the shape of the distribution is typical. Students will often have the height of the bar for 85+ taller than that for 75–84, confusing actual number of deaths with probability of death. There are fewer people in the 85+ category than in the 75–84 category, so fewer of them die.

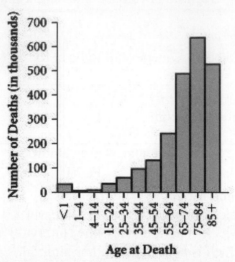

Source: *Statistical Abstract of the United States, 1997, Table 130.*

b. This distribution will be strongly skewed right. Most people get their driver's licenses at the earliest possible age or quite close to it.

c. The distribution of SAT scores for a large number of students should be approximately normal.

d. Selling prices of new cars should show a few very expensive models (like Corvettes) and a large number of relatively inexpensive (but not cheap!) ones (around $15,000 to $20,000). The distributions should be skewed right (toward the larger values.)

E7. a.

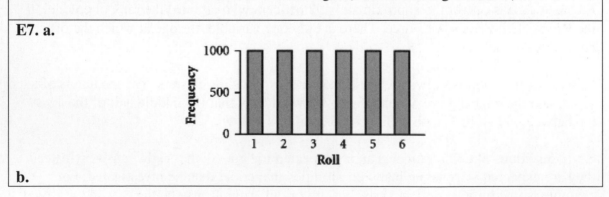

b.

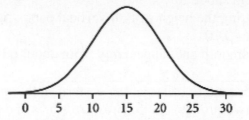

c. There are many possible answers. One such set of numbers is 1, 2, 3, 4, 9, 11, 13, 14, 17, 19, 21, 22, 23, 23, 24, 26, 27, 28, 29, 30. This boxplot represents one possible sketch.

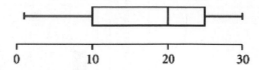

d. There are many possible answers. One such set of numbers is 0, 10, 20, 30, 80, 120, 150, 160, 170, 190, 210, 400, 500, 600, 700, 1300, 1500, 1800, 2400, 2900. This boxplot represents one possible sketch.

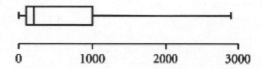

e.

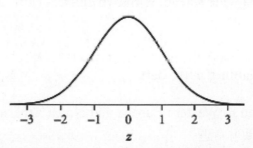

E9. a. Imagine connecting the dots in the scatterplot below by a normal bell curve:

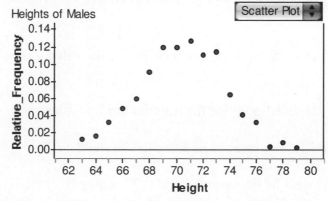

b. The mean is the balance point of the distribution, which occurs at about 71 in. The standard deviation is about 3 in.

c. This is obtained by adding the heights of all vertical bars to the left (and including)

19

the one at 74 inches, which is about 0.90.

d. This is obtained by adding the heights of all vertical bars strictly to the left of the one at 68 inches, which is about 0.10.

e. The distribution isn't smooth and ranges only from about 63 to 79 inches.

E11. a. The western states tend to be large, the eastern states small.

b. Dividing the data into two groups, one for *western* states and the other for *eastern* states, we have the following two histograms:

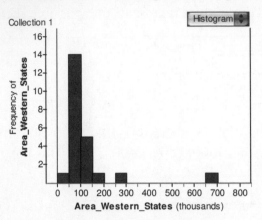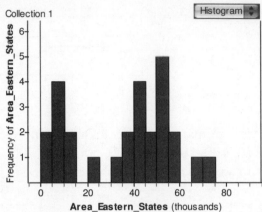

Yes; both groups cluster together, although eastern states spread out more.

E13. a. The distribution is wide spread, with two clusters (low GDP and media GDP) and one extremely large value.

b. Norway and Switzerland have the highest per capita GDP; they are not outliers because there is much variability in the data.

c. The higher GDPs belong to Western Europe and North America; the lower GPDs belong to Eastern Europe and Asia.

d. No, neighboring countries tend to have similar economies, producing a clustering effect.

E15. a. With a perfect uniform distribution on [0, 2], the value 1.0 would divide the values in half.

b. The values 0.5, 1.0, and 1.5 divide the distribution into quarters.

c. The values 0.5 and 1.5 enclose the middle 50% of the data.

d. About 15% of the data lie between 0.4 and 0.7 because the length of this interval is 0.3, and dividing it by the length of the entire interval, namely 2, yields 0.15.

e. The values 0.05 and 1.95 enclose the middle 95% of data values.

E17. a. Most tend to predict that domesticated animals have greater longevity due to the relative safety and good care provided by their habitat.

b.

```
           Domestic Wild

               4   0   13
              85   0   556778
           22220   1   0022222
               5   1   5555556
               0   2   000
                   2   5
                   3
                   3   5
                   4   1
                   1|5 represents 15 years
```

c. Both distributions appear to be slightly skewed to the right with possible outliers on the high side. The median of both distributions is 12, but the spread of the distribution for wild mammals is quite a bit larger. The middle half of the domestic mammals have an average longevity approximately between 8 and 12 years. The middle half of the wild mammals have an average longevity between 7.5 and 15.5 years.

E19. The bar charts are as follows:

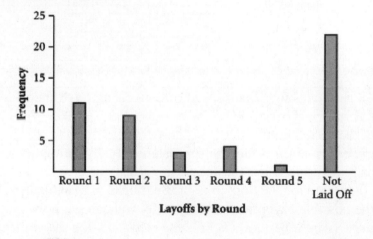

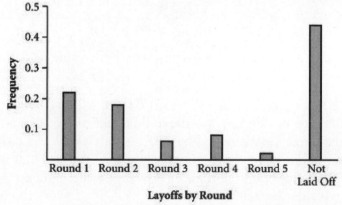

From the bar chart, it appears that Westvaco laid off the majority of workers in Rounds one and two. All together, Rounds 3, 4, and 5 make up about the same number of workers as in Round 2 alone.

21

E21. a. The heights of the first three bars are the number of nonpredators in the display that fall into the categories Domestic and Wild, and the total number of nonpredators. For example, the first bar shows that there are about 8 nonpredators that are domestic.

b. Looking at the middle set of bars, for predators, the second bar is taller than the first. Thus, a predator is more likely to be wild than domestic.

c. Looking at the first set of bars, for nonpredators, you can see that a nonpredator is also more likely to be wild than domestic because the second bar is taller than the first bar. However, for nonpredators, the first bar is a larger fraction of the second bar than is the case for predators. Thus, a predator is more likely to be wild than is a nonpredator.

Note: The way the bar chart is set up makes it easy to make the comparison asked for in part b but difficult to make the comparison in part c. You may wish to ask students to make a bar chart that makes it easy to answer part c. A two-way table like the one shown here can be helpful in summarizing the data on different categorical variables before making the bar chart.

	Nonpredator (0)	Predator (1)	Total
Domestic (0)	8	2	10
Wild (1)	19	10	29
Total	27	12	39

E23. a. Of the 18 mammals for which speeds are given, 12 have speeds that end in either 0 or 5.
b. Two-tenths of the 18 mammals, or 3.6
c. The most likely explanation is that the speeds are actually estimates for the wild mammals. Who is going to measure the speed of a grizzly bear in the wild? The speeds that don't end in 0 or 5 are for the dog, fox, giraffe, horse, pig, and squirrel. For these mammals, with the possible exception of the giraffe, you can see how speed could be measured accurately. (And it certainly is in horse races and dog races.)

E25. The seventh value is 10, as you can see from solving
$$25 = \frac{x + 24 + 47 + 34 + 10 + 22 + 28}{7}.$$

E27. a. The middle half of the speeds of domestic mammals are between 30 and 40. The spread for wild mammals is a bit larger—the middle half of the speeds are between 27.5 and 43.5. Half of the domestic mammals have speeds above 37 and half below. The median for wild mammals, again, is almost the same, at 36. Note that in each case the median is closer to the upper quartile than the lower quartile, indicating that the distribution of speeds may be skewed left.

b. The wild mammals are more likely to be predators than are the domestic animals (see E21), and the speeds of predators have the larger IQR.

E29. a. A – mound-shaped symmetric; B – skewed right;
C – skewed left; D - uniform

b. Keep in mind the main features of a boxplot, namely the maximum and minimum data values, and the three quartiles. Matching these distinct characteristics to the histograms yields: A – III; B- IV; C – II; D –I

E31. a.

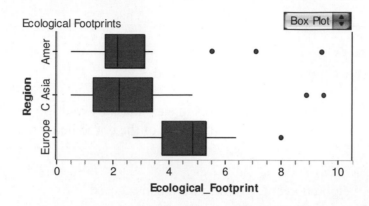

b-c. Distributions for all regions are skewed right, with at least one outlying country in each (United States, Canada and Uruguay; Kuwait and United Arab Emirates; Denmark). Central Asia and the Americas center around 2.5, with Europe centering around 5. Central Asia has the largest spread, followed closely by the Americas.

d. About 68%

E33. The third boxplot cannot be the plot for both classes combined because the minimum test score for the second period is about 10, and that would be the lowest for the combined set also. Students may come up with other possible reasons.

E35. a. II has the largest standard deviation, and III has the smallest.
b. II and III

E37. The set of heights of all female NCAA basketball players will have the larger mean because basketball players tend to be taller than other athletes, in general. The set of all female NCAA athletes will have the larger standard deviation because it will include tall, medium, and short athletes, whereas the set of all basketball players will include mostly tall athletes.

E39. a. With Seinfeld, the midrange is $\frac{(2.32+76.26)}{2} = 39.29$. Without Seinfeld, the midrange is $\frac{(2.32+58.53)}{2} = 30.425$. The midrange is not resistant and is extremely sensitive to outliers because it is computed using only the maximum and minimum.

b. The total of the ratings with the Seinfeld episode is $(101)11.187 = 1129.887$. Subtracting the Seinfeld rating of 76.26 leaves a sum for the remaining programs of 1053.627. So the mean rating is $\frac{1053.627}{100} \approx 10.54$.

E41. a. 3.11 grams
b. 0.04 gram
c. Yes, most of the weights are between 3.07 and 3.15.

E43. a. You can find the total of the scores in each class by multiplying the mean by the sample size, and this allows you to find the mean of the combined groups:
$$\overline{x} = \frac{30(75) + 22(70)}{52} = 72.9$$
This is called the weighted average of the two means.

b. You cannot find the median of the combined groups because to do so requires knowledge of the ordered arrangement of all the data values.

E45. $\sum(x - \overline{x}) = \sum x - \sum \overline{x} = \sum x - n\overline{x} = \left(\sum x\right) - \left(\sum x\right) = 0$

E47. a. Either could be used depending on the purpose of computing a measure of center. If, as is typical, there are a few expensive homes mixed in with many modestly priced homes, then the mean price will be larger than the median price. So real estate agents usually report the median price because it is lower and it tells people that half the prices are lower and half are higher. The tax collector would be interested in the mean price because the mean times the tax rate times the number of houses gives the total taxes collected.

b. As always, the answer depends on the purpose for computing a measure of center. Most likely, the reason here is to establish the total crop in Iowa for the year. In that case, it is best to find the mean yield per acre. This mean could be multiplied by the total acres planted in corn to approximate a total yield. An individual farmer probably would want to know the median as it gives the better indication of whether his or her yield was typical.

c. Again, the purpose of computing the measure of center determines which one you would use. Survival times are usually strongly skewed right. Telling a patient only the mean survival time would give too optimistic a picture. The smaller median would inform the person that half the people survive longer and half shorter. On the other hand, if you are the physician and must allocate your time by estimating the total number of hours you will be caring for your patients with this disease, the mean would be better. You would then multiply the mean number of survival days by the number of patients you have by an estimate of the number of hours each day that each patient takes.

E49. a. To make the histogram, students need only copy the histogram in the student text and then use the formula

$$C = \frac{5}{9}(F - 32)$$

to convert the numbers on the x-axis from °F to °C. The scale would then go from 36.67 to 58.89.

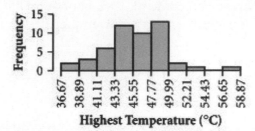

b. Note that the standard deviation is the tricky one: You just multiply by $\frac{5}{9}$. N stays the same. For each of the others, you subtract 32 and then multiply by $\frac{5}{9}$.

Variable	N	Mean	Median	StDev
Highest	50	45.61	45.56	3.72

Variable	Min	Max	Q1	Q3
Highest	37.78	56.67	43.33	47.78

c. Yes, there is an outlier on the high side. The IQR = 4.45, and 1.5(IQR) = 6.675. So, $Q_{3} + 1.5 \cdot$ IQR = 54.455, and the maximum is larger than that—so definitely an outlier. $Q_{1} - 1.5 \cdot$ IQR = 36.655, and the minimum is bigger—so none on the lower end.

E51. Let the mean of the original set of data be

$$\bar{x} = \frac{x_1 + x_2 + x_3 + x_4 + x_5}{5}$$

Then the mean of the transformed data is the result of the equation shown below.

$$\frac{(x_1 + c) + (x_2 + c) + (x_3 + c) + (x_4 + c) + (x_5 + c)}{5} = \frac{x_1 + x_2 + x_3 + x_4 + x_5 + 5c}{5}$$

$$= \frac{x_1 + x_2 + x_3 + x_4 + x_5}{5} + \frac{5c}{5} = \bar{x} + c$$

E53. a. Median: 23 or 24 cents. To estimate this, find the amount of change that corresponds to the 50th percentile.

b. Q_1 is about 10 cents and Q_3 is about 70 cents. To find Q_1 and Q_3 find the amount of change that correspond to the 25th and 75th percentiles, respectively. The IQR is 70 − 10 = 60 cents.

c. The set of data is skewed right. The larger increases toward the left side of the graph indicate larger numbers of students with smaller amounts of change. The large increases correspond to higher bars in a histogram. Or: 75% of the cases are between 0 and 70 cents and the top 25% is spread out from 70 cents to $2.50. So the distribution is skewed toward the larger values.

E55. Approximating from the histogram or from the cumulative relative frequencies (or percentiles), the 85% percentile is close to 32 mi/h. Rounding down to the nearest 5 mi/h., the speed limit should be set at 30 mi/h. (See the solution to E56 for the percentile plot.)

E57. The lower quartile is the 25th percentile, the median is the 50th and the upper quartile is the 75th. From the percentile plot with horizontal lines sketched across at these percentile values, you can see that the lower quartile is a little above 50, the median is a little above 55 and the upper quartile is about 60. So, (a) is the correct choice.

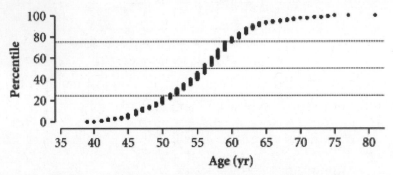

Note: The ages in months are obtained by simply multiplying the ages in years by 12. The shapes of the percentile plot will remain the same, so the median and quartiles also get multiplied by 12. So, the median of the ages in months is 56(12) or 672 months. The first and third quartiles are, respectively, 51(12) = 612 months and 60(12) = 720 months.

E59. **a.** 0.8413 or 84.13%; 0.9943 or 99.43%
b. 0.1587 or 15.87%; 0.0057 or 0.57%
c. 0.9332 or 93.32%
d. 0.6827 or 68.27% (0.6826 using Table A)

E61. a. 2   **b.** 1   **c.** 1.5   **d.** 3   **e.** −1   **f.** -2.5

E63. a.   **i.** 0.6340 (calculator: 0.6319)   **ii.** 0.0392 (calculator: 0.0395)
    iii. 0.3085 (calculator: 0.3101)

b. The middle 95% of scores range from, approximately,
$$505 - 1.96(111) \approx 287 \text{ to } 505 + 1.96(111) \approx 723.$$

E65. First, find the percentage of men who qualify.

$$z = \frac{x - mean}{SD} = \frac{62 - 70.4}{3.0} = -2.8$$
$$z = \frac{x - mean}{SD} = \frac{72 - 70.4}{3.0} \approx 0.53$$

The area between these z-scores is about 0.4974 + 0.2019 = 0.6993. About 70% of men aged 20 to 29 in the United States meet the height qualifications to be a flight attendant for United Airlines.

    Next, find the percentage of women who qualify.

$$z = \frac{x - mean}{SD} = \frac{62 - 65.1}{2.6} \approx -1.19$$
$$z = \frac{x - mean}{SD} = \frac{72 - 65.1}{2.6} \approx 2.65$$

The area between these two z-scores is about 0.3830 + 0.4960 = 0.8790. About 88% of women aged 20 to 29 in the United States meet the height qualifications to be a flight attendant for United Airlines, a higher percentage than men.

E67. a. 0.1587

b. 8.16

c. Solving $-1.34 = \frac{6 - mean}{3}$, you get mean = 10.02.

d. For P = 0.6, z = 0.25. Solving $0.25 = \frac{12 - 10}{SD}$, you get SD = 8.

E69. Make use of the standard rule of thumb for a symmetric normal-like distribution. Doing so, we have the following percentages presented in the order requested in the problem: 68%; 95%; 16%; 84%; 97.5%; 2.5%

E71. a. The z-score for the height 68 is $\frac{(68 - 70.4)}{3.0} = -0.80$. The area under the normal curve to the left of this point is 0.2119. Thus, about 21.19% of U.S. males between 20 and 29 are less than 68 inches tall.

b. From part a, 0.2119 of the men are below the height of 68 inches. Similarly, the z-score for a height of 67 inches is −1.13, and so 0.1292 of the men are below that height. The proportion in between is 0.2119 − 0.1292 = 0.0827. So, about 0.0827(19,000,000) = 1,571,300 are between the two heights.

c. A percentile of 90 corresponds to a z-score of about 1.28. Hence, the desired height is
$$x = mean + z \cdot SD = 70.4 + 1.28(3.0) = 74.24 \text{ inches}.$$
Equivalently, you could solve the equation:
$$1.28 = \frac{x - 70.1}{3.0}$$

E73. a. The distribution is probably skewed right because it's not possible for the length of a reign to be much more than 1 standard deviation below the mean.

b. The z-score for 0 is $\frac{(0 - 18.5)}{15.4} = -1.20$, so about 0.1151 of the reigns.

c. If all values in the distribution must be positive and two standard deviations or less below the mean is less than 0, the distribution isn't approximately normal.

E75. a.

```
            Stem-and-leaf of Number N = 51
              Leaf Unit = 1.0
                      16   0   0000000111122244
                      25   0   566778899
                     (5)   1   00013
                      21   1   5
                      20   2   244
                      17   2   66
                      15   3   03
                      13   3   579
                      10   4   2
                       9   4   58
                       7   5   14
                       5   5
                       5   6
                       5   6
                       5   7   2
                       4   7   69
              HI 95, 232

                      0I5 stands for 5 tornadoes
```

b. Minimum: 0 Q_1: 2   Median: 10 Q_3: 35
Maximum: 232

c. Outliers fall below $2 - 1.5(35 - 2) = -47.5$ or above $35 + 1.5(35 - 2) = 84.5$.
There are two outliers, Florida and Texas.

d.

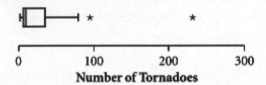

e. Both plots show the strong skewness in the data and both outliers: Florida and Texas.
In the stemplot, you can see that half the states have less than 10 tornadoes. The
cleanness of the boxplot makes it clear how much of an outlier Texas actually is. (Of
course, it is a very large state, which helps explain why it is an outlier.) However, you
can't see from the boxplot at all that so many states have at most 1 tornado. Because the
stemplot has a reasonable number of values in it and is consequently easy to read while
carrying almost all the complete values, it is reasonable to select it as the most
informative.

f. The distribution is strongly skewed right with two outliers and a wall at 0. The median
number of tornadoes is 10, with the middle 50% of states having between 2 and 35
tornadoes.

E77. a. Outliers would lie below $585 - 1.5(670 - 585) = 457.5$ or above $670 + 1.5(670 - 585) = 797.5$.

b. The bunching up between the lower quartile and the median suggest the distribution is probably skewed right.

E79. a. Developing countries have lower life expectancies than do developed countries. Thus, Region 1 must be Africa (developing countries, for the most part) and Region 3 must be Europe (developed countries). The Middle East, Region 2, has a mixture of developed and developing countries.

b. C is for Region 1 (Africa); B is for Region 2 (Middle East); A is for Region 3 (Europe).

E81. No, this is true only for normal distributions. For example, the set of values {2, 2, 4, 6, 8, 8} is symmetric and has mean and median both equal to 5. The standard deviation is about 2.76. Only two of the values, or about 33% of them, are within one standard deviation of the mean.

E83. a. The median number of deaths is 13. Half of the cities have fewer than 13 pedestrian deaths per year, and half have more.

b. The lower quartile is the average of the 10th and 11th data points, namely 8. The upper quartile is the average of the 30th and 31st data points, namely 24.5.
An outlier is a city whose number of deaths exceeds $24.5 + 1.5(24.5-8) = 49.25$ or is less than $8 - 1.5 (24.5-8) = -16.75$, which is impossible. There are three outliers, Phoenix, Los Angeles, and New York. One explanation is the large populations these cities have; these are the three largest cities in the United States.

c. Plots and explanations will vary. A stemplot is a very good choice. It reveals the three outliers and shows that the distribution is skewed right. It is given by:

0	0134556678888999
1	0002334455679
2	22788
3	06
4	58
5	9
6	
7	
8	
9	9
10	
11	
12	
13	
14	
15	7

d. New York 1.91; Miami 6.68. The rate is adjusted for population size and gives a more accurate picture of pedestrian safety.

E85. The mean grade is 2.98 with a standard deviation of 1.33.

E87. There are many possible responses. An example is {1, 1, 1, 1, 1, 2, 2, 10}, which has mean 2.375 and standard deviation of about 3.11. One standard deviation below the mean is less than 0.

E89. a.   i. 6.325 v. 6.667
        ii. 2.000 v. 2.010
       iii. 0.632 v. 0.633
b. No, as n gets larger, the difference between s and σ goes to 0.

E91. a. The state with the lowest per capita income in 1980 had an average income per person of $6,573.
b. There is at least one outlier on the high end for both 2000 and 2007, but none for 1980. There are no outliers on the low end for any of the years.
c. There was a noticeable improvement in position as the z-scores went from -1.0 in 1980 to -0.86 in 2007.
d. As the years progress, the histograms become more skewed toward the higher values and both the centers and spreads increase.
e. The z-score would work best for the 1980 data, as it is more mound-shaped and symmetric then the others.

E93. $176 \pm (1.645)30$ or 126.65 mg/dl and 225.35 mg/dl.

E95. a. Again, the histogram of batting averages looks quite normal in shape with center at about 0.260. The standard deviation is approximately 0.040.

b. The batting averages in both leagues have distributions that are approximately normal in shape. The American League has a higher mean (by about 0.010) and less spread.

c. The z-score corresponding to 0.300 in the National League is $z = 1$. The corresponding batting average, x, in the American League would still be 1 standard deviation above the mean, or
$$x = 0.270 + (1)(0.030) = 0.300$$
so that batter would be expected to have a similar batting average in the American League.

Chapter 3

Practice Problem Solutions

P1. Plot a shows a positive relationship that is strong and linear. There is fairly uniform variation across all values of x.

Plot b shows a negative relationship that is strong and linear, again with fairly uniform variation across all values of x.

Plot c shows a positive relationship that is moderate and linear with fairly uniform variation across all values of x. There is one point that lies a short distance from the bulk of the data.

Plot d shows a negative relationship that is moderate and linear with fairly uniform variation across all values of x. Again, there is one outlier.

Plot e shows a positive relationship that is strong and linear except for the outlier. As students will learn, the one outlier has dramatic influence on the strength of this relationship. There is fairly uniform variation across all values of x.

Plot f shows a negative relationship that is very strong and curved. Again, one point lies in the general pattern but far away from the remainder of the data, which accentuates the strong relationship. Another outlier lies below the bulk of the data on the left.

Plot g shows a negative relationship that is strong and curved. The two points at either end of the array accentuate the curvature. There is a bit more variability among values of y for smaller values of x than for larger values of x.

Plot h shows a positive relationship that is strong and curved. Again, the outlier on the extreme right accentuates the curved pattern and would have dramatic influence on where a trend line might be placed. The variability in y is fairly constant across all values of x.

P2. a. You may have to remind students that a scatterplot without labels and units on the axes is meaningless. Emphasize the importance of appropriate labeling.
Here is the scatterplot.

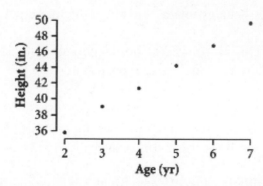

b. These data are not very interesting to describe. The x-axis shows ages 2 to 7 years, and the y-axis shows an median height of children at each age. The shape is linear, the trend is positive, and the strength is very strong. That is, the scatterplot shows a very strong positive linear trend. A typical child grows about 2.7 inches per year.

c. The linear trend could reasonably be expected to hold for another year. However, *median height* could not be expected to increase at this rate to age 50, as people typically stop growing around age 20.

P3. a. The relationship is a positive trend, has moderate strength, and slight curvature with more variability at larger values of x.

b. Trees of same age at another location may have similar pattern; older trees may not fit this pattern.

c. As trees age, growth slows and variability among tree sizes is more pronounced.

P4. a. The points (40, 91) and (240, 88.3) lie on or near the regression line, so the slope is about –0.0135. Each day, the eraser tended to lose around 0.0135 gm of weight, on average.

b. The y-intercept is 91.48 grams. This is a prediction of the weight of the eraser at the beginning of the year.

P5. a. About 0.8.

b. If one student has a hand length that is 1 in. longer than that of another student, the first student's hand tends to be 0.8 in. wider.

c. The y-intercept is 1.69. This is a prediction of the width of a hand with no length. This is not reasonable.

d. The students in the lower cluster did not spread their fingers. The slope would increase. In fact, if these points were removed, the regression line would move up slightly at the end for smaller hand lengths and move up a bit more at the end for longer hand lengths.

P6. a. The slope is 0.072. This is an estimate of average thickness per sheet (in mm).

b. The y-intercept is 2.7 mm. This is the predicted thickness of a stack comprised of 0 sheets; i.e., the thickness of the cover. This seems reasonable.

c. $\hat{y} = 2.7 + 0.072(175) = 15.3 mm$

d. The biggest residual occurs for 100 sheets; it is positive.

e. The residuals are $y - \hat{y}$. Evaluating for all data points, we have the following:

Number of sheets x	y	Regression Line $\hat{y} = 2.7 + 0.072x$	Residual $y - \hat{y}$
50	6.0	6.3	-0.3
100	11.0	9.9	1.1
150	12.5	13.5	-1.0
200	17.0	17.1	-0.1
250	21.0	20.7	0.3

The sum of the values in the rightmost column is 0.

32

P7. a.

b. The equation is $\hat{y} = 279.75 + 2.75x$. The slope and y-intercept are found using the table below.

Pizza	x	y	$x - \bar{x}$	$y - \bar{y}$	$(x - \bar{x}) \cdot (y - \bar{y})$	$(x - \bar{x})^2$
1	9	305	−2	−5	10	4
2	11	309	0	−1	0	0
3	13	316	2	6	12	4
Sum	33	930	0	0	22	8
Mean	11	310				

So, slope $= \dfrac{22}{8} = 2.75$, $\quad y-\text{intercept} = 310 - 2.75(11) = 279.75$.

P8. a. (*adult weight*) = -28.81 + 13.22 (*birth weight*)

b. For the llama, $y = 140$ and so $\hat{y} = -28.81 + 13.22(11) = 116.61$. So, the residual is $y - \hat{y} - 23.39$.

c.

Birth Weight	y	Regression Line $\hat{y} = 28.81 + 13.22x$	Residual $y - \hat{y}$
36	475	504.73	-29.73
37	434	517.95	-83.95
11	140	174.23	-34.23
11.5	100	180.84	-80.84
7.21	62	124.13	-62.13
5.74	50	104.69	-54.69

d. Square all values in the rightmost column in (c) and sum to obtain $\sum (y - \hat{y})^2 = 2586.2$.

e. $-28.81 + 13.22(\bar{x}) = -28.81 + 13.22(18.08) = 210.2 = \bar{y}$

P9. Yes, the regression equation of $\hat{y} = 279.75 + 2.75x$ and the mean of the response variable, 310, are the same as we computed by hand. The SSE is $0.5^2 + 1^2 + (-0.5)^2 = 1.5$ and is found in the Analysis of Variance table in row "Error," column "Sum of Squares."

P10. a. The slope is 1.13; on the average, the area increase by about 1.13 inches per year
b. The y-intercept is -3.70; it predicts the area of a tree at year 0 – this is not a practical interpretation.

P11. a. -0.5 **b.** 0.5 **c.** 0.95 **d.** 0 **e.** -0.95

P12. a. About 0.95   **b.** 0.99

P13. a. Note that
$$\bar{x}=0,\ \bar{y}=0.40,\ s_x=\sqrt{2.5}\approx1.581,\ s_y=\sqrt{0.8}\approx0.894.$$
Next, consider the following table of values:

x	$z_x=\frac{x-0}{1.581}$	y	$z_y=\frac{y-0.40}{0.883}$	$z_x\cdot z_y$
-2	-1.265	-1	-1.566	1.981
-1	-0.633	1	0.671	-0.4247
0	0	0	-0.447	0
1	0.633	1	0.671	0.4247
2	1.265	1	0.671	0.8488

So, $r=\frac{1}{5-1}\sum z_x\cdot z_y=0.7075$.

b. Similarly, note that
$$\bar{x}=0,\ \bar{y}=3.0,\ s_x=\sqrt{2}\approx1.414,\ s_y=\sqrt{1}=1.$$
Next, consider the following table of values:

x	$z_x=\frac{x-0}{1.414}$	y	$z_y=\frac{y-3.0}{1}$	$z_x\cdot z_y$
-2	-1.414	2	-1	1.414
0	0	2	-1	0
0	0	3	0	0
0	0	4	1	0
2	1.414	4	1	1.414

So, $r=\frac{1}{5-1}\sum z_x\cdot z_y=0.707$.

P14. 0.99; all but one of the products are positive.

P15. a. Positive

b. The point at the extreme upper right of the plot at about (5, 5) will make the largest positive contribution to the correlation because it is farthest away from the new origin $(\bar{x},\bar{y})$ and from the new coordinate axes $(x=\bar{x})$ and $(y=\bar{y})$ and so has a large $z_x\cdot z_y$.

c. In Quadrants I and III; 20 have a positive product.

d. In Quadrants II and IV; 7 have a negative product.

P16. The plot on the top has a strong curvature. A line would not be appropriate here. The plot on the bottom is linear. The cloud of points is roughly elliptical. A line would be appropriate for this plot.

P17. a. The correlation is 0.650:

$$b_1 = r \cdot \frac{s_y}{s_x} \quad \Rightarrow \quad 0.368 = r \cdot \frac{7}{12.37} \quad \Rightarrow \quad r = 0.650$$

b. The regression equation is *exam 2* = 48.94 + 0.368 *exam 1*. The predicted *exam 2* score is 78.38.

c. The regression equation is *exam 1* = −132.99 + 2.72 *exam 2*.

d.

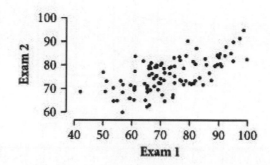

P18. a. The size of the city's population.
b. You should divide each number by the population of the city to get the number of fast–food franchises per person and the proportion of the people who get stomach cancer.

P19. An obvious lurking variable is the age of the child. Parents tend to give higher allowances to older children, and vocabulary is larger for older children than for younger.

P20. A careless conclusion would be that people are too busy watching television to have babies. The lurking variable is how affluent the people in the country are. More affluent people tend to have more televisions and have fewer children.

P21. a. The formula relating these quantities is
$$r^2 = \frac{\text{SST} - \text{SSE}}{\text{SST}} = \frac{480.25 - 212.37}{480.25} \approx 0.5578$$
(as given in the output) so $r = \pm 0.747$. Because the slope of the regression line is negative (the scatterplot goes downhill), $r = -0.747$.

b. The slope of −0.621 tells us that if one state has a high school graduation rate that is one percentage point higher than another, we expect its poverty rate to be lower by 0.621 percentage points, on average.

c. x is in percentage of high school graduates, y is in percentage of families living in poverty, b_1 is in percentage of families living in poverty per percentage of high school graduates, and r has no units.

P22. The plot should look similar to one of the two options shown here. The regression line is flatter than the line connecting the endpoints of the ellipse. This plot shows the regression effect as well. This time, it is the older sisters of the taller younger sisters who tend to be less tall than their younger sisters.

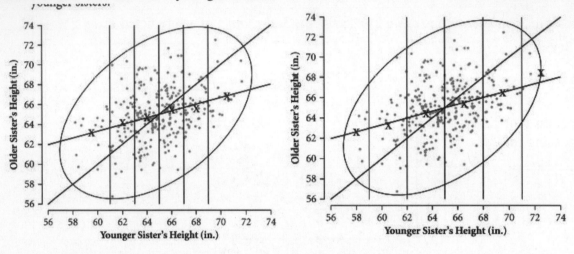

P23. There is evidence of regression to the mean. An ellipse around the cloud of points will have a major axis that is steeper than the regression line. The slope of the regression line is only about 0.4, much less than 1. Also, for the vertical strip containing exam 1 scores above 95, the mean exam 2 score is about 93. For exam 1 scores less than 70 the mean exam 2 score is about 76.

P24. **a.** Consider the following scatterplot with regression line:

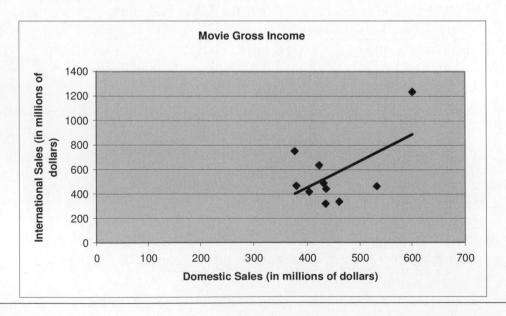

36

The cluster of points has a decreasing trend except for one extreme value in upper right.

b. *International* = -417 + 2.17*Domestic*; *r* = 0.56

c. *International* = 979 – 1.16*Domestic*; *r* = -0.4. Removing the point allows the line to fit the cloud of points better, but reduces the *r*.

P25. a. The student did not predict very well. The estimates were consistently low.

b. (180, 350) appears to be the most influential point. It is an outlier in both variables and is not aligned with the other points.

c. With the point (180, 350) the regression equation is $actual = 12.23 + 1.92 \cdot estimate$, and $r = 0.975$. Without this point, the equation is $actual = -27.10 + 3.67 \cdot estimate$, with $r = 0.921$. This point pulls the right end of the regression line down, decreasing the slope and increasing the correlation.

P26. a. The scatterplot with the regression line is shown next.

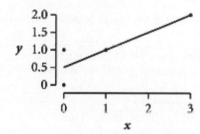

b. The table is as follows:

x	y	Predicted Value	Residual
0	0	0.5	−0.5
0	1	0.5	0.5
1	1	1	0
3	2	2	0

c. The residual plot is shown here.

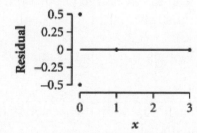

d. The residual plot straightens out the tilt in the scatterplot so that the residuals can be seen as deviations above and below zero rather than above and below a tilted line. The symmetry of the residuals in this example shows up better on the residual plot.

P27. **a.** Population density appears to increase linearly, with little variation around a straight line.

b. Residuals show that the rate of increase in population density is not constant. In the 1800s the rate is a little lower than average, from 1900 to 1930 the rate is higher than average, lower again from 1930 to 1960 and higher since 1960.

P28. **a.** A–IV; B–II; C–I; D–III

b. <u>Curve with increasing slope</u>: The residual plot will open upward, as in plot II.

<u>Unequal variation in the responses</u>: The residual plot has a fan shape, as in plot I.

<u>Curve with decreasing slope</u>: The residual plot will open downward, like an inverted cup. No plot in this example shows this pattern.

<u>Two linear patterns with different slopes</u>: The residual plot will be V–shaped, as in plot III. The pattern is more clear if you ignore the point with a residual of about 10.

c. Plot D and residual plot III show a scatterplot that looks as though it should be modeled by two different straight lines. The V shape can be seen in the scatterplot, but it may be more obvious to many in the residual plot because the overall linear trend (tilt of the V) has been removed.

Exercise Solutions

E1. a. Positive and strong: As eggs get bigger, both length and width increase proportionally.
b. Positive and moderate: Most students tend to score relatively high on both parts of the exam, middling on both parts, or relatively low on both parts of the exam.
c. Positive and strong: Trees produce one new ring each year.
d. Negative and moderate: People tend to lose flexibility with age.
e. Positive and strong: The number of representatives is roughly proportional to the population of a state.
f. Positive and weak: Large countries tend to have large populations, but there are notable exceptions such as Canada and Australia. Also, some small countries in area have very large populations, such as Indonesia.
g. Negative and strong (but curved): Winning times tend to improve (get smaller) over the years.

E3. a. The worst record for baggage handling during this period is Delta, and Northwest has the lowest on-time percentage among the airlines.
b. Airlines with a high percentage of on-time arrivals and a low rate of mishandled baggage would fall in the upper left of the plot. Thus, United is "best" for mishandled baggage; America West is best for percentage on-time arrivals.

c. False. American's baggage mishandling rate of 6.5 mishandled bags per thousand was not twice Southwest's rate of a little under 4.5. It appears to be more than twice because the scale on the *x*-axis starts at 3.75, not 0.

d. The relationship between the two variables is negative and moderate. The negative relationship shows that an airline that is "bad" on one variable tends to be "bad" on the other as well.

e. Yes, since an airline that is on-time a higher percentage of the time probably pays attention to other details as well (like handling baggage).

E5. a. Plots A, B, and C are the most linear. Plot D is not linear because of the seven universities in the lower right, which may be different from the rest. The first plot, of *graduation rate* versus *alumni giving rate,* gives some impression of downward curvature. However, if you disregard the point in the upper right, the impression of any curvature disappears.

Plots A, B, and C, all have just one cluster. However, Plot D, the plot of *graduation rate* versus *top 10% in high school,* has two clusters. Most of the points follow the upward linear trend, but the cluster of seven points in the lower right with the highest percentage of freshmen in the top 10% shows little relationship with the graduation rate.

Plots A and C have possible outliers. Plot A, the plot of *graduation rate* versus *alumni giving rate* has a possible outlier in the upper right. It is below the general trend and its *x*-value (but not its *y*-value) is unusually large. In Plot C, the plot of *graduation rate* versus *SAT 75th percentile,* the points toward the upper left and the middle right should be examined because they are farther from the general trend than the other points, although neither their *x*-values nor their *y*-values are unusual. The point in the lower right of Plot D should also be examined, along with the other six points nearby.

b. The plots of *graduation rate* versus *alumni giving rate* and *graduation rate* versus *SAT 75th percentile* have similar moderate positive linear trends. The plot of *graduation rate* versus *top 10% in high school* shows wide variation in both variables, with little or no trend. The plot of *graduation rate* versus *student/faculty ratio* is the only plot that shows a negative trend.

c. Among these four variables, it appears that the alumni giving rate is the best predictor of the graduation rate and SAT scores (as measured by the 75th percentile) is second best. However, both of these relationships are moderate and neither is a strong predictor of graduation rate. Ranking in high school class (as measured by the top 10%) is almost useless as a predictor of college graduation rate. The plot of *graduation rate* versus *alumni giving rate* owes part of the impression of a strong relationship to the point in the upper right. This plot shows some heteroscedasticity, with the graduation rate varying more with smaller alumni giving rates.

Even though the relationship between, say, the graduation rate and the student/faculty ratio is negative, that's not what makes the student/faculty ratio a poor predictor of the graduation rate. Given a specific student/faculty ratio, we can predict a graduation rate.

The problem is that there is a great deal of uncertainty about how close the actual rate would be to the predicted rate because the range of graduate rates is large for any given student/faculty ratio.

d. The relationships could change considerably when looking at all universities because these are highly rated universities, so the values of all variables tend to be "good." With a larger collection of universities, there may be more spread in the values of all variables and, most likely, more pronounced patterns.

Consider the scatterplot given below. On this plot, the **x**'s represent the top 25 universities and the closed circles represent the next 25 most highly rated universities. Note that if you look at either group, there is very little upward trend. However, putting the two groups together gives a stronger linear trend. If another group of 25 universities were added, the trend probably would be stronger. This phenomenon is sometimes called the *effect of a restricted range*.

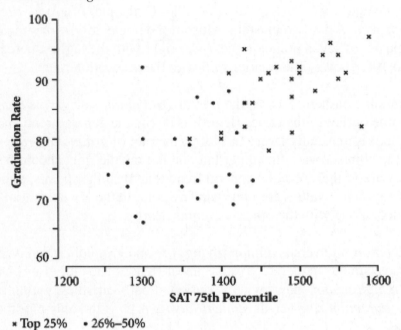

e. Graduation rates may increase as SAT scores increase because better prepared students may be more successful in college courses, so a university with a greater number of prepared students will, on average, graduate a higher percentage of students. Alumni giving rates may increase as graduation rates increase because the university has produced happy alumni. These data do not "prove" this claim, however, because there are other possible explanations. These types of observational studies cannot prove claims; the proof of a claim requires an experiment, which is one of the topics of the next chapter.

Note: These relationships hold for *universities* and that the data provide no evidence about whether the relationships hold for *individuals*.

E7. a. This plot does not help us decide this question. Because so many people who were hired in the early years (and even recently) would have retired, we do not know whether older people were hired then or not. To determine whether age discrimination in hiring may have existed, we need a plot of the age at hire of all people hired, not just those who remained at the time of layoffs.

b. From this plot, it appears that people hired earliest were more subject to layoff, not necessarily the older employees. Everyone with 30 years seniority was laid off, but not all of the older people were laid off. Perhaps, then, it was higher salaries because of seniority or obsolete job skills that resulted in a greater proportion of older employees being laid off.

E9. a. The scatterplot comprised of data which exhibit less variance is the one that provides a stronger relationship. In this sense, PPE does better, but neither does well.

b. Utah has a high graduation rate and low PPE.

c. The eastern states cluster to the left, high graduation rates and low student–teacher ratios; the Western states cluster to the right, high student–teacher ratios and large variation in graduation rates; the Midwestern states cluster in the upper middle while the Southern states cluster in the lower middle of the plot.

d. Mention that Delaware is at the median in graduation rate, in the upper quartile in PPE, and a little above the median on student-teacher ratio. Further reducing the student-teacher ratio can possibly help increase graduation rate.

E11. a. I – E; II – C; III – A; IV – D; V – B
b. I – A; II – E; III – B; IV – D; V – C

E13. a. The slope is about $(67 – 37)/(14 – 2)$, or 2.5 inches/year. Or, 2.47 based on the regression line.

On the average, boys tend to grow about 2.5 inches per year from the ages of 2 to 14.

b. The y-intercept of 31.6 would mean that an average newborn is 31.6 inches long. Because this is clearly too long, this extrapolation is not valid.

E15. a. The predictor variable is the arsenic concentration in the well water. The response variable is the concentration of arsenic in the toenails of people who use the well water.

b. There is a moderate positive linear relationship between arsenic concentration in the toenails of well water users and the arsenic concentration in the well water. There is a cluster around well water arsenic concentrations of 0 to 0.005 parts per million.

c. The residual for the person with highest concentration of arsenic in the well water is about 0.3 ppm.

d. This person has a concentration of about 0.076 ppm in their well water. The concentration of arsenic in this person's toenails is about 0.4 parts per million.
e. Seven of the 21 wells are above this standard.

E17. **a.** In the order listed, Bristol, Martinsville, Richmond, New Hampshire, Daytona.

b. The residual is -64.13. It tells you that the track's TQS is much smaller than would be predicted from data on all tracks.

c. Talladega has a positive residual, its data point lies above the line. Richmond has a negative residual, its data point lies below the line

---

E19. **a.** Observe that

$$\bar{x} = 2, \ \bar{y} = 26$$

$$b_1 = \frac{\sum(x-\bar{x})(y-\bar{y})}{\sum(x-\bar{x})^2} = \frac{(1-2)(31-26)+(2-2)(28-26)+(3-2)(19-26)}{(1-2)^2+(2-2)^2+(3-2)^2} = \frac{(-5)+0+(-7)}{1+0+1} = -6$$

$$b_0 = \bar{y} - b_1\bar{x} = 26 - {}^-6 \cdot 2 = 38$$

So, the equation is $\hat{y} = 38 - 6x$, where $\hat{y}$ is the predicted AQI and x is the number of years after 2000.

b. The AQI in Detroit tended to decrease 6 points per year on average.

c. The residuals for each year are calculated in the table below:

x	y	$\hat{y}$	Residual = $y - \hat{y}$
1	31	$38 - 6 \cdot 1 = 32$	-1
2	28	$38 - 6 \cdot 2 = 26$	2
3	19	$38 - 6 \cdot 3 = 20$	-1

The largest residual is for $x = 2$, which represents the year 2002. The residual is 2.

d. $SSE = (-1)^2 + 2^2 + (-1)^2 = 6$
e. $-1 + 2 + -1 = 0$

f. The equation for this line is y = 40 – 6x. (Since the residual for this point was 2, simply shift the line up two units.) SSE for this line would be $(-3)^2 + 0^2 + (-3)^2 = 18$. This is larger than the SSE for the least squares line, so according to the least squares approach the first line was better. Students should agree that the least squares line fits better because it passes through the middle of the set of points. The line through the point for 2002 is too high.

g. This equation would be y = 37 – 6x. The fitted value for 2002 would be $37 - 6 \cdot 2 = 25$. The only nonzero residual would be for this point, so SSE would be $3^2 = 9$.

h. As the plot below shows, the least squares line is a better indicator of the trend of all the points than either of the others because it represents the overall trend of the data. For both of the other lines, all points are on or to the same side of the line.

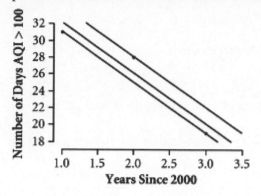

E21. a. The equation is $height = 31.6 + 2.43 \cdot age$. Here is the plot.

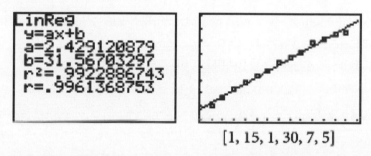

[1, 15, 1, 30, 7, 5]

b. The residual appears to be positive because the point appears to be above the line. The actual residual is 59.5 – 58.33 = 1.17, which is positive. (The residual is 1.21 with no rounding.)

c. The mean age is 8 years, and the mean height is 51 inches. Substitute into the regression equation to see that this point is on the regression line:

$$51 \neq 31.6 + 2.43 \cdot 8 = 51.04$$

This small discrepancy is due to rounding error. To avoid this, use the stored values for a and b from your calculator. For the TI-83 Plus and TI-84 Plus they can be found in the variables menu ($\boxed{\text{VARS}}$). Select **5:Statistics…** then right arrow to **EQ**. The **a** and **b** listed are the values from the most recently calculated regression equation. When those values are used, the right side of the equation comes out to 51 exactly.

d. This equation is very close to that for boys obtained earlier.

43

E23. a. *Age* is the explanatory variable and *area* is the response variable.

b. The line has slope close to 1; in comparing 10 and 20 year-old trees, there areas are also about 10 and 20 square inches.

c. 30 year-olds would have larger error because there is more variation in the residuals near 30.

d. Near 0 (0.38); its predicted area almost equals its actual area.

e. Age 38 and the area is about 20; it has a much smaller area compared to other trees of its age.

E25. a. $height = 31.5989 + 2.47418 \cdot age$. Answers will vary according to students' original estimates. It should be fairly close.

b. SSE = 2.4. This does seem reasonable from the scatterplot. The residuals are all quite small.

E27. For these proofs to make sense, you will need the following properties of finite sums:

(i) For a constant a, $\sum a = n \cdot a$, since that means adding up n instances of a,

(ii) The mean of a set of numbers is a constant, so $\sum \bar{x} = n \cdot \bar{x}$ where $\bar{x}$, in turn, is equal to the sum of the individual x values $\sum x_i$

a. A horizontal line has an equation of the form $y = a$. The residual for a point (x_i, y_i) would be $y_i - a$. Adding these up for n points, we obtain

$$\sum (y_i - a) = \sum y_i - \sum a = \sum y_i - na.$$

We want this sum to be zero.

$$\sum y_i - n \cdot a = 0$$

$$\sum y_i = n \cdot a$$

$$\frac{\sum y_i}{n} = \frac{\cancel{n} \cdot a}{\cancel{n}}$$

$$\frac{\sum y_i}{n} = a$$

And, of course, the left side of this equation is the mean of the y-values. This horizontal line has equation $y = \bar{y}$ so it passes through $(\bar{x}, \bar{y})$.

Conversely, if we start with the fact that the horizontal line passes through the point $(\bar{x}, \bar{y})$, the equation of the line is $y = \bar{y}$. The residual for a point (x_i, y_i) would be $y_i - \bar{y}$. Summing these residuals we obtain

$$\sum(y_i - \overline{y}) = \sum y_i - \sum \overline{y} = \sum y_i - n \cdot \overline{y} = \sum y_i - n\frac{\sum y_i}{n} = \sum y_i - \sum y_i = 0.$$

b. For a point (x_i, y_i), the predicted y-value is $a + b \cdot x_i$. So the residual is $y_i - a - b \cdot x_i$. We want the sum of these residuals to be zero.

$$\sum(y_i - a - b \cdot x_i) = 0$$
$$\sum y_i - n \cdot a - \sum b \cdot x_i = 0$$
$$\sum y_i = n \cdot a + b \cdot \sum x_i$$
$$\frac{\sum y_i}{n} = \frac{n \cdot a}{n} + \frac{b \cdot \sum x_i}{n}$$
$$\overline{y} = a + b \cdot \overline{x}$$

This means that the point $(\overline{x}, \overline{y})$ must satisfy the equation of the line if the residuals are to sum to zero.

Conversely, if we start with the fact that the line passes through $(\overline{x}, \overline{y})$, that means the equation $\overline{y} = a + b \cdot \overline{x}$ must be true. So $a = \overline{y} - b \cdot \overline{x}$. The residual for the point (x_i, y_i) would be $y_i - (\overline{y} - b \cdot \overline{x}) - b \cdot x_i = y_i - \overline{y} + b \cdot \overline{x} - b \cdot x_i$. The sum of these residuals then is

$$\sum y_i - n \cdot \overline{y} + n \cdot b \cdot \overline{x} - b \cdot \sum x_i = n \cdot \overline{y} - n \cdot \overline{y} + n \cdot b \cdot \overline{x} - n \cdot b \cdot \overline{x} = 0$$

c. As shown in parts a and b, *any* line that passes through the point $(\overline{x}, \overline{y})$ makes the sum of the residuals equal to zero.

E29. a. The plot is shown here. This association shows a positive trend that is moderately strong. Even though a few points lie relatively far from the pattern, fat content could be used as a reasonably good predictor of calories. The equation of the regression line is $\hat{y} = 195 + 10.05x$, where $\hat{y}$ is the predicted number of calories and x is the number of grams of fat. The slope means that for every additional gram of fat, the pizzas tend to have 10.05 additional calories, on average.

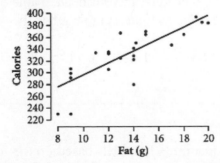

b. The plot is shown next. There is possibly a very weak positive association between *fat* and *cost*. The equation of the regression line is $\hat{y} = 10.7 + 2.41x$, where $\hat{y}$ is the predicted number of grams of fat and x is the cost. The slope means that for every additional dollar in cost, the number of grams of fat tends to increase by 2.41 grams, on average. (You will be able to check to see if this association is "real" in a subsequent chapter.)

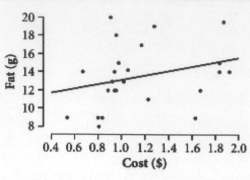

c. The plot is shown here. There appears to be no linear association between calories and cost.

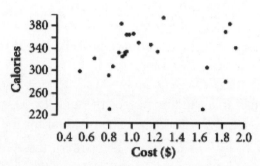

d. In the analysis of the pizza data, *calories* has a moderately strong positive association with *fat*—the two tend to rise or fall together. This makes sense because fat has a lot of calories. *Fat* has a weak positive association with *cost*. There appears to be no association between *cost* and *calories*.

E31. The correlations of the scatterplots are
    a. 0.66   **b.** 0.25   **c.** –0.06   **d.**   0.40   **e.**   0.85   **f.**   0.52
    g. 0.90   **h.** 0.74

E33. This problem isn't as much work as it looks like. Notice that for the first four tables, the means are all 0 and the standard deviations are all 1, so they have already been standardized. You can find the average product in your head. Table e has the same correlation as table a, table f has the same as table c, table g has the same as table b, and table h has the same as table c.
    a. 1   **b.**   0.87   **c.** –0.5   **d.** –1   **e.** 1   **f.** –0.5
    g. 0.5   **h.** –0.5

E35. a. About 0.95.

b. All but three of the points lie in quadrants I and III (based on the 'origin' $(\overline{x}, \overline{y})$.) In Quadrant I, z_x and z_y are positive and in Quadrant III, both , z_x and z_y are negative. In either of these quadrants the products $z_x \cdot z_y$ are positive. Thus, the correlation is positive.

c. The point in the lower left corner of Quadrant III makes the largest contribution. It is the most extreme point for both the *x* and *y* values, giving it the largest (in absolute value) *z*–score for both variables.

d. The point just below $(\overline{x}, \overline{y})$ makes the smallest contribution. z_x and z_y are both near zero, so the product of these *z*–scores will be quite small.

E37. a. No because the units will be different. For example, for the group that measures in chirps per second and uses temperature for *x*, the units of the slope will be chirps per second per degree temperature. For the group that measures in chirps per minute, the units will be chirps per minute per degree temperature. So the slope for the second group should be 60 times that of the first group. For a group that measures in chirps per minute and uses chirps for *x*, the units of the slope will be degrees temperature per chirps per minute. Even if they use the same units, groups that interchange *x* and *y* will get different slopes (chirps per minute per degrees Centigrade, or degrees Centigrade per chirps per minute).

b. Yes. The correlation is the same no matter what the units or what you use for *x* and for *y*. That is because *r* is equal to the average product of *z*–scores, which have no units and $z_x \cdot z_y = z_y \cdot z_x$.

E39. a. True. This is a direct result of the formula

$$b_1 = r \frac{s_y}{s_x}$$

If $s_x < s_y$, the factor on the right, s_y/s_x, will be greater than 1, which makes b_1 greater in absolute value than *r*.

b. The formula

$$1.6 = 0.8 \cdot \frac{s_y}{s_x}$$

can be true only if $s_x = 25$ and $s_y = 50$.

c. $b_1 = r \dfrac{s_{estimated}}{s_{actual}} \implies 0.36 = r \cdot \dfrac{4.12}{0.93} \implies r = 0.081$

d. $b_1 = r \dfrac{s_{actual}}{s_{estimated}} \implies b_1 = 0.081 \cdot \dfrac{0.93}{4.12} \implies b_1 = 0.0183$

E41. a. A large brain helps animals live smarter and therefore longer. The lurking variable is overall size of the animal. Larger animals tend to develop more slowly, from gestation to "childhood" through old age, and larger animals tend to live longer than smaller ones.

b. If we kept the price of cheeseburgers down, college would be more affordable. The lurking variable is inflation over the years—all costs have gone up over the years.

c. The Internet is good for business. The lurking variable is years. Stock prices generally go up due to inflation over the years. The Internet is new technology, and so the number of Internet sites also is increasing over the years.

E43. a.

$$r^2 = \frac{\text{SST} - \text{SSE}}{\text{SST}} = \frac{0.088740 - 0.016304}{0.088740} \approx 0.8163$$

So $r = 0.903$. The value of r^2 above is equal to the "R–sq" in the regression analysis.

b. The largest residual occurs for the point (55, 0.318). Its value is
$$y - \hat{y} = 0.318 - [0.202 + 0.00306(55)] = 0.318 - 0.3703 = -0.0523.$$

c. Yes. Hot weather causes people to want to eat something cold. But the correlation alone does not tell us that.

d. Degrees Fahrenheit, pints per person, pints per person per degree Fahrenheit, and no units.

e. "MS" was computed by dividing "SS" by "DF."

E45. The center of this cloud of points should not be modeled by a straight line. This can be considered as a plot with curvature, or as a plot with two very influential points in the upper right corner. In either case, some adjustments should be made to the data before attempting to fit a line to it. One possibility is to transform the data by techniques you will learn in Section 3.5. Because r^2 is a measure of how closely the points cluster about the regression line, it would not make sense in this context.

E47. Scoring exceptionally well, for example, on a test involves more than just studying the material. There is a certain amount of randomness involved, too—the instructor asked questions about what the student knew, the student was feeling well that day, the student was not distracted, and so on. It's unlikely that this combination of knowing the material and good luck will happen again on the next test for this same student. The student probably will get a lower, but still high, score on the next test even if he or she doesn't slack off. However, it would appear as if doing well the first time and getting praised prompted the student to relax and study less. On the second test, the student's place at the top of the class may be taken by another student who knew just as much for the first test but was also effected by randomness on the first test—perhaps unlucky in the questions the instructor chose or unlucky in another way at the time.

At the other end, a student who scores exceptionally poorly on the first test also has a bit of randomness involved—bad luck this time. Whether or not he or she is praised, the student scoring exceptionally poorly probably won't have all of the random factors go against him or her on the next test and the student's score will tend to be higher.

E49. a. The scatterplot appears here (with the regression line). A line is not a good model because the cloud of points is not elliptical with one extremely influential point.

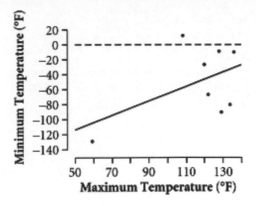

b. The regression equation is $\hat{y} = -161.90 + 0.954x$, and $r = 0.49$.

c. With Antarctica removed, the slope of the regression line changes from positive (0.954) to negative (−1.869) and the correlation becomes negative, $r = -0.45$. The plot appears as shown here. (Notice that a new potential influential point has appeared: Oceania.) Without a plot of the data, you might come to the following incorrect conclusion: In general, continents tend to be "warm" or "cold"; that is, continents with higher maximums also tend to have higher minimums. In fact, there is little relationship.

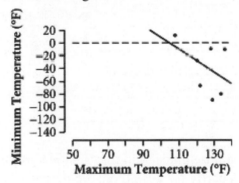

E51. a. The plot with the regression line and the regression summary are shown here. The equation of the line is $\hat{y} = -1.63 + 0.745x$.

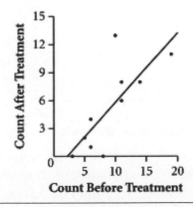

49

```
Dependent variable is:    y
No Selector
13 total cases of which 3 are missing
R squared = 58.4%   R squared (adjusted) = 53.2%
s = 3.177 with 10 - 2 = 8 degrees of freedom

             Sum of
Source       Squares    df   Mean Square   F-ratio
Regression   113.348    1    113.348       11.2
Residual     80.7516    8    10.0939

                        s.e. of
Variable   Coefficient  Coeff      t-ratio   prob
Constant   -1.63057     2.299      -0.709    0.4984
x          0.745223     0.2224     3.35      0.0101
```

Here is the completed table:

x	y	Predicted Values	Residual
11	6	6.567	−0.567
8	0	4.331	−4.331
5	2	2.096	−0.096
14	8	8.803	−0.803
19	11	12.529	−1.529
6	4	2.841	1.159
10	13	5.822	7.178
6	1	2.841	−1.841
11	8	6.567	1.433
3	0	0.605	−0.605

b. The residual plot shown here is a little unusual in that it shows more variability in the middle than at either end. But this is partly because there are more cases in the middle.

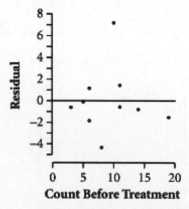

c. The disinfectant appears to be unusually effective for the person with the large negative residual, the point (8, 0) on the original scatterplot. It is seemingly ineffective for the person with the large positive residual, the point (10, 13) on the scatterplot.

50

E53. **a.** Here is what the residual plot for these data actually looks like.

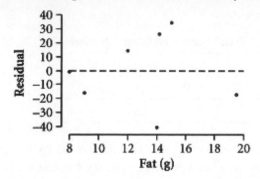

b. The largest positive residual belongs to Pizza Hut's Stuffed Crust, which has more calories than would be predicted from a simple linear model. The largest negative residual belongs to Pizza Hut's Pan Pizza. None appear so far away that it would be called an exception, or outlier. In fact, you can check this by making a boxplot of the residuals, as shown here.

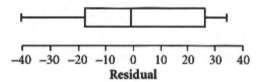

c. The complete data set shows a moderately strong positive trend with a slope of about 14.9 calories per gram of fat and correlation of 0.908. The most influential data point would be the one farthest away from the main cloud of points on the x–axis (Domino's Deep Dish). Removing Domino's Deep Dish yields a slope of around 18 calories per gram of fat and correlation of 0.893. None of the other points have nearly as much influence on the slope.

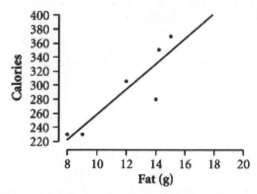

E55. **A.** I **B.** IV **C.** III **D.** II   A linear model would be appropriate for C and D. Both C and D show a random scatter of points around the residual, but the slope of the regression line is almost zero for plot C, and there appears to be no correlation. Plot B does not have a random scatter around the line; the pattern appears to be cyclical. This is typical of a situation in which something changes approximately linearly over time. What happens next usually depends on what just happened, causing this up–and–down pattern in the residuals.

E57. **a.** The regression equation is $\hat{y} = 366.67 + 16x$. The completed table is shown here.

Aircraft	Seats	Cost	Predicted	Residual
ERJ–145	50	1100	1166.67	−66.67
DC9	100	2100	1966.67	133.33
MD–90	150	2700	2766.67	−66.67

b. The plot of the residuals versus x (seats) is given first followed by the plot of the residuals versus $\hat{y}$. The only difference between the two plots is the scaling on the horizontal axis.

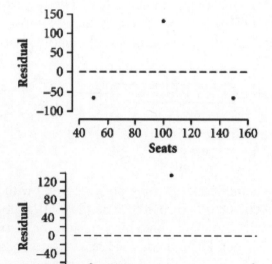

E59. Scatterplots of the original data, without and with the regression line, follow the commentary.

To estimate the recommended weight for a person whose height is 64 inches, add the fitted weight (given to be 145 pounds) to the residual of about 1.2 to get 146.2. The slope of the regression line must be about $(187 − 145)/(76 − 64) = 3.5$. For the second height, 65 inches, the fitted weight would be $145 + 3.5(1) = 148.5$. The residual is about 0.9 pounds. Thus, the recommended weight is about $148.5 + 0.9 = 149.4$ pounds. You could continue point by point to get the next plot, but a rough sketch can be obtained by making use of the linear patterns in the residuals (and hence in the original scatterplot). The points on the scatterplot must form a straight line up to a height of 71, where the weight must be about $145 + 7(3.5) − 1.5 = 168.0$.

The remainder of the points must form (approximately) another straight line up to a height of 76, where the weight must be about $145 + 12(3.5) + 2.2 = 189.2$.

52

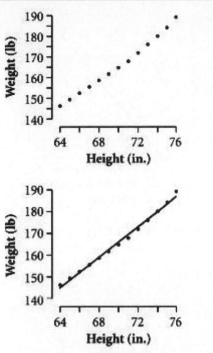

It is difficult to see the strong V shape of the residuals in a scatterplot drawn on the actual scale of the data.

E61. a. If Leonardo is correct, the data should lie near the lines:

$$arm\ span = height$$

$$kneeling = \frac{3}{4}\ height$$

$$hand = \frac{1}{9}\ height$$

Looking at the next plots, these rules appear to be approximately correct.

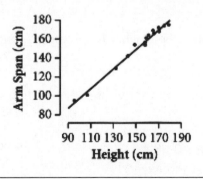 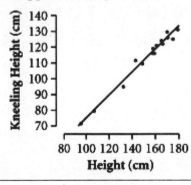

53

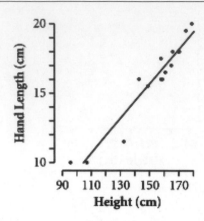

The least squares regression equation for predicting the arm span from the height is:
$$\hat{y} = -5.81 + 1.03x.$$

The least squares regression equation for predicting the kneeling height from the height is:

$$\hat{y} = 2.19 + 0.73x.$$

The least squares regression equation for predicting the hand length from the height is:
$$\hat{y} = -2.97 + 0.12x.$$

b. For the first plot, the slope is 1.03. This means for every 1 cm increase in *height,* there tends to be a 1.03 cm increase in *arm span.* Leonardo predicted a 1 cm increase.

For the second plot, the slope is 0.73, which means that for every 1 cm increase in *height,* there tends to be a 0.73 cm increase in *kneeling height.* Leonardo predicted a 0.75 cm increase.

For the third plot, the slope is 0.12, which means that for every 1 cm increase in *height,* there tends to be a 0.12 cm increase in *hand length.* Leonardo predicted a $\frac{1}{9} \approx 0.11$ cm increase.

Leonardo's claims hold reasonably well. The slopes are about what he predicted, and the *y*-intercepts are close to 0 in each case.

c. In each case, the points are packed tightly about the regression line and so there is a very strong correlation. The correlations are

arm span and *height:* 0.992 (strongest)
kneeling height and *height:* 0.989
hand length and *height*: 0.961 (weakest)

E63. **a.** Yes, the student who scored 52 on the first exam and 83 on the second lies away from the general pattern. This student scored much higher on the second exam than would have been expected. This point will be influential because the value of *x* is extreme on the low side and the point lies away from the regression line. The point sticks out on the residual plot. There is a pattern in the rest of the points; they have a positive correlation.

b. The slope should increase, and the correlation should increase. In fact, the slope increases from 0.430 to 0.540, and the correlation increases from 0.756 to 0.814.

c. The residual plot appears next. The residuals now appear scattered, without any obvious pattern, so a linear model fits the points well when point (52, 83) is removed.

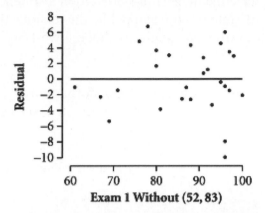

d. Yes, there is regression to the mean in any elliptical cloud of points whenever the correlation is not perfect. For example, the student who scored lowest on Exam 1 did much better on Exam 2. The highest scorer on Exam 1 was not the highest scorer on Exam 2. A line fit through this cloud of points would have slope less than 1.

E65. a. Each value should be matched with itself.

b. Match each value with itself, except match 0.5 (or –0.5) with 0: (–1.5, –1.5) (–0.5, –0.5) (0, 0), (0, 0.5), (0.5, 0), (1.5, 1.5) for a correlation of .950,

c. (–1.5, 0) (–0.5, 0.5) (0, 1.5), (0, –1.5), (0.5, –0.5), (1.5, 0) has a correlation of –.1.

d. Match the biggest with the smallest, the next biggest with the next smallest, etc.: (–1.5, 1.5) (–0.5, 0.5) (0, 0), (0, 0), (0.5, –0.5), (1.5, –1.5).

E67. a. True. Both measure how closely the points cluster about the "center" of the data. For univariate data, that center is the mean; for bivariate data, the center is the regression line.

b. True. Refer to E66 for examples.

c. False. For example, picture an elliptical cloud of points with major axis along the y-axis. The correlation will be zero, but there will be a wide variation in the values of y for any given x.

d. True. Intuitively, a positive slope means that as x increases, y tends to increase. This is equivalent to a positive correlation. Similar statements can be made for a negative slope and a zero slope. Alternatively, the fact that this statement is true can be seen from the relationship

$$b_1 = r \frac{s_y}{s_x}$$

Because the standard deviations are always positive, b_1 and r must have the same sign.

E69. a. Public universities that have the highest in-state tuition also tend to be the universities with the highest out-of-state tuition, and public universities that have the lowest in-state tuition also tend to be the universities with the lowest out-of-state tuition. This relationship is quite strong.

b. No, the correlation does not change with a linear transformation of one or both variables. However, if you were to take logarithms of the tuition costs, the correlation would change.

c. The slope would not change with the first transformation. Consider the formula for the slope:

$$b_1 = r \frac{s_y}{s_x}$$

The correlation remains unchanged with the change of units. The standard deviations would each be $\frac{1}{1000}$ as large as previously, but the factor of $\frac{1}{1000}$ would be in both the numerator and denominator and so would cancel out. But if you were to take logarithms of the tuition costs, the slope would change because the proportion s_y/s_x would be different.

E71. You can compute the values of r using the formula

$$b_1 = r \frac{s_y}{s_x}$$

Solving for r, the formula becomes

$$r = b_1 \frac{s_x}{s_y}$$

These correlations are A: 0.5; B: 0.3; C: 0.25. So from weakest to strongest, they are ordered C, B, A.

E73. For example, stocks that do the best in one quarter may not be the ones that do the best in the next quarter. As another example, the best 20 and worst 20 hitters this year in major League Baseball are not likely to repeat this kind of performance again next year.

E75. a. The next scatterplot shows a very strong linear relationship between the expenditures for police officers and the number of police officers per state. This makes sense. There is one outlier and influential point, California, which has by far the largest population of any state listed. The correlation is 0.976, and the equation of the regression line is

$$expPolice = -403 + 73.7(number\ of\ police)$$

So, for every additional thousand police officers, costs tend to go up by \$73,700,000, or \$73,700 each. If the influential point of California is removed, the slope of the line decreases but the correlation increases.

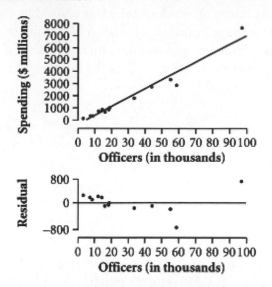

b. The scatterplot again shows a very strong linear relationship, with California as an outlier and influential point. The larger the population of the state, the more police officers. This time the correlation is 0.987, and the equation of the regression line is

$$number\ of\ police = 1.5 + 2.91\ population$$

That is, for every increase of 1 million in the population, the number of police officers tends to go up by 2910. If the influential point of California is removed, the slope of the line increases a little and the correlation decreases a little.

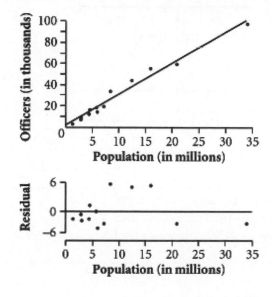

c. The scatterplot shows that there is a moderate positive but possibly curved relationship. For this scatterplot, it is not appropriate to compute the correlation or equation of the regression line. But, in general, the larger the number of police, the higher the rate of violent crime. (Note that this is the rate per 100,000 people in the state, not the number of violent crimes.) Almost equivalently, the larger the population of a state, the higher the violent crime rate. It is not at all obvious why this should be the case. Why would larger states (more police = more population) tend to have higher *rates* of violent crime? (Because of the strong relationship between the number of police officers per state and the population, a scatterplot of the crime rate versus the number of police officers per 100,000 population looks about the same.)

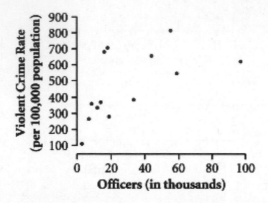

A log-log transformation of the number of police straightens these data quite well, as shown here in the scatterplot and residual plot.

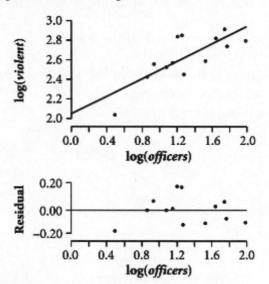

E77. **a.** The scatterplot shows that *graduation rate* and *student-teacher ratio* are not correlated: $r = 0.311$. The equation of the regression line is
$$Graduation\ Rate = -0.92\ Student\text{-}Teacher\ Ratio + 89.4$$

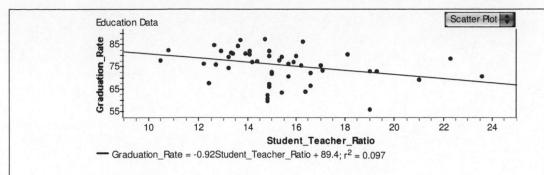

Graduation_Rate = -0.92Student_Teacher_Ratio + 89.4; $r^2 = 0.097$

b. The scatterplot shows that *graduation rate* and *PPE* are not correlated: $r = 0.275$. The equation of the regression line is

$$\text{Graduation Rate} = 0.00101 \text{ PPE} + 65.4$$

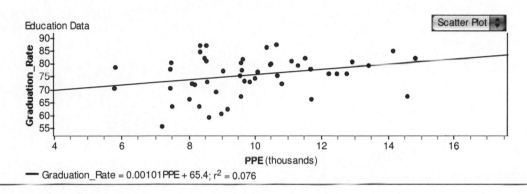

Graduation_Rate = 0.00101PPE + 65.4; $r^2 = 0.076$

Chapter 4

Practice Problem Solutions

P1. a. The population is the collection of all the households in your community; the units are the individual households.
b. Sampling is quicker, cheaper, and easier than conducting a census.

P2. This is a two-stage cluster sample, not a simple random sample. Any sample with more than 10 students from a single dorm, for example, cannot occur by this process.

P3. a. No
b. No; two students with last names starting with different letters cannot both be in the sample.
c. No; two students sitting in different rows cannot both be in the sample.
d. Yes
e. No; a group of six girls cannot all be in the sample.

P4. Design c is not valid because the groups proposed as strata are not mutually exclusive.

P5. a. Compute the respective percentages of 1200: 516; 384; 300

b. Age since 40 seems to be an established cutoff point for likelihood of getting receding gums.

P6. $84(0.65) + 69(0.35) \approx 0.79$

P7. a. Choose a random start between persons 1 and 20, and take every 20th person.
b. Choose a random start between persons 1 and 5, and take every 5th person.

P8. Choose a random start between 1 and 17, and take every 17th person.

P9. a. Pick a random sample of zip codes. The sample consists of all households on the carrier routes along in each zip code in the sample.

b. Start as in part b, then take a random sample of households within each "zip-code cluster" – maybe certain streets within the carrier route within each zip code.

P10. a. Use pages as clusters. You can concentrate on only a small number of pages in the book, but some sections of the text may contains more capitalized words than other parts (for instance, an author index). A sample containing such a non-representative page would through off the results.

b. Take an SRS of pages. Then take an SRS of lines from each of those pages.

c. Take an SRS of characters from each line selected in part b.

P11. **a.** Al. The shots are closely spaced but off center.
b. Cal. The shots are widely spaced and off center.
c. Bal. The shots are closely spaced and cluster around the center.
d. Dal. The shots are widely spaced but cluster around the center.
e. Al. His shots are consistent. You just need to get him to adjust his direction. That would be easier than getting one of the others to be more consistent.

P12. **a.** Size bias. Larger farms are more likely to be selected than smaller farms.

b. This is a judgment sample. In his attempt to get a diverse sample, the professor may have inadvertently selected too few valedictorians from the most typical high schools. He may also have missed some important groups. The students may also see this as a convenience sample because the professor only used high schools in Illinois.

c. Voluntary response bias. Teachers whose students did well are more likely to report. Overall, the teachers from the AP discussion list reported that 426 out of 535, or 80%, passed (got a 3 or better). When the official results were released, the pass rate was 62.1%.

d. Size bias. Longer strings are more likely to be selected.

e. Voluntary response bias. Responders tend to have stronger opinions than nonresponders. Even though Ann Landers received a very large number of responses, a large sample offers no guarantees about bias. Voluntary response samples are so likely to be biased that you should not trust them.

P13. The estimate will be too high. The sample will catch almost all those who came to Boston to see the museum, but only a tiny fraction of those who came to Boston for other reasons because those in the latter group are not likely to go to the museum at all. This is an example of a convenience sample.

P14. Question I is more likely to draw "yes" responses, which would show support for some control of gun ownership. Question II is more likely to draw "no" responses, which would indicate a lack of support for controls.

P15. This is incorrect response bias where people gave an answer that they thought made them look knowledgeable. They probably considered Princeton's overall reputation and assumed that any program there must be strong.

P16. **a.** The subjects are the children; the response variable is having/not having leukemia; and the treatments are living near to or not living near to a major power line.

b. No, it's merely observational. Samples of children are taken from a collection of areas, some near to a major power line and some not, and they are assessed as to whether or not they have leukemia.

c. It is very difficult to prove a causal relationship. They can list many other factors that could contribute to the children developing leukemia.

P17. **a.** Number the twelve test sites and randomly select four of them for each type of glass. Collect the measures for the amount of energy lost at each site and compare the results from each of the three types of glass.

b. The subjects, or experimental units, are the twelve site-generator combinations. The treatments are the three types of glass. Each treatment is replicated four times.

c. If each site had six generators, each treatment would be randomly assigned two generators at each site. The experimental units would then be the 72 site-generator combinations, but you are "blocking" on sites so that each treatment gets studied fairly on each site. You will learn about blocking in the next section.

P18. **a.** The lurking variable is the person's age. Older people are more likely to do both.

b. I causes II. Experiments have shown that drinking more milk, which is high in calcium, results in stronger bones. Of course, there are other causes of strong bones, such as heredity and exercise.

c. II causes I. People who go to school longer tend to earn more money. However, this statement is based on observational studies, not on randomized experiments, so occasionally the cause and effect relationship is questioned. Some people believe that family background is a lurking variable: People from more well-off families tend to go to school longer and to earn more money, but it isn't the schooling that caused the higher income, it was their family's support in getting established. However, most people believe that although family background certainly contributes, the number of years of schooling is another cause of higher income.

P19.

		Motivation	
		High	**Low**
Took Course?	**Yes**	Higher SAT	No evidence
	No	No evidence	Lower SAT

P20. The variable "time spent on the activity" could be confounded by the use of activities other than those assigned, in the study or outside of it.

P21. **a.** The factors are type of lighting (brightness of the room) and type of music. For the type of lighting, the levels are low, medium, and high. For the type of music, the levels are pop, classical, and jazz.

b. Answers will vary. Possibilities are heart rate, blood pressure, or self-description of anxiety level.

P22. **a.** Students should not believe this. Their explanations will vary but the actual explanation is in P23. The pipe- and cigar-smokers are a bit older on average.

b. Observational study

c. The factor is smoking behavior. The levels are nonsmoking, cigarette smoking, and pipe or cigar smoking. The response variable is the number of deaths per 1000 men per year.

P23. Older men have a higher death rate; the pipe- and cigar-smokers are older than the others. Here are the age-adjusted death rates per 1000 men: nonsmokers 20.3, cigarette-smokers 28.3, and pipe- and cigar-smokers 21.2. The new factor is the age. The discussion loosely follows Paul R. Rosenbaum's *Observational Studies* (New York: Springer-Verlag, 1995, page 60). (These data come from William G. Cochran, "The Effectiveness of Adjustment by Subclassification in Removing Bias in Observational Studies," *Biometrics* 24 (1968): 205–213. The original study was by Best and Walker.)

P24. **a.** This study is observational.

b. The factor is the legal age for driving. The levels are the age groups. The response variable is the highway death rate by state.

c. The "treatments," legal driving age, may be confounded with driver education, as states with higher age limits may generally be more restrictive about getting a driver's license and so also require driver's education courses. Also, if it is the case that the legal driving age tends to be lower in western states, the geography of the states would be confounded with age because western states with many miles of high-speed highways tend to have higher death rates than congested eastern states.

P25. **a.** The response variable is death or survival; the explanatory variable is low birth weight.

b. This is an observational because the data had already been obtained prior to the study and are simply being analyzed.

c. Low birth weight appears to be strongly associated with neonatal death.
d. The following could be confounded with birth weight: other diseases or physical impairments of the baby; health of the mother; difficulty of delivery.

e. Within each group, the percentages are nearly equal and most of the deaths come from the premature births; thus, it is unclear if the deaths are mainly due to low birth weight, or premature birth (or some other cause).

P26. For his control group, Dr. Mayo used adults who died of nonrespiratory causes. He was right to use a control group, but he failed to isolate the effect of interest because when it comes to thymus size, adults and infants are not comparable. If Mayo had used infants who died of nonrespiratory causes as his control group, he would have seen that they all had large thymus glands and would not have concluded that the thymus was responsible for the breathing problems. The placebo effect could have been a factor because Dr. Mayo was a trusted physician. It was not possible for this study to be blind or double-blind because it was obvious to everyone which infants had had surgery.

P27. **a.** There are quite a few problems with this study. It is possible that students were pointing home not from some magnetic sense, but because they could feel how much "turning" they had done or because they knew the roads. It is also possible that they could tell direction because they knew the direction of the sun even though they were blindfolded.

b. The group that has the magnets.

c. Whether the students were assigned randomly to the treatments. Students may also mention that the study was not double-blind.

d. Answers will vary, but the second design is certainly better than the first.

e. Because the subjects could tell whether or not they had the magnet because of the magnet's attraction to the walls of the van, this study was not blind. It was not double-blind either because the experimenter who evaluated how well the students pointed home was the same person who chose which kind of bar they wore.

P28. treatments: the two text books   units: the 10 classes
sample size: 10, with 5 for each treatment.

P29. The experimental units are carnation plants, and the ones used for the experiment should be as nearly alike as possible. Randomly assign the new product to about half of a number of plants, and leave the others growing under standard conditions. There must be at least two plants in each group, and preferably many more. The plants receiving the standard treatment can be used as a control; the goal is to see whether the new product is better than the standard treatment at producing large blooms. Putting the plants in their places before randomly assigning treatments helps ensure that the two treatments aren't confounded with variables such as the amounts of heat, light, water, and temperature the various plants receive.

P30. **a.** The first factor is the brand of paper towel, with levels Brand A and Brand B. The second factor is wetness, with levels dry and wet. The response variable is the number of pennies it can hold before breaking.

b. Randomize the assignment of wet or dry to each set of ten towels (five from each brand to each treatment) and the order of testing. There are four treatments: Brand A wet, Brand A dry, Brand B wet, Brand B dry. Write each treatment name on five slips of paper, then draw to determine the order. The experimental units are the 20 time-towel combinations in which the tests will be performed.

c. This is an experiment, although the results may not generalize beyond the towels tested unless it can be established that they are essentially the same as all towels of these brands.

P31. **a.** Students should list reasons why dorms that get the same soap still vary in the number of visits to the infirmary. Some possible reasons include the following: Some dorms have a less healthy atmosphere than others (e.g., they are too stuffy or contain contaminated drinking fountains), some dorms may be for athletes or others that tend to be healthier on average, or some dorms may have students who just happen to be sicker more often than usual.

b. To equalize variation between the treatments.

P32. **a.** randomized paired comparison with repeated measures

b. randomized paired comparison with matched pairs

c. The choice to use randomized paired comparison with matched pairs was likely made because there may be a residual effect of one or both of the drugs—that is, it does not clear out of the bloodstream in the time allowed between treatments.

P33. The most striking difference would likely occur in the factor "whether the person can roller skate." One should block on the answer to this question so that prior experience with skating doesn't confound the results.

P34. a. There is a lot of variability in how well students memorize, so some sort of blocking seems desirable. Your students may suggest a randomized paired comparison design with repeated measures. First, find the seniors in the school who are willing to participate. Then randomly select half to go into a room that has the radio playing and the other half to go into a room that is quiet. The students would be given a familiar task to study, such as a list of new vocabulary words. Afterward, they would be tested on the meaning of the words. The number that they remember is the response variable. Then, the students would switch rooms and be given another new list of vocabulary words. If your students feel that this would take too much time, they might suggest matching students by GPA or, if that is impossible, a completely randomized design.

b. For this experiment, the treatments are soup with MSG and soup without MSG. The response variable is the amount of soup eaten by a customer who orders soup. Even though there is a lot of variation in how much soup people will eat, blocking seems difficult to do. (However, it might be possible to block by estimating a person's weight.) In this case a completely randomized design seems best. For each customer who orders soup that night, a treatment is randomly selected, perhaps by flipping a coin each time.

P35. a. The question is difficult to answer from the plots. The plots show that the number of eyelid flicks per minute cover about the same range with both treatments, and there is no way to tell which dots represent measures from the same patient.

b. Here it is clear that the average number of flicks per minute is higher for Drug A, indicating that Drug A works better than Drug C for reducing drowsiness. If they were equal, the points would lie on the line.

c. You could subtract the number of flickers per minute for Drug C from that for Drug A. If the differences are generally greater than zero, Drug A is more effective at reducing drowsiness.

d. This is a randomized paired comparison with repeated measures. By using the same unit for each treatment, the within-treatment variability is reduced, allowing the between treatment variability to become apparent.

P36. a. Green bears went farther. The display below shows the average distances for each team for both red bears and green bears, using the data in the student text. On the surface, it looks as though color has a real effect on launch distance.

Team	Red	Green	Difference
1	26.2	32.4	6.2
2	54.4	89.8	35.4
3	31.8	60.2	28.4
4	30.0	34.8	4.8
5	31.2	41.2	10.0
6	102.6	115.4	12.8

b. The greater distances traveled by the green bears is due to confounding of launch order with bear color. As students got more practice, they were able to launch the bear farther. This table shows how the variables are confounded.

		Practice Doing Launches	
		Less	More
Color of Bear	Red	Shorter distances	No evidence
	Green	No evidence	Longer distances

c. The plots show a clear time trend: Later launches tend to travel farther, because most people get better with practice. (If you do this experiment and your class gets results that do not show a time trend, it is worth spending a few minutes trying to figure out why. For example, several practice launches before starting to record data could minimize confounding of the launch order with bear color.)

d. Randomize the order in which the bears of different colors are launched. For example, put five bears of each color in a box and mix them up. For each launch, draw out a bear without looking. Don't return that bear to the box. A more risky strategy is to use practice launches until the learning effect appears to go away.

Exercise Solutions

E1. a. This is a stratified random sample with strata of grocery store owners and restaurant owners.

b. Perhaps a systematic sample of customers coming into the businesses would work.

E3. Stratification by gender is likely to be the best strategy. This could potentially avoid cost bias in the sense that females are more likely to spend a higher amount on hair cuts than males.

E5. Consider the farms as the five strata, and take a random sample of, say, 10 acres from each farm.

E7. Systematic sampling with random start in both cases; in the second case this leads to stratified sampling.

E9. a. Too high. Your polltakers will never get in touch with people who are away from home between 9:00 a.m. and 5:00 p.m., so these people will eventually be dropped from the sample. Because these are the people who are less likely to have children under the age of 5 at home (those with small children are more likely to stay home to take care of them), the sample will contain an artificially high percentage of those who do have children under the age of 5 at home.

b. At dinnertime, adults are more likely to be at home.

c. This is a case of sampling bias. The problem is with the design of the survey: calling between 9 a.m. and 5 p.m. The people not at home did not get the opportunity to respond.

E11. a. A lower percentage of abstainers responded to the mailed survey, perhaps due to lack of interest; among the drinkers, a higher percentage were found to be frequent excessive drinkers in the follow-up, perhaps more accurately brought out by the interview.

b. Yes. This is not surprising because excessive drinkers may be reticent to respond to such a survey for fear of ramifications if the results were to ever be non-anonymous.

E13. a. Among adults aged 18 to 24, 39.1% are not at risk.

b. The young and the old have the higher percentages at risk, due perhaps to risky life styles among the young and illness among the old.

c. It is self-reported data on sensitive issues.

E15. This is a convenience sample, resulting in sample selection bias. Students who take a statistics course are more likely to like math than students in general, unless it is a required class. The estimated response will tend to be too high in the first case and about right in the second case.

E17. From respondents 40 years old or older you would expect to get an estimate that is too high. 40-year-olds are older than average. As people get older, they tend to visit more and more states. (They can't visit fewer!) From the residents of Rhode Island you might get an estimate that is too high as well, as Rhode Island is a small state close to many other small states. Compare this result to what you might expect from asking people who live in Texas or Montana.

E19. a. A convenience sample seems likely here.

b. No. The results of the survey are a bit suspect because they were obtained using self-reported data in a survey on guitars conducted by a guitar manufacturer.

c. Less than 67% because those interested in learning to play the guitar may have no interest at all in video games. In fact, it is arguable that those serious about playing an instrument would not tend toward the video game version; rather, those who are less talented may be more apt to pursue the video game version as a compromise measure.

E21. The estimate will be too high. There are two reasons. First, families with no children have no chance of being in the sample. Second, families with many children will be over-represented. In Activity 4.1a, a person who stays 5 days is five times as likely to be chosen as a person who stays 1 day. Similarly, families with 5 children would be five times more likely to be represented in the sample than families with only 1 child.

E23. From the Gallup poll, 36% were in favor of U.S. air strikes. From the ABC poll, 65% were in favor. Note that the ABC News poll mentions the involvement of allies of the United States and also makes it sound as though the air strikes will hit only objects and not soldiers. Students might reasonably think that the Gallup poll will produce a higher favorable response, because it mentions why the United States would want to conduct air strikes, i.e., it justifies air strikes. It can be hard to predict in advance which direction the psychology of questionnaire bias will go.

E25. There is no right answer here, but some answers are better than others. A good way to potentially balance the characteristics is to take a random sample. This method may not perfectly balance the characteristics, but has a better chance of producing something close to a balanced sample than most other methods.

E27. Observational study. Grade level is the 'treatment' and this cannot be assigned randomly to students.

E29.   **a.** Randomization: yes, five plants are selected at random from the ten.

Replication: yes, each treatment is applied to five plants
Control: yes, five plants are replanted in their normal environment.

b. All ten needed to be dug up so that the variables *shade* and *being dug up* are not confounded.

E31.   **a.** dormitories   **b.** 20   **c.** experimental

E33.   **a.** Factors are feeding method, education program, and hospital stay, each at two levels; all possible combinations lead to 8 treatments

b. Researchers may have randomized treatments as follows: List the n mothers who volunteer to participate in the study; form a box of n chips of 8 different colors (one for each treatment) with the numbers of chips of each color being as equal as possible; draw a chip at random for each mother.

c. No; one cannot assign birth weight to subjects.

E35.   A reason for using death *rate* instead of total number of deaths is that the sizes of the populations are different.

There is a possibility that climate is confounded with age. Florida has a much higher proportion of older people than does Alaska because Florida is a state where a great number of people move after retiring.
This is an observational study.

E37.   No, not unless the subjects are in random order to begin with. Suppose the subjects in poorest health are all at the end of the list. Then, all of those in poorest health probably are going to be assigned to the same treatment group, a serious confounding effect.

E39.   The modified SAT study is a completely randomized experiment, where students are randomly assigned to two groups, one group for each treatment.

Experimental Units: high school students who want to take the special course
Factor and levels: the only factor is course-taking behavior, and the levels are the same
        as the treatments
Response: SAT score
Blocks: none
Treatments: course, or no course (control)

E41. This is a randomized paired comparison (repeated measures) design.

Block: patient (more precisely, two weeks of a patient's time)
Treatments: low phenylalanine diet or regular diet
Experimental Unit: one week of a patient's time
Design: the design is a randomized paired comparison design with repeated measures
Response: dopamine level

E43. Treatments were assigned to units using a block design. In particular, the design was a randomized block with repeated measures on the same subject.

Response: rate of finger tapping
Unit: one day of a subject's time
Treatments: caffeine, theobromine, or placebo
Block: subject (more precisely, three days of a subject's time)

E45. The treatments were assigned completely at random in this completely randomized design.

Response: quantity eaten (relative to body weight)
Unit: hornworm
Treatments: diet of regular food or diet of 80% cellulose
Blocks: none

E47. a. Sample. The measurement process is destructive, and the size of the population makes a census too expensive.

b. Census. The population size is small, the information is easy to get, and the measurement process is not destructive.

c. Sample. The population is so large that a census would be too costly and time-consuming.

E49. a. The class is likely to be reasonably representative.

b. The class will not be representative: Its average age will be much lower than that of the population.

c. The class is likely to be reasonably representative.

d. Students who choose to take a statistics course as an elective are more likely to prefer science to English than students who do not take statistics. So in this case, it would not be reasonably representative. If your classmates took statistics because it is required for a variety of majors, they could be reasonably representative.

e. Blood pressure increases with age. The average blood pressure for the class will almost surely be much lower than for the population.

E51. Yes; actors and actresses listed in the almanac are better known, and so better paid than those who are not listed.

E53. The frame will not include most owners who bought their 2008 models used from either a private party or from dealers of other makes. It will include people who have sold their 2008 model. These owners may be the ones with the higher repair bills.

E55. Your net allows tiny fish to escape. Dragging it behind a boat will miss fish that are far below the surface.

E57. *New York Times* readers tend to have higher incomes and more years of education in comparison with the average New Yorker.

E59. Stage 1: Choose an SRS of states.
Stage 2: Choose an SRS of congressional districts from each state chosen in stage 1.
Stage 3: Choose an SRS of precincts from each congressional district chosen in stage 2.
Stage 4: Choose an SRS of voters from each precinct chosen in stage 3.

E61. a. This is an observational study. Although subjects were selected at random, they weren't randomly assigned to a diet. Thus, this isn't an experiment.

b. Almost all of the people who follow a Mediterranean diet will be Greek and almost all of the people who follow a typical American diet will be American. Thus it will be impossible for the researcher to tell whether it was the diet or some other difference in American and Greek lifestyles that accounts for the difference in percentage of heart attacks. If it is some other difference in lifestyle, then a young American who switches to a Mediterranean Diet may not decrease his or her chance of getting a heart attack.

c. Lifestyle differences between the two countries may include more physical activity and less stress in Greece and so these factors are confounded with diet.
The American Heart Association page on "Mediterranean Diet" says
> "The incidence of heart disease in Mediterranean countries is lower than in the United States. Death rates are lower, too. But this may not be entirely due to the diet. Lifestyle factors (such as more physical activity and extended social support systems) may also play a part. Before advising people to follow a Mediterranean diet, we need more studies to find out whether the diet itself or other lifestyle factors account for the lower deaths from heart disease."

E63. a. This is an experiment as subjects were randomly assigned to treatments, there are a sufficient number of subjects, and there are four different treatments. The factors are presence of mother, with levels present and absent, and presence of siblings, with levels present and absent. The response is the difference in the mouse's weight from the initial weight to the weight 90 days later. The number of units is the total number of baby mice that were divided into the four groups.

b. This is a survey, with the population being all of the different possible 10 ft. by 10 ft.

squares on the hill. The explanatory variables are the amounts of the ten different nutrients. The response variable is the number of species of plants. The sample size is 4.

c. This is an observational study as there is no random selection of types of fruit or of particular fruit within a type. The factors are the different types of fruit. (There are no levels.) The response variable is float or not. The observational unit is a fruit and there is one observed unit for each type of fruit.

E65. a. The factor is location, with level being urban or rural. A block is one pair of twins.

b. This is an observational study. The treatment is already built into the units.

c. As the dotplots below show, the variation in response is greatest in rural twins and least in the differences. This study does demonstrate that the lungs of the urban twins took longer to clear than did the lungs of the rural twins.

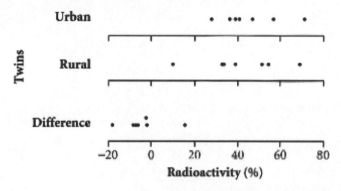

d. Twins were used because they are genetically identical. This will reduce any variability caused by genetic differences.

E67. You should include the following points in their design: a description of the treatments, a rationale for any blocking, a description of the blocking scheme, a description of the random assignment of treatments to units *within* blocks, the protocol for the experiment such as making it double-blind if possible, and the response variable to be measured and compared. (Have the students keep their notes so they can refer back to this example after Chapter 9.)

Sample response: The two treatments will be the two types of exercise bike. I will assign four bikes of each type to the eight units or spaces on the floor of the gym. Bikes nearer the door and bikes in front of the televisions may be more (or less) likely to be used than the other bikes. Thus, I will form four blocks of spaces with similar locations in the gym: 1 and 4, 2 and 3, 5 and 8, 6 and 7. Then *for each block,* I will flip a coin, and if it is heads, a bike of the first type goes into the even-numbered space and a bike of the second type into the odd-numbered space. If it is tails, a bike of the second type goes into the even-numbered space and a bike of the first type goes into the odd-numbered space. There is no reason to make this experiment blind or double-blind because we want the customers to observe the different choices and the number of hours of use is recorded automatically for each bike.

Chapter 5

Practice Problem Solutions

P1. **a.** Out of the 27 days listed on which the National Weather Service forecast a low temperature of 30°F, 13 of them actually had a low temperature of approximately 30°. Based on this, the best estimate of the probability would be 13/27, or about 0.48. Also, for 25°F, the best estimate of the probability would be 8/27, or about 0.30.

b. There appears to be a "cold" bias. When the prediction was incorrect the actual temperature was colder than predicted 10 times compared to only four times the temperature was warmer than predicted.

P2. **a.** There are 32 possible outcomes:

5 Heads	4 Heads	3 Heads	2 Heads	1 Head	0 Heads
HHHHH	HHHHT	HHHTT	HHTTT	HTTTT	TTTTT
	HHHTH	HHTHT	HTHTT	THTTT	
	HHTHH	HTHHT	THHTT	TTHTT	
	HTHHH	THHHT	HTTHT	TTTHT	
	THHHH	HHTTH	THTHT	TTTTH	
		HTHTH	TTHHT		
		THHTH	HTTTH		
		HTTHH	THTTH		
		TIITIHI	TTHTH		
		TTHHH	TTTHH		

b. Yes, all outcomes are equally likely.

c. For these, simply count the number of outcomes satisfying each condition and divide by 32. In the order presented, these probabilities are: 1/32, 5/32, 10/32, 10/32, 5/32, 1/32

P3. **a.** One way to do this is shown here:
H1, H2, H3, H4, H5, H6, T1, T2, T3, T4, T5, T6
b. Answers will vary. For the sample space shown in part A the outcomes are equally likely.
c. 1/12

P4. **a.** 28, 35   28, 39   28, 47
     28, 55   35, 39   35, 47
     35, 55   39, 47   39, 55
     47, 55
b. Yes.
c. 1/10
d. 6/10

P5. a. Label each person using the first letter of their first name. The sample space consists of the outcomes AB, AC, AD, BC, BD, CD.

b. Count all outcomes that contain an A, and divide by 6 to conclude that the probability is 3/6, or ½.

P6. a. Yes, you can list a sample space and it would look similar to the 32 outcomes in P1, with T representing being right-handed and B representing being left-handed.

b. You can not determine the probability without additional information about the percentage of students in the school who are right-handed.

P7. a. Since the proportion that were heads after the first two spins was zero, the first and second spins must have been tails. The third was heads because the proportion that were heads went up. The fourth was tails and the 50th was tails.
b. About 0.44.

P8. a.

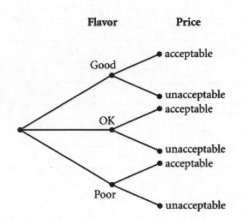

b. Six.
c. It is impossible to tell the probabilities without knowing the quality of the ice cream and price, but it is unlikely that the outcomes are equally likely.

P9. a. There are 3 · 7, or 21, different pairs of "levels of exercise" and "different medications" that you could end up with.

b. There is only one element of the sample space that corresponds to this event, so its probability is $\frac{1}{21}$.

c. Label the medications as "a, b, c, d, e, f, g" and the levels of exercises as "A, B, C". The possible outcomes, in array form, are:

	A	B	C
a	aA	aB	aC
b	bA	bB	bC
c	cA	cB	cC
d	dA	dB	dC
e	eA	eB	eC
f	fA	fB	fC
g	gA	gB	gC

d. Using the same labeling scheme as in (c), the possible outcomes, in tree form, are:

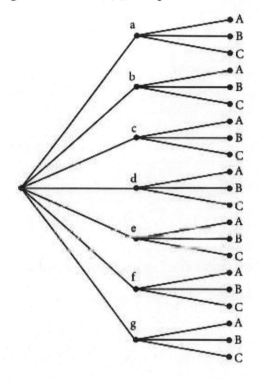

P10. **a.** $\frac{32.3+4.9}{298.6} \approx 0.125$

b. i. These are not disjoint since there are females 85 or older in the sample.

ii. These are disjoint since one cannot be ≤ 17 and ≥ 85 years of age simultaneously.

iii. These are not disjoint since a person in the "≥ 85 years of age" category is also in the "≥ 65 years of age" category.

P11. **a.** Use the information in the rightmost column since gender is not specified:
$$\frac{6841+3958}{17,231} \approx 0.627$$

b. Since the gender specified is male, use the leftmost column:
$$\frac{3171+1782}{7505} \approx 0.660$$

c. Since the gender specified is female, use the rightmost column:
$$\frac{3670+2176}{9726} \approx 0.60$$

P12. **a.** No **b.** Yes **c.** Yes **d.** No

P13. **a.** Yes, because selecting a junior and selecting a senior are disjoint events.
b. 33% + 27% = 60%

P14. **a.**

		Second Roll			
		1	**2**	**3**	**4**
First Roll	**1**	1, 1	1, 2	1, 3	1, 4
	2	2, 1	2, 2	2, 3	2, 4
	3	3, 1	3, 2	3, 3	3, 4
	4	4, 1	4, 2	4, 3	4, 4

b. P(6 or 7) = P(6) + P(7) = 3/16 + 2/16 = 5/16.
c. P(doubles or 7) = P(doubles) + P(7) = 4/16 + 2/16 = 6/16
d. Because these events are not disjoint. The outcome 3,3 is both a double and a 6.

P15. **a.** No. If they were mutually exclusive the cell in the column for alcohol involved and the row for speed related would be 0.

b. $\frac{10,928}{231,459} + \frac{87,086}{231,459} - \frac{4,436}{231,459} \approx 0.404$

P16. **a.**
$$P(doubles \ or \ sum \ of \ 4) = P(doubles) + P(sum \ of \ 4) - P(doubles \ and \ sum \ of \ 4)$$
$$= \frac{6}{36} + \frac{3}{36} - \frac{1}{36}$$
$$= \frac{8}{36}$$

b.
$$P(doubles \ or \ sum \ of \ 7) = P(doubles) + P(sum \ of \ 7) - P(doubles \ and \ sum \ of \ 7)$$
$$= \frac{6}{36} + \frac{6}{36} - \frac{0}{36} = \frac{12}{36}$$

c.
$$P(5 \ on \ first \ die \ or \ 5 \ on \ second \ die) = P(5 \ on \ first \ die) + P(5 \ on \ second \ die)$$
$$- P(5 \ on \ first \ die \ and \ 5 \ on \ second \ die)$$
$$= \frac{6}{36} + \frac{6}{36} - \frac{1}{36} = \frac{11}{36}$$

P17. $P(2 \ heads) = P(heads \ on \ first) + P(heads \ on \ second) - P(heads \ on \ both)$
$$= \frac{1}{2} + \frac{1}{2} - \frac{1}{4} = \frac{3}{4}$$

P18. $P(not \ doubles \ or \ sum \ of \ 8) = P(not \ doubles) + P(sum \ of \ 8)$
$$- \ P(not \ doubles \ and \ sum \ of \ 8)$$
$$= \frac{30}{36} + \frac{5}{36} - \frac{4}{36} = \frac{31}{36}$$

P19. **a.** $P(F) = P(\text{the person was female}) = \frac{470}{2201}$

b. $P(F \mid S) = P(\text{the person was female given that the person survived}) = \frac{344}{711}$

c. $P(\text{not } F) = P(\text{the person wasn't female}) = \frac{1731}{2201}$

d. $P(\text{not } F \mid S) = P(\text{the person wasn't female given that the person survived}) = \frac{367}{711}$

e. $P(S \mid \text{not } F) = P(\text{the person survived given that the person was not female}) = \frac{367}{1731}$

P20. **a.** $P(\text{worker is paid at or below minimum wage}) = \frac{1675}{73,756} \approx 0.023$

b. $P(\text{worker is paid at or below minimum wage} \mid \text{worker is white}) = \frac{1420}{61,061} \approx 0.023$

c. White workers have approximately the same likelihood of being paid at or below minimum wage as the population in general.

d. Find $P(\text{worker is black}) = \frac{9965}{73,756} \approx 0.135$

e. Find $P(\text{worker is black} \mid \text{worker is paid at or below minimum wage}) = \frac{205}{1675} \approx 0.122$

f. A worker that is paid at or below minimum wage is less likely to be black than a worker selected at random from all hourly workers.

P21. **a.** $P(\text{2nd draw is red} \mid \text{1st draw is red}) = \frac{2}{4} = \frac{1}{2}$

b. $P(\text{2nd draw is red} \mid \text{1st draw is blue}) = \frac{3}{4}$

c. $P(\text{3rd is blue} \mid \text{1st is red and 2nd is blue}) = \frac{1}{3}$

d. $P(\text{3rd is red} \mid \text{1st is red and 2nd is red}) = \frac{1}{3}$

P22. **a.** $P(club \mid black) = \frac{\text{number of black clubs}}{\text{total number of black cards}} = \frac{13}{26} = \frac{1}{2}$

b. $P(jack \mid heart) = \frac{\text{number of jack of hearts}}{\text{total number of black cards}} = \frac{1}{13}$

c. $P(heart \mid jack) = \frac{\text{number of jacks of hearts}}{\text{total number of jacks}} = \frac{1}{4}$

P23. **a.**

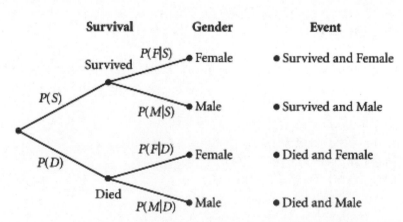

b. i. $P(S) = \frac{711}{2201}$ ii. $P(F \mid S) = \frac{344}{711}$ iii. $P(S \text{ and } F) = \frac{344}{2201}$

c. $P(S) \cdot P(F \mid S) = P(S \text{ and } F)$. This formula matches the computation in the text.

d. $P(M \text{ and } S) = P(M) \cdot P(S \mid M)$ and $P(M \text{ and } S) = P(S) \cdot P(M \mid S)$

P24. Without replacement, $P(HH) = \frac{13}{52} \cdot \frac{12}{51} = \frac{1}{17}$

With replacement, $P(HH) = \frac{13}{52} \cdot \frac{13}{52} = \frac{1}{16}$

P25. $P(W\ chosen\ 1st) = \frac{2}{4}$

$P(W\ chosen\ 2nd \mid W\ chosen\ 1st) = \frac{1}{3}$

$P(WW) = P(W\ chosen\ 1st) \cdot P(W\ chosen\ 2nd \mid W\ chosen\ 1st) = \frac{2}{4} \cdot \frac{1}{3} = \frac{1}{6}$

P26. $P(sum\ of\ eight\ and\ doubles) = P(doubles) \cdot P(sum\ of\ eight \mid doubles) = \frac{6}{36} \cdot \frac{1}{6} = \frac{1}{36}$

Note that this is not $P(sum\ of\ eight) \cdot P(doubles)$, which is $\frac{5}{36} \cdot \frac{6}{36} = \frac{5}{216}$.

P27. $P(doubles \mid sum\ of\ 8) = P(doubles\ and\ sum\ of\ 8)\ /\ P(sum\ of\ 8) = \dfrac{\frac{1}{36}}{\frac{5}{36}} = \dfrac{1}{5}$.

$P(sum\ of\ 8 \mid doubles) = P(doubles\ and\ sum\ of\ 8)\ /\ P(doubles) = \dfrac{\frac{1}{36}}{\frac{6}{36}} = \dfrac{1}{6}$.

These probabilities are not the same.

P28.

	Right-Handed	Left-Handed	Total
Blue Eyes	8	2	10
Brown Eyes	16	4	20
Total	24	6	30

$$P(right-handed \mid brown\ eyes) = \frac{P(right-handed\ \ and\ \ brown\ eyes)}{P(brown\ eyes)} = \frac{^{16}/_{30}}{^{20}/_{30}} = \frac{16}{20}$$

P29. You can interpret the two percentages given by Gallup as the approximate conditional probabilities that a randomly selected Republican woman, W, or Republican man, M, respectively, would have voted for Dole, D. That is,

$$P(D \mid W) = 0.16 \qquad and \qquad P(D \mid M) = 0.07$$

You can write the probability of a Republican voting for Dole, $P(D)$ in terms of two disjoint events as follows:

$$
\begin{aligned}
P(D) &= P[(D\ and\ W)\ or\ (D\ and\ M)] \\
&= P(D\ and\ W) + P(D\ and\ M) \\
&= P(W) \cdot P(D \mid W) + P(M) \cdot P(D \mid M) \\
&= 0.40 \cdot 0.16 + 0.60 \cdot 0.07 = 0.106
\end{aligned}
$$

P30. a.

		Test Result		
		Positive	Negative	Total
Disease?	**Yes**	2,985	$0.005 \cdot 3,000$ $= 15$	3,000
	No	$0.06 \cdot 97,000 =$ 5,820	91,180	97,000
	Total	8,805	91,195	100,000

b. $\frac{8,805}{100,000} = 0.088$

c. $\frac{0.0582}{0.088} \approx 0.66$ (Note that this result is the reason why many people do not believe that there should be universal, mandatory testing for the HIV virus, prostate cancer, and other such diseases. When the incidence of the disease is relatively low in the population, the number of false positives can be much larger than the number of cases of the disease.)

P31. a.

PPV = P(Fluid present | ultrasound positive) = $\frac{14}{29} \approx 0.483$

NPV = P(No fluid present | ultrasound negative) = $\frac{290}{299} \approx 0.970$

Sensitivity = P(ultrasound positive | fluid present) = $\frac{14}{23} \approx 0.609$

Specificity = P(ultrasound negative | fluid not present) $- \frac{290}{305} \approx 0.951$

b. No. The sensitivity is only 0.609. The test fails to find fluid is almost 40% of the women who have fluid. Even if positive, the more accurate test will be needed anyway because about half of the positives are false positives.

P32. a. $\frac{1}{2}$

b. 1

c. We cannot tell because the probability depends on the probability model used for selecting the coin; either coin could have produced the observed head.

d. Same as c.

e. 1

f. 0

P33. $P(didn't\ survive) = \frac{1490}{2201} \approx .677$ is not equal to $P(didn't\ survive \,|\, male) = \frac{1364}{1731} \approx .788$ So, the events *didn't survive* and *male* aren't independent. In fact, no two events described in this table are independent.

P34. Parts a and c give pairs of independent events, but part b does not.

a. $P(heart \,|\, jack) = \frac{1}{4} = \frac{13}{52} = P(heart)$

b. $P(heart \,|\, red\ card) = \frac{1}{2} \neq \frac{1}{4} = P(heart)$

c. $P(getting\ a\ 7 \,|\, heart) = \frac{1}{13} = \frac{4}{52} = P(getting\ a\ 7)$

P35. a. Let O denote the event that the person has type O blood and *not O* denote that the person does not have type O blood.

		Second Person		
		O	*Not O*	**Total**
First	*O*	0.1764	0.2436	0.4200
Person	*Not O*	0.2436	0.3364	0.5800
	Total	0.4200	0.5800	1.0000

b. *P(exactly one of the people has type O)* $= 0.2436 + 0.2436 = 0.4872$

c.

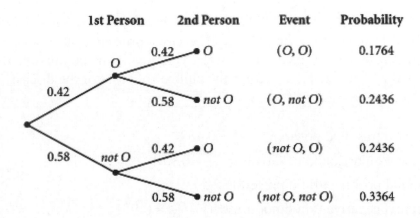

P36. a. P(*at least one has type O blood*) $= 1 - $ P(*none of them have type O blood*)
$$= 1 - 0.58^{10} \approx 0.9957$$

b. P(*at least one doesn't have type O blood*) $= 1 - $ P(*all of them have type O blood*)
$$= 1 - 0.42^{10} \approx 0.9998$$

P37. a. P(both freshmen must take remedial reading) = P(first freshman must take remedial reading) $\cdot$ P(second freshman must take remedial reading) $= 0.11 \cdot 0.11 = 0.0121$

b.

		First Freshman		
		Needs Remedial Reading	**Doesn't Need Remedial Reading**	**Total**
Second Freshman	**Needs Remedial Reading**	0.0121	0.0979	**0.11**
	Doesn't Need Remedial Reading	0.0979	0.7921	**0.89**
	Total	**0.11**	**0.89**	**1.00**

P(*exactly one freshman needs remedial reading*) =
   P(*first freshman needs remedial reading* and
      second freshman does not need remedial reading) +
   P(*second freshman needs remedial reading* and
      first freshman does not need remedial reading) = 0.0979 + 0.0979 = 0.1958.

c.

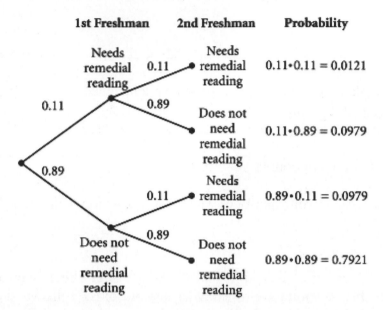

Follow the two branches that represent one needs remedial reading and the other does not, and add the probabilities for those two branches. 0.0979 + 0.0979 = 0.1958.

Exercise Solutions

E1. a. There are 16 possible outcomes:

4 Starbucks	3 Starbucks	2 Starbucks	1 Starbucks	No Starbucks
SSSS	MSSS	MMSS	MMMS	MMMM
	SMSS	MSMS	MMSM	
	SSMS	MSSM	MSMM	
	SSSM	SMMS	SMMM	
		SMSM		
		SSMM		

b. Since the sample space contains 16 elements, the probability that all 4 people choose Starbucks is $\frac{1}{16}$.

c. The probability that all four select Starbucks just by chance is small (0.0625), but not small enough to be convincing. (From Chapter 1, to be convinced that the tasters weren't just guessing, the probability should be less than 0.05.)

E3. **a.** $\frac{30}{36}$ **b.** $\frac{4}{36}$ **c.** $\frac{8}{36}$ **d.** $\frac{6}{36}$

     e. $\frac{11}{36}$ **f.** $\frac{1}{36}$ **g.** $\frac{9}{36}$ **h.** $\frac{3}{36}$

     i. $\frac{2}{36}$; these outcomes are (6, 1) and (1, 6).

E5. **a.** Disjoint and complete; the probabilities for 0, 1, and 2 fours are $\frac{9}{16}, \frac{6}{16}$, and $\frac{1}{16}$, respectively.

b. Not disjoint but complete; *the first roll is a four* and *the second roll is a four* can occur on the same pair of rolls.

c. Disjoint but not complete; only outcomes with fours are included.

d. Disjoint but not complete; *the sum equal to 2* is not included.

e. Disjoint but not complete; *the second die is a four* is not included.

E7. **a.** The first two rolls were not doubles, but the third roll was.

b. Six rolls were doubles. There were six places where the proportion went up.

c. 6/50 or 0.12.

d. As the number of rolls increases, one additional roll has a smaller effect on the cumulative proportion. To illustrate this, say you have rolled twice and one of the rolls was a double. The cumulative proportion is 0.5. The next roll will result in a cumulative proportion of 2/3 if the result is another double or 1/3 if the result is not a double. Either way it is a change of about 0.167. Now imagine there have been 100 rolls and 50 of them were doubles for a cumulative proportion of 0.5. The next roll would result in a cumulative proportion of 50/101 or 51/101. Either way this is a much smaller change in the cumulative proportion, which would result in a shorter segment.

E9. a. Five different backgrounds and four different types give 5 · 4, or 20, possible treatments.

c. With two levels of brightness added, the number of possible treatments is now 5 · 4 · 2, or 40.

b.

d.

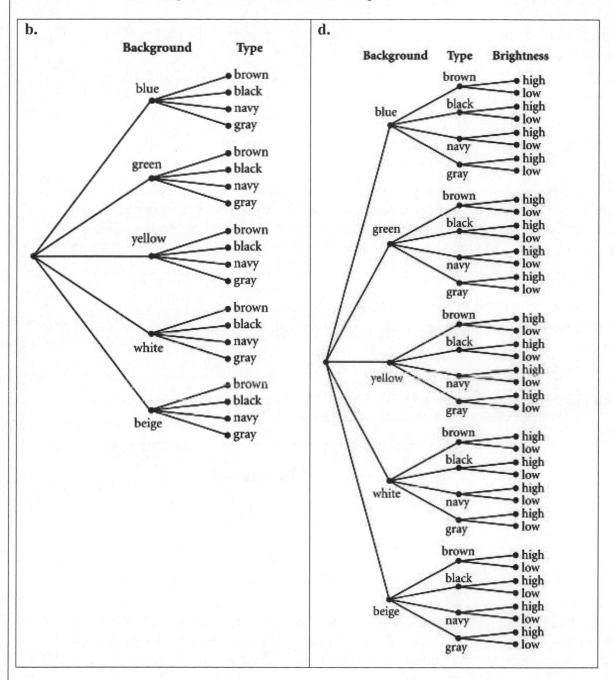

E11. a. Yes. Both candidates have an equal chance of selecting the white marble.

b. If they both draw a white marble on the same round they would tie again. The probability of this happening on any given round would be 1/9 · 1/9 = 1/81.

c. Adding more non-white marbles to the bags would reduce the chance of a tie, but would also increase the expected number of draws before a winner is declared.

E13. **a.** There are 2^6, or 64, equally likely outcomes; the probability of getting heads all six times is $\frac{1}{64}$.

b. There are $6 \cdot 6 \cdot 6 \cdot 6 \cdot 6 \cdot 6$, or 46,656, equally likely outcomes; the probability of getting a 3 all six times is $\frac{1}{46,656}$.

c. No. There are two choices for each person (school, no school) and so 2^{1200} possible outcomes. They are not equally likely (far fewer people are in school than are out of school), so you cannot find the probability without further information.

E15. a and c

E17. **a.** The categories are not disjoint because the numbers in the three categories add up to more than the total number of people who fish in the US. Some people fish in more than one category of water. The categories are complete because the two kinds of bodies of water are fresh and salt water. The Great Lakes are separated out as a separate category of fresh water.

b. Yes. $\frac{7.7}{30} \approx 0.257$

c. No because you don't know how many people who fish in the Great Lakes also fish in other freshwater.

d.

$$P(\text{fresh or salt}) = P(\text{fresh}) + P(\text{salt}) - P(\text{fresh and salt})$$
$$30,000,000 / 30,000,000 = 26,400,000 / 30,000,000 + 7,700,000 / 30,000,000 - P(\text{fresh and salt})$$
$$P(\text{fresh and salt}) = (34,100,000 - 30,000,000)/30,000,000 = 0.1367$$

E19. **a.**

	Owns a Laptop?		
	Yes	**No**	**Total**
Yes	26	46	72
Receives financial aid **No**	112	32	144
Total	138	78	216

b. The probability that a randomly selected student doesn't receive financial aid or doesn't own a laptop is $(144 + 78 - 32) / 216 \approx 88.0\%$. Alternately, the complement of "doesn't receive financial aid or doesn't own a laptop" is "receives financial aid and has a laptop. $1 - 26 / 216 \approx 88.0\%$.

E21. a.

Genetic Mutation B?

Genetic Mutation A?		Yes	No	Total
	Yes	55	25	80
	No	15	5	20
	Total	70	30	100

The proportion who have A or have B is $0.55 + 0.25 + 0.15 = 0.95$, or 95%

b. $P(A \text{ or } B) = P(A) + P(B) - P(A \text{ and } B) = 0.80 + 0.70 - 0.55 = 0.95.$

E23.

	Employed	Unemployed	Not in Labor Force	Total
High School Graduates	5,142,530	796,730	1,303,740	7,243,000
High School Dropouts	1,995,980	527,240	1,242,780	3,766,000
Total	7,138,510	1,323,970	2,546,520	11,009,000

The probability that a randomly selected person aged 16 to 24 who isn't enrolled in college or high school is employed is $7,138,510 / 11,009,000 \approx 0.648$.

E25. Jill has not computed correctly because it is not true that these two events are mutually exclusive. It is common to use the language that "two events are mutually exclusive if they can not happen at the same time." As you can see from this example, language can be misleading. These two events are not mutually exclusive even though they can not happen at the same instance in time. However, they can happen in the same situation. The first flip can be a head and the second flip can be a head, or *HH,* so the two events are not mutually exclusive.

E27. Each of the statements $P(A \text{ and } B) = 0$, $P(B \text{ and } C) = 0$, and $P(A \text{ and } C) = 0$ must be true in order for $P(A \text{ or } B \text{ or } C) = P(A) + P(B) + P(C)$.

(Note that technically we also need $P(A \text{ and } B \text{ and } C) = 0$, but this is implied by each of these three statements.)   To understand why this is true, examine this Venn diagram.

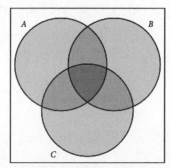

87

E29. a. $\frac{60,838}{232,556} \approx 0.262$

b. $\frac{7,798}{37,490} \approx 0.208$

c. $\frac{7,798}{60,838} \approx 0.128$

d. $\frac{29,692 + 30,888}{171,718} \approx 0.353$

e. Age 35 to 44 has $\frac{12,902}{42,301} \approx 0.305$, or 30.5% volunteers.   Age 16 to 24 has 20.8% volunteers.

E31. a.

		Receives additional medical attention?		
		Yes	**No**	**Total**
Age	**Under 90**	0.44	0.48	0.92
	90 and older	0.08	0	0.08
	Total	0.52	0.48	1.00

b.   P(90 or older | receives additional medical attention) $= \frac{0.08}{0.52} \approx 0.15$

c.   P(90 or older and does not receive additional medical attention) $= 0$

E33. a. $\frac{3}{5}$ **b.** $\frac{2}{5}$ **c.** $\frac{1}{2}$ **d.** $\frac{1}{2}$ **e.** $\frac{3}{4}$ **f.** $\frac{1}{4}$

Here is the tree diagram for drawing marbles:

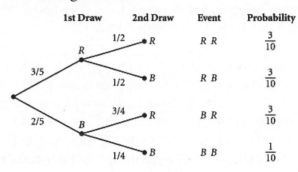

E35.   a. $4/52 \cdot 4/52 \approx 0.0059$

b. $4/52 \cdot 3/51 \approx 0.0045$

c. $4/52 \cdot 4/52 \approx 0.0059$

d. $4/52 \cdot 4/51 \approx 0.0060$

e. $1 \cdot 13/52$ (The first card can be anything, and the second card can be any of the 13 cards of the same suit.)

f. $1 \cdot 12/51 \approx 0.2353$

E37. This problem is easiest to understand if you first make a table.

		Rain Predicted?		
		Yes	No	Total
Rain?	Yes	6	18	24
	No	4	72	76
	Total	10	90	100%

The desired probability is 0.24, or 24%.

E39. **a.** Here are the expected results for 20 contaminated samples:

		Technician's Decision		
		Positive	Negative	Total
Contamination	Present	18	2	20
	Absent	8	72	80
	Total	26	74	100

b. The false positive rate is $\frac{8}{26} \approx 0.31$. Equivalently, we say PPV $= \frac{18}{26} \approx 0.69$.

c. The false negative rate is $\frac{2}{74} \approx 0.03$. Equivalently, we say NPV $= \frac{72}{74} \approx 0.97$.

d. With 50 contaminated samples out of 100, the expected results change as shown here.

		Test Result		
		Positive	Negative	Total
Contamination	Present	45	5	50
	Absent	5	45	50
	Total	50	50	100

The false positive rate is $\frac{5}{50} = 0.10$, or equivalently PPV $= 0.90$. The false negative rate is $\frac{5}{50} = 0.10$, or equivalently NPV $= 0.90$. It is important to note that the false positive rate goes down when the population being tested has a higher proportion of contaminated cases.

E41. **a.** No. The false positive rate is computed using

$$\frac{\text{number of false positive tests}}{\text{number of positive tests}}$$

You are given only the total number of tests, not the total number of positive tests. If all of the 28,436 tests were positive, the false positive rate would be $\frac{107}{28,436} \approx 0.0038$ or less than half of one percent. If only 107 tests were positive, the false positive rate would be 107/107 = 100%.

b. This is a new test so it is unlikely that false negatives have been reported yet. People who had negative tests were told that they tested negative for H.I.V. and so may not know that the test gave a false result until symptoms of AIDS appear. On the other hand, false positives will be reported right away because anyone who tests positive on this test will be given further studies to determine the extent of the H.I.V. infection.

c. Let x be the number of false negatives. Then

$$\frac{x + 107}{28,436} = 0.01$$

$$x = 177.36 \approx 177$$

E43. A table (or a tree diagram) might help here.

		Type of Coin		
		Fair Coin	Two-Headed	Total
Flip Result	**Heads**	$\frac{1}{4}$	$\frac{1}{2}$	$\frac{3}{4}$
	Tails	$\frac{1}{4}$	0	$\frac{1}{4}$
	Total	$\frac{1}{2}$	$\frac{1}{2}$	1

a. $\frac{1}{2}$ **b.** 1 **c.** $\frac{1}{3}$ **d.** $\frac{2}{3}$ **e.** 1 **f.** 0

E45. a. $P(doubles) = \frac{1}{6}$; $P(doubles \mid sum\ of\ 8) = \frac{1}{5}$. Not independent.

b. $P(sum\ of\ 8) = \frac{5}{36}$; $P(sum\ of\ 8 \mid 2\ on\ first) = \frac{1}{6}$. Not independent.

c. $P(sum\ of\ 7) = \frac{1}{6}$; $P(sum\ of\ 7 \mid 1\ on\ first) = \frac{1}{6}$. Independent.

d. $P(doubles) = \frac{1}{6}$; $P(doubles \mid sum\ of\ 7) = 0$. Not independent.

e. $P(1\ on\ first) = \frac{1}{6}$; $P(1\ on\ first \mid 1\ on\ second) = \frac{1}{6}$. Independent.

E47. a. $P(left\text{-}handed) = 12 / 100$ $P(left\ eye\ dominant) = 37 / 100$
$P(left\ eye\ dominant \mid left\text{-}handed) = 6/12$ $P(left\text{-}handed \mid left\ eye\ dominant) = 6 / 37$

b. No. $P(left\ eye\ dominant) \neq P(left\ eye\ dominant \mid left\text{-}handed)$.
c. No. $P(left\ eye\ dominant \mid left\text{-}handed) \neq 0$.

E49. a. *BB, BG, GB,* and *GG*
b. two boys; two girls
c. $(0.51)(0.51)$, or 0.2601, assuming births are independent
d. They may mean that their probability of getting a girl is higher than the population percentage for girls, and higher than the percentage for boys, if girls "run in the family." Under these conditions, *GG* would be the event with the highest probability.

E51. a. Because these two events are independent, we can use the Multiplication Rule for Independent Events:

$$P(type\ O\ and\ Rh+) = (0.42)(0.05) = 0.021$$

b. $P(type\ O\ or\ Rh+) = P(type\ O) + P(Rh+) - P(type\ O\ and\ Rh+) = 0.42 + 0.05 - 0.021 = 0.449$

c.

		Rh-Positive?		
		Yes	**No**	**Total**
Type O?	**Yes**	2.1	39.9	42
	No	2.9	55.1	58
	Total	5	95	100%

E53. a. Assuming that all batches are equally likely to be chosen, we seek P (J and J) = 0.25.
b. Suggest that each highway get 1 batch of cement H and 1 batch of cement J. Since the two sections comprising the highway are identical, using one type of cement on the two sections will yield results as to how each type of cement performs on two different highways.

E55. a. P(*all 12 require remedial reading*) = $0.11^{12} = 3 \cdot 10^{-12}$.
b. P(*at least one requires remedial reading*) = 1 − P(*none require remedial reading*)
$$= 1 - 0.89^{12} = 0.753.$$

E57. The results of the surgeries must be assumed to be independent in order to find the probability both surgeries will fail, with the given information. This typically is not a reasonable assumption, because the two surgeries are to be performed on the same person at the same time. If the surgeries can be assumed to be independent, as perhaps with minor surgery performed on different parts of the body, $P(A\ fails) = 0.15$ and $P(B\ fails) = 0.10$ and the probability that they both fail is the product 0.015.

E59. A two-way table will help. Begin by filling in the marginal totals for left and right-handed pitchers. Then calculate the probabilities in the cells under "Yes" as shown below. After that, adding and subtracting will allow you to fill in the rest of the table.

		Successful Hit?		
		Yes	**No**	**Total**
Pitcher?	**Left-handed**	$0.15 \cdot 0.20 = 0.03$	0.17	0.20
	Right-handed	$0.23 \cdot 0.80 = 0.184$	0.616	0.80
	Total	0.214	0.786	1.00

He hits successfully 21.4% of the time.

E61. Let's call these three conditions I, II, and III.
I: $\quad P(A) = P(A \mid B)$ $\qquad$ II: $\quad P(B) = P(B \mid A)$ $\qquad$ III: $\quad P(A \text{ and } B) = P(A) \cdot P(B)$

The proof that I $\Rightarrow$ II is as follows:
Suppose $P(A \mid B) = P(A)$. Then by the definition of conditional probability,
$$P(B \mid A) = \frac{P(A \text{ and } B)}{P(A)} = \frac{P(B) \cdot P(A \mid B)}{P(A)} = \frac{P(B) \cdot P(A)}{P(A)} = P(B)$$
By switching A and B, this proof also shows that II $\Rightarrow$ I.

<u>The proof that I $\Rightarrow$ III is as follows:</u>
First write the definition of conditional probability:

$$P(A \mid B) = \frac{P(A \text{ and } B)}{P(B)}$$

Because I is assumed to be true, substitute $P(A)$ for $P(A \mid B)$ to get

$$P(A) = \frac{P(A \text{ and } B)}{P(B)}$$

Multiplying both sides by $P(B)$ gives

$$P(A) \cdot P(B) = P(A \text{ and } B)$$

which is III.

<u>The proof that III $\Rightarrow$ I is as follows:</u>
Because III is true, we have

$$P(A \text{ and } B) = P(A) \cdot P(B)$$

Dividing both sides by $P(B)$ gives

$$P(A) = \frac{P(A \text{ and } B)}{P(B)}$$

But from the definition of conditional probability, we know that

$$P(A \mid B) = \frac{P(A \text{ and } B)}{P(B)}$$

Comparing the last two equations, we see that $P(A) = P(A \mid B)$. So I is true.

Because we have shown that II $\Leftrightarrow$ I and I $\Leftrightarrow$ III, we get by transitivity that II $\Leftrightarrow$III, and the three conditions are equivalent.

E63. a. Think of a table with A and B intersecting. The symbol $\overline{A}$ denotes *not A*.

	B	$\overline{B}$
A		
$\overline{A}$		

From the table, $P(A) = P(B \text{ and } A) + P(\overline{B} \text{ and } A)$, which, using the Multiplication Rule, is equal to $P(B) \cdot P(A \mid B) + P(\overline{B}) \cdot P(A \mid \overline{B})$.

b. Use this to find $P(\textit{Dodgers win})$:

$$P(win) = P(win \mid day) \cdot P(day) + P(win \mid not\ day) \cdot P(not\ day) = \frac{11}{21} \cdot \frac{21}{78} + \frac{30}{57} \cdot \frac{57}{78} = \frac{41}{78}$$

c. $\quad P(B \mid A) = \frac{P(A \text{ and } B)}{P(A)} = \frac{P(B) \cdot P(A \mid B)}{P(A)}$

But from part a, $P(A) = P(A \mid B) \cdot P(B) + P(A \mid \overline{B}) \cdot P(\overline{B})$.

d.

$$P(win \mid day) = \frac{P(day \mid win) \cdot P(win)}{P(day \mid win) \cdot P(win) + P(day \mid not\ win) \cdot P(not\ win)} = \frac{\dfrac{11}{41} \cdot \dfrac{41}{78}}{\dfrac{11}{41} \cdot \dfrac{41}{78} + \dfrac{10}{37} \cdot \dfrac{37}{78}} = \frac{11}{21}$$

E65. a. $4 \cdot 6$, or 24

b.

	Regular Die					
	1	**2**	**3**	**4**	**5**	**6**
1	1, 1	1, 2	1, 3	1, 4	1, 5	1, 6
2	2, 1	2, 2	2, 3	2, 4	2, 5	2, 6
3	3, 1	3, 2	3, 3	3, 4	3, 5	3, 6
4	4, 1	4, 2	4, 3	4, 4	4, 5	4, 6

(left column label: **Tetrahedral Die**)

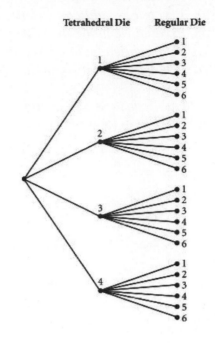

c. $\frac{4}{24}$

d. $\frac{2}{24}$

e. These events are not disjoint because you can roll (2, 2). These events are not independent events because $P(sum\ of\ 4) = \frac{3}{24}$, but $P(sum\ of\ 4 \mid doubles) = \frac{1}{4}$.

f. These events are not disjoint because you can roll (2, 5). These events are independent because $P(2\ on\ tetrahedral) = \frac{1}{4} = P(2\ on\ tetrahedral \mid 5\ on\ the\ regular\ die)$.

E67. $P(favorite\ song\ plays\ at\ least\ once) = 1 - P(favorite\ song\ does\ not\ play) = 1 - (\frac{8}{9})^7 \approx .56$.

E69. Since both were observed at random, the events

man washes his hands

and

woman washes her hands

are independent.

a. $P(both\ the\ man\ and\ the\ woman\ wash\ their\ hands) = 0.75 \cdot 0.90 = 0.675$

b. $P(neither\ wash\ their\ hands) = 0.25 \cdot 0.10 = 0.025$

c. $P(at\ least\ one\ washes\ their\ hands) = 1 - P(neither\ wash\ their\ hands) = 0.975$.

d. $P(man\ washes\ his\ hands \mid woman\ washes\ her\ hands) = P(man\ washes\ his\ hands) = 0.75$

E71. a. $P(two\ teachers) = \frac{2}{4} \cdot \frac{1}{3} = \frac{1}{6} \approx .167$

b. $P(two\ teachers) = \frac{5,000,000}{10,000,000} \cdot \frac{4,999,999}{9,999,999} = .249999975$

c. The draws in both (a) and (b) are technically dependent, but the large population size in (b) makes independence a reasonable assumption. The probability is essentially $(0.5)(0.5) = 0.25$ for the large population.

E73. a. i. $P(Republican) = 16/42 \approx 0.381$
  ii. $P(voted\ Republican) = 13/42 \approx 0.310$
  iii. $P(voted\ Republican\ and\ Republican) = 12/42 \approx 0.286$
  iv. $P(voted\ Republican\ or\ Republican) = 17/42 \approx 0.405$
  v. $P(Republican\ |\ voted\ Republican) = 12/13 \approx 0.923$
  vi. $P(voted\ Republican\ |\ Republican) = 12/16 \approx 0.75$

b. No. $P(Republican) = 0.381$. $P(Republican\ |\ Voted\ Republican) = 0.923$. These are not equal.
c. No. There was one Republican who voted for a Democrat.

E75. a.

	A	Not A	Total
B	$\frac{1}{2} \cdot \frac{1}{4} = \frac{1}{8}$	$\frac{1}{4} \cdot \frac{3}{4} = \frac{3}{16}$	$\frac{5}{16}$
Not B	$\frac{1}{8}$	$\frac{9}{16}$	$\frac{11}{16}$
Total	$\frac{1}{4}$	$\frac{3}{4}$	1

b. No. $P(B) = 5/16 \neq P(B\ |\ A)$

E77. The table summarizes the results of the study:

		Answer to Single Question		
		Yes	No	Total
Clinically Depressed?	**Yes**	37	6	43
	No	8	28	36
	Total	45	34	79

The sensitivity of the single question is $\frac{37}{43}$, or about 86%. This means that 86% of the depressed patients will be identified as depressed by the single question.

The specificity of the single question is $\frac{28}{36}$, or about 78%. This means that 78% of the people who aren't depressed will be identified as not depressed by this question.

The positive predictive value is $\frac{37}{45}$, or about 82%. This means that 82% of the people identified as depressed by the single question actually are depressed.

The negative predictive value is $\frac{28}{34}$, or about 82%. This means that 82% of the people identified as not depressed by the single question are not depressed.

This is a reasonably good test, but the sensitivity is not as high as it should be. The number of false positives, 8, isn't much of a problem as these people will receive further testing. The number of false negatives, 6, is the bigger problem as these people won't receive the counseling they need. If depression is indeed a serious impediment to recovery from a stroke, a good screening test should have higher sensitivity than 86%.

E79. **a.** There are 99,999 possible zip codes: 00001, 00002, . . . , 99999. The average number of people per zip code is 300,000,000 / 99,999 ≈ 3000.

b. There are 99,999 possible zip codes and 9,999 possible plus 4 codes. Thus, there are 99,999 · 9,999 = 999,890,001 possible zip plus 4 codes. The average number of people per zip plus 4 code is 300,000,000 / 999,890,001 ≈ 0.3. We each could have almost three zip plus 4 codes.

Chapter 6

Practice Problem Solutions

P1. Define X to be the smaller of the two numbers when rolling two six-sided dice. Then the probability distribution of the random variable X is given by this table:

x	p	x	p
1	$\frac{11}{36}$	4	$\frac{5}{36}$
2	$\frac{9}{36}$	5	$\frac{3}{36}$
3	$\frac{7}{36}$	6	$\frac{1}{36}$

P2. The four possible outcomes are:
   None are spam
   The first selected message is the only spam
   The second selected message is the only spam
   The first and second selected messages are spam

The corresponding probabilities are:

   P(None are Spam) $= 0.3 \cdot 0.3 = 0.09$
   P(*The first selected message is the only spam*) $= 0.7 \cdot 0.3 = 0.21$
   P(*The second selected message is the only spam*) $= 0.3 \cdot 0.7 = 0.21$
   P(*The first and second selected messages are spam*) $= 0.7 \cdot 0.7 = 0.49$

Define X to be the number of selected messages that are spam. Then the probability distribution of the random variable X is given by this table:

X	P(X)
0	0.09
1	0.42
2	0.49

P3. a. There are 10 possible samples:

BCM, BCA, BCS, BMA, BMS, BAS, CMA, CMS, CAS, MAS

b. Each sample has a probability of 1/10 of being selected.
c. Define X to be the number of committee members who live in the dorms. Then the probability distribution of the random variable X is given by this table:

x	p	x	p
0	$\frac{1}{10}$	2	$\frac{3}{10}$
1	$\frac{6}{10}$	3	$\frac{0}{10}$

P4. $0(0.09) + 1(0.42) + 2(0.49) = 1.4$

P5. a. $\mu = \sum x \cdot P(x) = 0(0.524) + 1(0.201) + 2(0.179) + 3(0.070) + 4(0.026) = 0.873$

b. On average, a family in the US has 0.873 children.

c. The expected number of children when 10 families are chosen at random is $10(0.873)$ = 8.73.

P6. The possible outcomes when flipping a coin five times are:

5 Heads	4 Heads	3 Heads	2 Heads	1 Head	0 Heads
HHHHH	HHHHT	HHHTT	HHTTT	HTTTT	TTTTT
	HHHTH	HHTHT	HTHTT	THTTT	
	HHTHH	HTHHT	THHTT	TTHTT	
	HTHHH	THHHT	HTTHT	TTTHT	
	THHHH	HHTTH	THTHT	TTTTH	
		HTHTH	TTHHT		
		THHTH	HTTTH		
		HTTHH	THTTH		
		THTHH	TTHTH		
		TTHHH	TTTHH		

Each of these outcomes is equally likely with probability 1/32. Define X to be the number of heads when flipping a coin five times. Then the probability distribution of the random variable X is given by this table and graph:

x	p	x	p
0	$\frac{1}{32}$	3	$\frac{10}{32}$
1	$\frac{5}{32}$	4	$\frac{5}{32}$
2	$\frac{10}{32}$	5	$\frac{1}{32}$

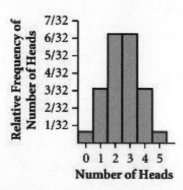

The expected value is 2.5.

P7. a.

Sum of Two Tetrahedral Dice	Probability
2	$\frac{1}{16}$
3	$\frac{2}{16}$
4	$\frac{3}{16}$
5	$\frac{4}{16}$
6	$\frac{3}{16}$
7	$\frac{2}{16}$
8	$\frac{1}{16}$
Total	$\frac{16}{16} = 1$

The probability that the sum is 3 is $\frac{2}{16}$.

b. The expected value of this distribution is

$$2\left(\tfrac{1}{16}\right)+3\left(\tfrac{2}{16}\right)+4\left(\tfrac{3}{16}\right)+5\left(\tfrac{4}{16}\right)+6\left(\tfrac{3}{16}\right)+7\left(\tfrac{2}{16}\right)+8\left(\tfrac{1}{16}\right)=5.0.$$

P8. a. $\mu_X = 600 \cdot \frac{1}{500} + 0 \cdot \frac{499}{500} = \$1.20$ **b.** $\mu_X = 1000 \cdot \frac{1}{500} + 400 \cdot \frac{2}{500} + 0 \cdot \frac{497}{500} = \$3.60$

P9.

Outcome	Payout	Probability
No burglary	$0	0.9768
Burglary	$5000	0.0232

The expected payout per policy is 0(0.9768) + 5000(0.0232) = $116, so you should charge $116 to break even.

P10.

x	p	x · p
0	0.45	0
1	0.09	0.09
2	0	0
3	0.04	0.12
4	0.15	0.60
5	0.27	1.35
Sum	1	2.16

Observe that

$$\mu = \sum x \cdot P(x) = 2.16, \qquad \sigma = \sqrt{\sum (x-\mu)^2 \cdot P(x)} \approx 2.221.$$

The mean score on the problem was 2.16, but it would be typical for a student to get a score 2.221 points higher or lower than that. (But not less than 0.)

P11. Answers will vary depending on the choice of center. Here we use the midpoint of the first two age groups, and 35 for the last group.

Age	Proportion
19.5	0.442
26	0.314
35	0.244

Observe that

$$\mu = \sum x \cdot P(x) = 25.323, \qquad \sigma = \sqrt{\sum (x-\mu)^2 \cdot P(x)} \approx 6.163 \, .$$

Using the midpoint as the center for the first two intervals and 35 for the last interval, the mean age of students at UTSA is 25.232, but it would be typical for a student to be 6.163 years older or younger than that.

P12. **a.** 1,691,290 tickets out of 8,054,375, or 0.210, are winners. This is about equal to $1/4.76 \approx 0.210$.

b. On average, you will lose 70¢ for each ticket purchased. The variability in the outcomes is quite large: $23.69, which indicates there is a small chance of winning a very big prize.

P13. **a.** The probabilities do not add to one. So, it must be the case that the probability of not being able to rent the cleaner on a typical Saturday is $1 - (0.4 + 0.3 + 0.1) = 0.2$. When this happens, the rental income is $0. The expected rental income for the store is then:

$$\mu_x = \sum x \cdot P(x) = 30 \bullet 0.4 + 50 \bullet 0.3 + 60 \bullet 0.1 + 0 \bullet 0.2 = \$33$$

b.

$$\sigma_x = \sqrt{\sum (x-\mu_x)^2 \cdot P(x)}$$

$$= \sqrt{(30-33)^2 \cdot 0.4 + (50-33)^2 \cdot 0.3 + (60-33)^2 \cdot 0.1 + (0-33)^2 \cdot 0.2} \approx \$19.52$$

The possible deviations from the mean (in absolute value) are $3, $17, $27, and $33. No deviation is very close to $19.52, but as typical deviation, it seems reasonable.

P14. **a.** Observe that

$$\mu_x = \sum x \cdot P(x) = 0(0.524) + 1(0.201) + 2(0.179) + 3(0.070) + 4(0.026) = 0.873$$

$$\sigma_x = \sqrt{\sum (x-\mu_x)^2 \cdot P(x)}$$

$$= \sqrt{\begin{array}{l}(0-0.873)^2 (0.524) + (1-0.873)^2 (0.201) + (2-0.873)^2 (0.179) + (3-0.873)^2 (0.070) \\ + (4-0.873)^2 (0.026)\end{array}}$$

$$= 1.096$$

On average, a randomly selected family will have 0.873 children, give or take 1.096

children. (The fact that the standard deviation is larger than the mean is an indication that the distribution is skewed right.)

b. On average, 30 randomly selected families will have 30(0.873) = 26.19 children, give or take 6.003 children.

P15. a. The expected yearly savings is 52($35) = $1,820. The standard deviation for yearly savings is $\sqrt{52}\,(\$15) = \$108.17$.

b. The expected yearly savings in this case is 52($35 + $10) = $2,340. The standard deviation for yearly savings in this case is the same as in (a) because adding a constant to a random variable does not change its standard deviation.

P16. a. The expected value and standard deviation of this distribution are
$$\mu = 2\left(\tfrac{1}{16}\right)+3\left(\tfrac{2}{16}\right)+4\left(\tfrac{3}{16}\right)+5\left(\tfrac{4}{16}\right)+6\left(\tfrac{3}{16}\right)+7\left(\tfrac{2}{16}\right)+8\left(\tfrac{1}{16}\right)=5.0$$
$$\sigma^2 = (2-5)^2\left(\tfrac{1}{16}\right)+(3-5)^2\left(\tfrac{2}{16}\right)+(4-5)^2\left(\tfrac{3}{16}\right)+(5-5)^2\left(\tfrac{4}{16}\right)$$
$$+(6-5)^2\left(\tfrac{3}{16}\right)+(7-5)^2\left(\tfrac{2}{16}\right)+(8-5)^2\left(\tfrac{1}{16}\right)=2.5.$$

b. The expected value and standard deviation of this distribution are
$$\mu = 1\left(\tfrac{1}{4}\right)+2\left(\tfrac{1}{4}\right)+3\left(\tfrac{1}{4}\right)+4\left(\tfrac{1}{4}\right)=2.5$$
$$\sigma^2 = (1-2.5)^2\left(\tfrac{1}{4}\right)+(2-2.5)^2\left(\tfrac{1}{4}\right)+(3-2.5)^2\left(\tfrac{1}{4}\right)+(4-2.5)^2\left(\tfrac{1}{4}\right)=1.25.$$

c. The mean roll is 2.5. So, the sum of the means of two such rolls is 5. This coincides with the mean of the sum from (a).

d. This is similar to (c). The variance of each roll is 1.25, so that the sum for two rolls is 2.5. This coincides with the variance from (a).

P17. a. For a duplex, the mean and standard deviation are
$$\mu = 0(0.008)+1(0.058)+2(0.176)+3(0.287)+4(0.278)$$
$$+5(0.153)+6(0.040)=3.388$$
$$\sigma^2 = (0-3.388)^2(0.008)+(1-3.388)^2(0.058)+(2-3.388)^2(0.176)+(3-3.388)^2(0.287)$$
$$+(4-3.388)^2(0.278)+(5-3.388)^2(0.153)+(6-3.388)^2(0.040)=1.579$$

For a single household, the mean and standard deviation are
$$\mu = 0(0.087)+1(0.331)+2(0.381)+3(0.201)=1.696$$
$$\sigma^2 = (0-1.696)^2(0.087)+(1-1.696)^2(0.331)+(2-1.696)^2(0.381)+(3-1.696)^2(0.201)$$
$$=0.78758$$

b. This is similar to P16 c and d. Indeed, from part (a), the mean number of vehicles for a single household is 1.696, so that for two such households, the mean is 1.6962=3.392, which coincides with the mean for a duplex. Similarly, the variance for each household is 0.78758, so that for two households the variance is 2*0.78758=1.575, which also coincides with the results from part (a) for a duplex.

P18. Let X = number of heads from 8 flips of a coin.

a. $P(X = 3) = \binom{8}{3}(0.5)^8 = 0.21875$

b. Since 25% of 8 is 2 heads, we compute $P(X = 2) = \binom{8}{2}(0.5)^8 = 0.109375$.

c. $P(X = 7 \text{ or } X = 8) = 0.03125 + 0.00391 \approx 0.03516$

P19. a. $\binom{5}{1}\left(\frac{6}{36}\right)^1\left(\frac{30}{36}\right)^4 \approx 0.4019$

b. $\binom{5}{3}\left(\frac{6}{36}\right)^3\left(\frac{30}{36}\right)^2 \approx 0.0322$

c. $P(\text{at least one } 7) = 1 - P(\text{no } 7) = 1 - \left(\frac{30}{36}\right)^5 \approx 0.5981$

d. $P(\text{at most one } 7) = P(\text{no } 7) + P(\text{one } 7) = \binom{5}{0}\left(\frac{6}{36}\right)^0\left(\frac{30}{36}\right)^5 + \binom{5}{1}\left(\frac{6}{36}\right)^1\left(\frac{30}{36}\right)^4 \approx 0.8038$.

P20. a. The probability of touching either side of the forehead is the same, namely 0.5.

b. Since such trials are assumed to be independent, the probability that she touches the mark every time is $\binom{12}{12}(0.5)^{12}(0.5)^0 \approx 0.0002441$, or about a 0.02% chance.

c. Happy isn't picking the side to touch at random; she is favoring the side with the mark so she must understand where the mark is on her forehead.

P21. Let X = number with blood type A and p = probability of having type A blood.

a. This is a binomial experiment with $n = 10$ and $p = 0.40$. As such, we have

$$P(X = 0) + P(X = 1) = 0.04636.$$

b. Yes, since $P(X \geq 2) = 0.95364$.

P22. Let X = number of dropouts. Choosing a sample of 5 people at random in the given age group, we have a binomial random variable with $p = 0.11$.

a. The distribution is as follows:

X	Probability
0	$\binom{5}{0}(0.11)^0(0.89)^5 \approx 0.5584$
1	$\binom{5}{1}(0.11)^1(0.89)^4 \approx 0.3451$
2	$\binom{5}{2}(0.11)^2(0.89)^3 \approx 0.0853$
3	$\binom{5}{3}(0.11)^3(0.89)^2 \approx 0.0105$
4	$\binom{5}{4}(0.11)^4(0.89)^1 \approx 0.0006515$
5	$\binom{5}{5}(0.11)^5(0.89)^0 \approx 0.00001605$

b. The histogram is as follows. It is strongly skewed right, towards larger values

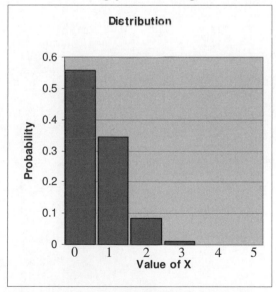

c. $\mu = 5(0.11) = 0.55,$ $\sigma^2 = 5(0.11)(0.89) = 0.4895$, so that $\sigma = 0.6996$

d. No. If 11% is the correct percentage, the probability that none are dropouts is quite high, 0.5584.

P23. Let X = number of drivers wearing seatbelts. Choosing a sample of 4 people at random, we have a binomial random variable with $p = 0.84$.

a. The distribution is as follows:

X	Probability
0	$\binom{4}{0}(0.16)^4(0.84)^0 \approx 0.000655$
1	$\binom{4}{1}(0.16)^3(0.84)^1 \approx 0.0138$
2	$\binom{4}{2}(0.16)^2(0.84)^2 \approx 0.1084$
3	$\binom{4}{3}(0.16)^1(0.84)^3 \approx 0.3793$
4	$\binom{4}{4}(0.16)^0(0.84)^4 \approx 0.4979$

The histogram is as follows. It is skewed left, towards the smaller values.

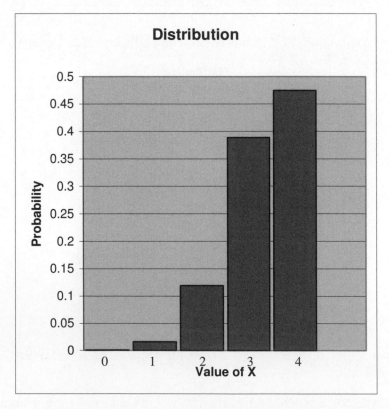

b. Observe that $\mu = 4(0.84) = 3.36$, So, on average, 3.36 of the 4 randomly selected drivers will be wearing seatbelts.

c. The standard deviation is $\sigma = \sqrt{4(0.16)(0.84)} \approx 0.7332$, so it would be typical to find 3 or 4 drivers wearing seatbelts, which are the values within 0.7332 of the mean.

d. Definitely; the probability that no drivers in a random sample of 4 drivers were wearing seatbelts is only 0.000655 if NHTSA is correct that 84% of drivers use seatbelts regularly.

Exercise Solutions

E1. a. The outcomes are pairs of selections of distinct concertos. Assuming each digit in a pair refers to the number of the concerto played, the 10 outcomes are as follows:
$$12, \ 13, \ 14, \ 15, \ 23, \ 24, \ 25, \ 34, \ 35, \ 45$$

Let X = number of minutes played in pair of concertos. The values of X corresponding to each outcome are as follows:

Pair	Value of X	Pair	Value of X
12	68	24	63
13	73	25	70
14	71	34	68
15	78	35	75
23	65	45	73

The distribution for X is as follows:

X	Probability
63	$\frac{1}{10}$
65	$\frac{1}{10}$
68	$\frac{2}{10}$
70	$\frac{1}{10}$
71	$\frac{1}{10}$
73	$\frac{2}{10}$
75	$\frac{1}{10}$
78	$\frac{1}{10}$

The expected value of X is:
$$\mu = 63\left(\tfrac{1}{10}\right) + 65\left(\tfrac{1}{10}\right) + 68\left(\tfrac{2}{10}\right) + 70\left(\tfrac{1}{10}\right) + 71\left(\tfrac{1}{10}\right) + 73\left(\tfrac{2}{10}\right) + 75\left(\tfrac{1}{10}\right) + 78\left(\tfrac{1}{10}\right) = 70.4\,\text{min.}$$
So, on average, 2 concertos played back to back will provide 70.4 minutes of music.

b. $P(X \leq 74) = 1 - \left[P(X = 75) + P(X = 78)\right] = 0.8$.

E3. a. Let N = no seat belt and S = wears seat belt.
If we choose a sample of three people, the 8 possible outcomes are:

NNN, SNN, NSN, NNS, SSN, SNS, NSS, SSS

Let X = number of drivers who DO NOT wear a seat belt.
This is a binomial random variable with $p = 0.84$. The distribution of X is as follows:

X	Probability
0	$\binom{3}{0}(0.16)^0(0.84)^3 \approx 0.593$
1	$\binom{3}{1}(0.16)^1(0.84)^2 \approx 0.339$
2	$\binom{3}{2}(0.16)^2(0.84)^1 \approx 0.065$
3	$\binom{3}{3}(0.16)^3(0.84)^0 \approx 0.004$

b. No, the probability that none of the three were wearing seat belts is 0.593, which is quite likely if 16% of pickup truck occupants wear seat belts.

E5. a. Let X = number of speeders who get another speeding ticket within the year.
Let Y = yes, they get another ticket and N = no, they do not get another ticket.
If choosing a random sample of three drivers from this population, the possible outcomes are:

YYY, NYY, YNY, YYN, NNY, NYN, YNN, NNN

The distribution for X is as follows:

X	Probability
0	$\binom{3}{0}(0.11)^0(0.89)^3 \approx 0.705$
1	$\binom{3}{1}(0.11)^1(0.89)^2 \approx 0.261$
2	$\binom{3}{2}(0.11)^2(0.89)^1 \approx 0.032$
3	$\binom{3}{3}(0.11)^3(0.89)^0 \approx 0.001$

b. You cannot tell because you need to know something about the people who were speeding and might have gotten a ticket but didn't. If more than 11% of those people got at least one speeding ticket in the next year, then you would have some evidence that tickets deter speeding.

E7. a. Possible outcomes for number of TVs in Units 1 and 2, respectively, in a duplex:

Outcome (unit 1, unit 2)	Probability
0,0	$0.012 \cdot 0.012 = 0.000144$
0,1	$0.012 \cdot 0.274 = 0.003288$
0,2	$0.012 \cdot 0.359 = 0.004308$
0,3	$0.012 \cdot 0.218 = 0.002616$
0,4	$0.012 \cdot 0.095 = 0.001140$
0,5	$0.012 \cdot 0.042 = 0.000504$
1,0	$0.274 \cdot 0.012 = 0.003288$
1,1	$0.274 \cdot 0.274 = 0.075076$
1,2	$0.274 \cdot 0.359 = 0.098366$
1,3	$0.274 \cdot 0.218 = 0.059732$
1,4	$0.274 \cdot 0.095 = 0.026030$
1,5	$0.274 \cdot 0.042 = 0.011508$
2,0	$0.359 \cdot 0.012 = 0.004308$
2,1	$0.359 \cdot 0.274 = 0.098366$

Outcome (unit 1, unit 2)	Probability
2,2	$0.359 \cdot 0.359 = 0.128881$
2,3	$0.359 \cdot 0.218 = 0.078262$
2,4	$0.359 \cdot 0.095 = 0.034105$
2,5	$0.359 \cdot 0.042 = 0.015078$
3,0	$0.218 \cdot 0.012 = 0.002616$
3,1	$0.218 \cdot 0.274 = 0.059732$
3,2	$0.218 \cdot 0.359 = 0.078262$
3,3	$0.218 \cdot 0.218 = 0.047524$
3,4	$0.218 \cdot 0.095 = 0.027010$
3,5	$0.218 \cdot 0.042 = 0.009156$
4,0	$0.095 \cdot 0.012 = 0.001140$
4,1	$0.095 \cdot 0.274 = 0.026030$
4,2	$0.095 \cdot 0.359 = 0.034105$
4,3	$0.095 \cdot 0.218 = 0.027010$
4,4	$0.095 \cdot 0.095 = 0.009025$
4,5	$0.095 \cdot 0.042 = 0.003990$
5,0	$0.042 \cdot 0.012 = 0.000504$
5,1	$0.042 \cdot 0.274 = 0.011508$
5,2	$0.042 \cdot 0.359 = 0.015078$
5,3	$0.042 \cdot 0.218 = 0.009156$
5,4	$0.042 \cdot 0.095 = 0.003990$
5,5	$0.042 \cdot 0.042 = 0.001764$

Define X to be the number of color television sets per duplex. Then the probability distribution of the random variable X is given by this table

X	p	X	p
0	0.000144	6	0.138750
1	0.006576	7	0.084176
2	0.083692	8	0.027337
3	0.201964	9	0.007980
4	0.250625	10	0.001764
5	0.209592		

b. $E(X) = 0 \cdot 0.000144 + 1 \cdot 0.006576 + 2 \cdot 0.083692 + 3 \cdot 0.201964 + 4 \cdot 0.250625$
$+ 5 \cdot 0.209592 + 6 \cdot 0.138750 + 7 \cdot 0.084176 + 8 \cdot 0.027337 + 9 \cdot 0.007980$
$+ 10 \cdot 0.001764 = 4.5602.$

The expected value is 4.5602 color televisions per duplex.

c. The expected value of number of color televisions per household is:
$$\mu = 0(0.012) + 1(0.274) + 2(0.359) + 3(0.218) + 4(0.095) + 5(0.042) = 2.236 .$$
On average, a household has 2.236 color televisions.

E9. 13.7% of the houses, or about 69 houses, would be expected to have 3 vehicles. Each of these would need to park one vehicle on the street. 5.8% of the houses, or about 29 of them would have four vehicles, two of which would need to be parked on the street. This means about 127 vehicles would be parked on the street. With three spaces per house, the 29 houses with four vehicles would each have one vehicle on the street.

E11. **a.** $134.50 profit if there is no burglary; 5000 – 134.50, or $4865.50 loss if there is a burglary.

b. $134.50 + $5000 / 1000 = $139.50. At the $134.50 rate the expected gain of the company is 0. Therefore, they need to add $5.00 per customer to the rate charged.

c. Examples include whether there is a burglar alarm; whether there are good locks on the doors; whether the house contains expensive items; what proportion of the time someone is at home.

E13. **a.** $\mu_X = \$0.185$

b. 1,000,000(0.50 – 0.185), or $315,000. Alternatively, the state would take in $500,000 for 1,000,000 tickets and would pay out prizes worth $185,000 for a profit of $315,000.

c. 37% of the income was returned in prizes.

E15. **a.** Buy service contract: Expected cost is $250
Don't buy service contract: Expected cost is
$150(1)(0.30) + $150(2)(0.15) + $150(3)(0.10) + $150(4)(0.05) = $165

b. The advantage of buying the service plan is that you if the computer happens to be a lemon and require at least two repairs, you will have saved money, given that unlimited repairs are free. The advantage of not buying the service contract is that you expect to pay less than if you were to buy the service contract.

E17. **a.** The expected number of vehicles is 200 • 1.696 = 339.2.

b. The standard deviation can be calculated by $\sqrt{200} \cdot 0.88745926 = 12.551$. It would not be unusual to be off by 12 or so vehicles.

E19. **a.** The expected total weekly tips is
$$200(0.1) + 300(0.3) + 400(0.4) + 500(0.2) = \$370.$$
The standard deviation for total weekly tips is
$$\sigma^2 = (200 - 370)^2 (0.1) + (300 - 370)^2 (0.3) + (400 - 370)^2 (0.4) + (500 - 370)^2 (0.2) = \$8100$$
$$\sigma = \$90$$
b. The expected weekly salary is $60 + $370 = $430. The standard deviation for the weekly salary remains $90 since adding a constant to a random variable does not change its standard deviation.

c. Assuming that you keep 80% of the tips, the expected weekly salary would be $60 + 0.80($370)= $356. The corresponding standard deviation is 0.80(s.d. in (a)) = $72.

E21. **a.** Begin with possible combinations of working times:

Main Pump	Backup Pump	Total Working Time	Probability
1	1	2	$0.1 \cdot 0.2 = 0.02$
1	2	3	$0.1 \cdot 0.8 = 0.08$
2	1	3	$0.3 \cdot 0.2 = 0.06$
2	2	4	$0.3 \cdot 0.8 = 0.24$
3	1	4	$0.6 \cdot 0.2 = 0.12$
3	2	5	$0.6 \cdot 0.8 = 0.48$

Define X to be the number of months the two pumps together work. Then the probability distribution of the random variable X is given by this table:

x	p	x	p
2	0.02	4	0.36
3	0.14	5	0.48

b. $\mu_x = 4.3$ months, $\sigma_x = 0.781$

c. $\mu_{\text{Main}} = 2.5$ months, $\mu_{\text{Backup}} = 1.8$ months. The sum of these is 4.3 months.

E23. **a.**

Difference	Probability	Difference	Probability
-3	$\frac{1}{24}$	2	$\frac{4}{24}$
-2	$\frac{2}{24}$	3	$\frac{3}{24}$
-1	$\frac{3}{24}$	4	$\frac{2}{24}$
0	$\frac{4}{24}$	5	$\frac{1}{24}$
1	$\frac{4}{24}$		

b. Using the formula $\mu = \sum x \cdot P(x)$, we find this distribution has mean $\mu = 1$. (It is also apparent by symmetry that the mean is 1.) The mean of the distribution of the outcomes of the six-sided die is 3.5, and the mean of the distribution of the outcomes of the four-sided die is 2.5. Thus, the mean of the sampling distribution of the difference is equal to $\mu_1 - \mu_2 = 3.5 - 2.5 = 1$.

Using the formula $\sigma^2 = \sum (x - \mu)^2 \cdot P(x)$, we find this distribution has variance $\sigma^2 = 4.167$. The variance of the distribution of the outcomes of the six-sided die is 2.917, and the variance of the distribution of the outcomes of the four-sided die is 1.25. Thus, the variance of the sampling distribution of the difference is indeed equal to $2.917 + 1.25 = 4.167$.

E25. a. The expected total yearly earnings is 52($356) = $18,512.

The standard deviation for yearly earnings is $\sqrt{52}\,(\$72) = \$519.20$.

b. Note that 18,512 + 2.87(519.20) = 20,000. So, a yearly salary of $20,000 is nearly 3 standard deviations from the mean. So, it would be very unusual to earn more than $20,000 in one year.

E27. Let X = number who correctly identify gourmet coffee.
The situation described is binomial with $n = 6$ and $p = 0.5$ (since equally likely to tell gourmet coffee from ordinary coffee).
The distribution of X is as follows:

X	Probability
0	$\binom{6}{0}(0.5)^6 \approx 1(0.015625) = 0.015625$
1	$\binom{6}{1}(0.5)^6 \approx 6(0.015625) = 0.09375$
2	$\binom{6}{2}(0.5)^6 \approx 15(0.015625) = 0.234375$
3	$\binom{6}{3}(0.5)^6 \approx 20(0.015625) = 0.3125$
4	$\binom{6}{4}(0.5)^6 \approx 15(0.015625) = 0.234375$
5	$\binom{6}{5}(0.5)^6 \approx 6(0.015625) = 0.09375$
6	$\binom{6}{6}(0.5)^6 \approx 1(0.015625) = 0.015625$

The histogram is as follows. It is symmetric and mound shaped.

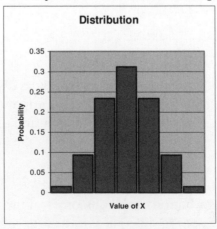

0   1   2   3   4   5   6

110

b. On average, the expected number of people who correctly identify the gourmet coffee is np = 6(0.5) = 3.

c. While you would get 3 people on average who correctly identify the gourmet coffee, you could well get $\sqrt{np(1-p)} = \sqrt{6(0.5)(0.5)} = 1.225$ people more or fewer.

d. Yes, if the people can't tell the difference, the probability is only 0.015625 that all 6 will choose correctly.

E29. Let X be the number of people in the sample of 20 residents that do not have health insurance.

a. $P(X \geq 3) = 1 - P(X < 3) = 1 - P(X = 0) - P(X = 1) - P(X = 2)$

$$1 - \left(\binom{20}{0}(0.15)^0(0.85)^{20} + \binom{20}{1}(0.15)^1(0.85)^{19} + \binom{20}{2}(0.15)^2(0.85)^{18} \right) = 0.5951$$

b. $E(X) = np = 20 \cdot 0.15 = 3$ people, and $\sigma_X = \sqrt{np(1-p)} = \sqrt{20 \cdot 0.15 \cdot 0.85} = 1.5969$ people.

E31. To get 60% correct, you have to get 3 or more on the 5-question quiz and 12 or more on the 20-question quiz. Looking at the graphs in the display, the probability of getting 3 or more correct on a 5-question quiz is 0.5. On a 20-question quiz, the probability of getting 12 or more correct is about 0.25. It would be much better to have a 5-question quiz.

E33. a. Because of the $15 spent for the tickets, in order to gain $10 or more, you must win $25 or more. The only way you can do this is to win on 3 or more of the tickets:
$$P(3 \ or \ more \ wins) = P(X \geq 3) = 1 - P(X \leq 2) \approx 1 - 0.9429 = 0.0571.$$

b. $E(earnings) = 15(0.06)(10) - 15(1) = -\$6.00$

c. $\sigma_W = \sqrt{np(1-p)} = \sqrt{15 \cdot 0.06 \cdot 0.94} \approx 0.9198$, where W is the number of winning tickets. The earnings $E = 10W - 15$, so the $\mu_E = 10\mu_W \approx 10 \cdot 0.9198$, or $9.20.

E35. a. $P(at \ least \ one \ alarm \ sounds) = 1 - P(all \ alarms \ fail) = 1 - (0.3)^3 = 0.973$

b. $P(X \geq 1) = 1 - P(X = 0) = 1 - (0.3)^6 = 0.99927$. No, the two probabilities are nearly the same.

c. Solving $1 - (0.3)^n = 0.99$ gives $n \approx 3.825$, which rounds up to $n = 4$. This equation can be solved by trial and error, by graphing, or by logarithms.

E37. a. If X denotes the number of defective DVDs in the sample, then
$$P(X \ is \ at \ least \ 1) = P(X \geq 1) = 1 - P(X = 0) = 1 - (1 - p)^n,$$
where p is the probability of observing a defective DVD. This probability, $P(X \geq 1)$, is given as 0.5. With $p = 0.1$, the equation becomes $1 - (0.9)^n = 0.5$ or $n \approx 6.5788$, which

rounds up to $n = 7$.

b. With $p = 0.04$, the equation is $1 - (0.96)^n = 0.5$ or $n \approx 16.980 \approx 17$.

E39. a. Observe that

$$\mu = 0(1-p) + 1(p) = p$$

$$\sigma^2 = (0-p)^2(1-p) + (1-p)^2(p)$$

$$= p^2(1-p) + p(1-p)^2$$

$$= p(1-p)(p+(1-p)) = p(1-p)$$

$$\sigma = \sqrt{p(1-p)}$$

b. Observe that

$$\sigma^2_{X_1+X_2+\cdots+X_n} = \sigma^2_{X_1} + \sigma^2_{X_2} + \cdots + \sigma^2_{X_n}$$

$$= p(1-p) + p(1-p) + \cdots + p(1-p)$$

$$= np(1-p)$$

So, $\sigma_{X_1+X_2+\cdots+X_n} = \sqrt{np(1-p)}$.

E41. a. There are $(12)(12) = 144$ possible outcomes. Fortunately, we do not have to list them because there are only three ways to get a sum of 3 or less: 1, 1; or 2, 1; or 1, 2. Thus the probability is 3/144.

b. Each of the outcomes 1, 2, 3, 4, . . . , 12 has a $\frac{1}{12}$ chance of occurring. By symmetry, the mean is 6.5. The standard deviation is 3.452.

c. The mean of the sampling distribution of the sum of two rolls of a 12-sided die is

$$\mu_{sum} = 2(6.5) = 13$$

and the standard deviation is

$$\sigma_{sum} = \sqrt{2 \cdot 3.452^2} = 4.882.$$

E43. a. You are looking for the distribution of the minimum lifelength of the two pumps, since the system stops when the first pump stops. The combinations can be displayed in a two-way table. Since the pumps operate independently, the probabilities can be calculated by multiplying the individual probabilities.

		Pump 1		
		1 month	2 months	3 months
	1 month	0.01	0.03	0.06
Pump 2	2 months	0.03	0.09	0.18
	3 months	0.06	0.18	0.36

Define X to be the minimum lifelength of Pumps 1 and 2. Then the probability distribution of the random variable X is given by this table:

x	p
1 month	0.19
2 months	0.45
3 months	0.36

b. The lifelength of System II is the maximum lifelength of the two pumps. Using the same table, replacing 'Pump 1' and 'Pump 2' with 'Pump 3' and 'Pump 4,' respectively:

Define X to be the maximum lifelength of Pumps 3 and 4. Then the probability distribution of the random variable X is given by this table:

x	p
1 month	0.01
2 months	0.15
3 months	0.84

c. The expected lifelength of System I is 2.17 months and the expected lifelength for System II is 2.83 months. You should recommend System II because it has a larger expected lifelength.

E45. The chance that at least one donation out of ten will be type B is
$$1-(0.9)^{10} \approx 1-0.349 \approx 0.651$$
You are assuming that the probability that the first donation checked is type B is 0.1, that the probability that any subsequent donation is type B is 0.1, and that the probability that a donation is type B is independent of the type of the donations previously checked.

E47. a. Let X = number of women who have the BS degree. Then, X is binomial with $n = 20$ and $p = 0.33$. $P(X > 10) = 0.0350$; assuming the women are selected randomly and independently and that the probability is 0.33 for each one.

b. Let X = number of men who have the BS degree. Then, X is binomial with $n = 20$ and $p = 0.26$. $P(X > 10) = 0.0132$; assuming the men are selected randomly and independently and that the probability is 0.26 for each one.

c. Assuming that the proportion of all people with bachelor's degrees is about 29.5% (which is the average of the probabilities used in (a) and (b).

Let X = number of people with BS degree. Then, X is binomial with $n = 40$ and $p = 0.295$. $P(X > 20) = 0.0019$.

E49. **a.** On a TI-83 Plus or TI-84 Plus calculator, enter the command **binompdf(130,0.9)**, press $\boxed{\text{STO}^{\rightarrow}}$ $\boxed{\text{2nd}}$ **[L1]** $\boxed{\text{ENTER}}$, which stores the probabilities of the different numbers of no shows in list **L1.** The first $P(X)$ in the list below, 0.84793, is a cumulative probability.

Number of Passengers Who Show	Payout x	Probability p
≤120	0	0.84793
121	100	0.06399
122	200	0.04248
123	300	0.02487
124	400	0.01263
125	500	0.00546
126	600	0.00195
127	700	0.00055
128	800	0.00012
129	900	0.00002
130	1000	0.00000

b. $E(payout) = \sum x \cdot P(x) \approx \$31.807$.

E51. **a.**

GAME B	
Value of Prize	**Probability**
$0.69	0.25
$0.00	0.75

GAME C	
Value of Prize	**Probability**
$1.44	0.125
$0.00	0.875

GAME D	
Value of Prize	**Probability**
$1.99	0.0625
$0.00	0.9375

The expected winnings for each game are:

Game A: $0.55(0.50) = \$0.275$
Game B: $0.69(0.25) = \$0.1725$
Game C: $1.44(0.125) = \$0.18$
Game D: $1.99(0.0625) = \$0.124375$

Game A yields the highest expected earnings.

b. Some may not want the drink or might just want to try for the bigger prize.

114

E53. The probability of rolling an even number in one roll of a six-sided die is 3/6 or 0.5. Let X = number of even numbers rolled from seven rolls of a die.

$$P(X = 2) = \binom{7}{2}(0.5)^7 = 0.1641$$

$$P(X \geq 4) = \binom{7}{4}(0.5)^7 + \binom{7}{5}(0.5)^7 + \binom{7}{6}(0.5)^7 + \binom{7}{7}(0.5)^7 = 0.5$$

Chapter 7

Practice Problem Solutions

P1. a. The smallest number of moons would be 1 by describing the moons of Mercury, Venus, and Earth. The largest number of moons would be 146 moons, by describing the moons of Uranus, Saturn, and Jupiter.

b. I. Randomly select three planets.
II. Calculate the sum of the numbers of moons.
III. Repeat steps I and II many times.
IV. Display the distribution of the sums.

c. Answers will vary. One sample plot is shown below:

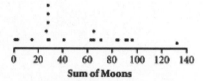

This can be done fairly easily on a TI-83 Plus or TI-84 Plus with a short program.

```
FnOff
{0,0,1,2,63,56,27,13}→L₁
ClrList L₃
For(I,1,20)
rand (9)→L₂
SortA(L₂,L₁)
L₁(1)+L₂(2)+L₃(3)→L₃(I)
End
Plot1(Histogram,L₃)
ZoomStat
```

The For(I,1,20) begins a loop. The first time through, I = 1. Every command is executed until the line End is reached. At that point, the loop is repeated, this time with I = 2. This program takes the algorithm for selecting without replacement and repeats it twenty times, recording the result in the appropriate position in L3.

P2. Histogram C corresponds to I, given that all the values are skewed to the left. Likewise, Histogram A corresponds to II for similar reason. Histogram B corresponds to III since the values all tend to fall near the center of the given distribution in Display 7.11.

P3. a. The most you could be paid is 10% of Elvis Presley's and Charles Schulz's earnings, which is $8,500,000. The least you could be paid is $2,700,000.

b. The possible pairs, along with your possible earnings, are

52 + 33	$8,500,000
52 + 20	$7,200,000
52 + 18	$7,000,000
52 + 15	$6,700,000
52 + 12	$6,400,000
33 + 20	$5,300,000
33 + 18	$5,100,000
33 + 15	$4,800,000
33 + 12	$4,500,000
20 + 18	$3,800,000
20 + 15	$3,500,000
20 + 12	$3,200,000
18 + 15	$3,300,000
18 + 12	$3,000,000
15 + 12	$2,700,000

The sampling distribution is shown below.

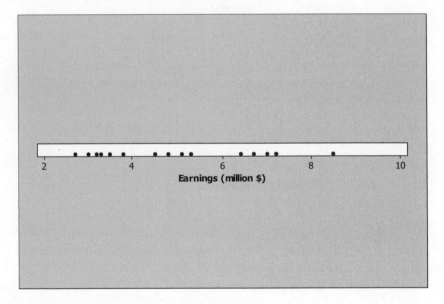

c. Only three of the 15 combinations were $7 million or more, so the probability is 3/15, or 0.2,

P4. a. The mean is $\dfrac{0+1+2+\cdots+5}{6} = 2.5$ and the standard deviation is

$$\sqrt{\frac{(0-2.5)^2 + (1-2.5)^2 + (2-2.5)^2 + \cdots + (5-2.5)^2}{6}} \approx 1.71.$$

b. Label the cars as 0, 1, 2, 3, 4, and 5, which corresponds to their respective number of defects. The 15 possible 2-samples, with their mean number of defects, are:

2-sample	Mean number of defects
0,1	0.5
0,2	1
0,3	1.5
0,4	2.0
0,5	2.5
1,2	1.5
1,3	2.0
1,4	2.5
1,5	3.0
2,3	2.5
2,4	3.0
2,5	3.5
3,4	3.5
3,5	4.0
4,5	4.5

c. The mean of the means listed in column 2 of the table in (b) is obtained by summing all 15 values and dividing by 15. This yields 2.5, which coincides with the population mean.

d. The standard deviation is given by

$$\sqrt{\frac{\sum_{x}(x-2.5)^2}{15-1}} \approx 1.12 < 1.71,$$

where the sum is taken over the values in the right column in the table in (b).

P5. a. The range is $527 - 56 = 471$ sq. mi.
b.

Sample of Three Parks	Range of the Areas (sq mi)
A, B, C	$527 - 56 = 471$
A, B, R	$378 - 56 = 322$
A, B, Z	$229 - 56 = 173$
A, C, R	$527 - 119 = 408$
A, C, Z	$527 - 119 = 408$
A, R, Z	$378 - 119 = 259$
B, C, R	$527 - 56 = 471$
B, C, Z	$527 - 56 = 471$
B, R, Z	$378 - 56 = 322$
C, R, Z	$527 - 229 = 298$

c.

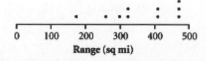

119

d. No. The mean of the sampling distribution of the sample range is 360.3 sq. mi. This is much less than the actual range of 471 sq. mi. You could tell this without calculating because the greatest possible range for a sample occurs when the population maximum and minimum are contained in the sample. All other samples will have a smaller range.

e. There is much variation. There is a difference of 278 between the greatest sample range and the smallest sample range, and a standard error of 96.68 sq. mi.

P6. a.   **A.** II   **B.** I   **C.** III

b. The mean of all sampling distributions is 30, which can easily be seen from the symmetry of the distributions.

c. Distribution A has the smallest standard error since its values deviate from 30 by the least amount.

P7. a. Plot A is the population, starting out with only five different values and a slight skew. Plot C is for a sample size of 4, having more values, a smaller spread, and a more nearly normal shape. Plot B is for a sample size of 10, which has even more values, an even smaller spread, and a shape that is closest to normal of all the distributions. The mean appears to be about 1.7 for each distribution. The standard deviation of plots A, B, and C appear to be about 1, 0.3, and 0.5, respectively.

b. The mean will be the 1.7 for each distribution, which is consistent with the estimates.

c. For a sample size of 4, $\sigma_{\bar{x}} = 1 / \sqrt{4} = 0.5$. For $n = 10$, $\sigma_{\bar{x}} = 1 / \sqrt{10} \approx 0.316$. These match the estimates very closely.

d. The population distribution is roughly mound-shaped with a slight skew. The two sampling distributions however are approximately normal with the distributions becoming more nearly normal as the sample size increases.

P8. The shape will be approximately normal because the population's distribution is approximately normal; the mean will be the mean of the population, or 0.266, and the standard error will be

$$\frac{\sigma}{\sqrt{n}} = \frac{0.037}{\sqrt{15}} \approx 0.009553 .$$

P9. Because the weights are normally distributed, we can use the normal approximation with all sample sizes.

a. The z-score is

$$z = \frac{0.148 - 0.15}{0.003} = -0.667$$

which gives a probability of 0.2525. Alternatively, on a TI-84 the result of the command

120

normalcdf(–9999999,0.148,0.15,0.003) is approximately 0.2525.

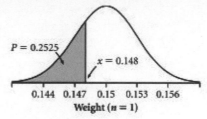

b. The z-score is

$$z = \frac{0.148 - 0.15}{0.003 / \sqrt{4}} = -1.3333$$

which gives a probability of 0.0912.

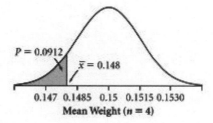

c. The z-score is

$$z = \frac{0.148 - 0.15}{0.003 / \sqrt{10}} = -2.1082$$

which gives a probability of 0.0175.

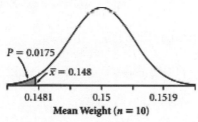

P10. a. The sampling distribution should be approximately normal because this is a very big sample size. With random samples of 1000 families, the potential values of the sample mean are centered at 0.9—the mean number of children per family—and have a standard error of 1.1 divided by the square root of 1000, or about 0.035.

b. The middle 95% of the sample means would lie between $0.9 \pm 1.96(0.035)$ or 0.8314 to 0.9686. So, the distribution of potential values of the sample mean is concentrated very tightly around the expected value of 0.9.

c. The z-score for a mean of 0.8 children is

$$z = \frac{0.8 - 0.9}{1.1 / \sqrt{1000}} \approx -2.87 \, .$$

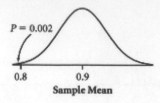

The probability that the mean is 0.8 or less is about 0.002.

d. We must compute $P(0.8 < \overline{x} < 1.0)$. The corresponding z-score for 0.8 is -2.87 (from (c)), and the z-score for a mean of 1.0 children is

$$z = \frac{1.0 - 0.9}{1.1 / \sqrt{1000}} \approx 2.87 .$$

So,

$$P(0.8 < \overline{x} < 1.0) = P(-2.87 < z < 2.87) = 0.9960 ,$$

or 0.9959 using Table A.

P11. **a.** Since the distribution of the population is roughly symmetric, a sample size of 15 should be large enough to assume the sampling distribution of the total will be approximately normal. The mean of the sampling distribution, $\mu_{\overline{x}}$, would be $1.7 \cdot 15 = 25.5$, and the standard error of the sum, σ_{sum}, is $\sigma \sqrt{n} = 1 \cdot \sqrt{15} \approx 3.873$. The z-score is

$$z = \frac{30 - 25.5}{3.873} \approx 1.16$$ and the probability is $P = 0.1226$.

Alternatively, **normalcdf**(30, 1E99, 1.7•15, $\sqrt{(15)}$) = 0.1226. Or, using the means, **normalcdf**(30/15, 1E99, 1.7, $1/\sqrt{15}$) = 0.1226.

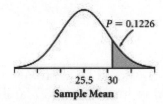

b. The two z-scores are

$$z = \frac{sum - n\mu}{\sigma\sqrt{n}} = \frac{25 - 20(1.7)}{1.0\sqrt{20}} \approx -2.012$$

$$z = \frac{sum - n\mu}{\sigma\sqrt{n}} = \frac{30 - 20(1.7)}{1.0\sqrt{20}} \approx -0.894,$$

so the probability is approximately 0.1635 (or 0.1645 using Table A).

P12. **a.** $\sigma_{\overline{x}} = \frac{4.7}{\sqrt{5}} \approx \$2.102 \text{ million}$. The z-score for a mean salary of \$1.56 million is

$$z = \frac{1.5 - 4.6}{2.102} \approx -1.475 .$$

So, $P(\bar{x} < 1.5) = P(z < -1.475) = 0.0708$.

b. This probability corresponds to the leftmost 2 bars, the sum of which is approximately $0.005 + 0.02 = 0.025$.

c. No, the sampling distribution isn't approximately normal for a sample size of 5, so using the normal distribution to approximate it won't be very accurate.

P13. The analysis would not change because the sample size in both cases is a very small fraction of the population size.

P14. a. This sampling distribution can be considered approximately normal because both $np = 53$ and $n(1 - p) = 47$ are 10 or greater. The sampling distribution has a mean of 0.53 and a standard error of

$$\sigma_{\hat{p}} = \sqrt{\frac{p(1-p)}{n}} = \sqrt{\frac{0.53(1-0.53)}{100}} \approx 0.05$$

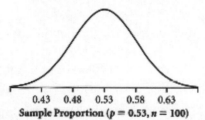

0.43 0.48 0.53 0.58 0.63
Sample Proportion ($p = 0.53$, $n = 100$)

b. No. As you can see from the sampling distribution in part a, the probability of getting 9 (or 9%) or fewer women just by chance is almost 0. That there are so few women in the U.S. Senate cannot reasonably be attributed to chance alone, and so we should look for other explanations (for example, that fewer women go into politics, that women are not able to raise as much money for their campaigns, or that voters are reluctant to elect a woman senator).

P15. a. First, note that $p = 0.15$ and $\sigma_p = \sqrt{\frac{(0.15)(0.85)}{400}} \approx 0.018$. Also, the sampling distribution of $\hat{p}$ is approximately normal because $np = 0.15(400) > 10$ and $n(1-p) = 0.85(400) > 10$. Hence, the reasonably likely proportions are

$$0.15 \pm 1.960(0.018),$$

or 0.115 to 0.185. This means that there is a 95% chance it is within 3.5% of 15%.

b. First, note that $p = 0.15$ and $\sigma_p = \sqrt{\frac{(0.15)(0.85)}{4000}} \approx 0.0056$. Hence, the reasonably likely proportions are

$$0.15 \pm 1.960(0.0056),$$

or 0.139 to 0.161. This means that there is a 95% chance it is within 1.1% of 15%.

P16. a. Here, $n = 435$ and $p = 0.38$. This sampling distribution can be considered approximately normal because both $np = 165.3$ and $n(1 - p) = 269.7$ are 10 or greater. The sampling distribution has a mean of 0.38 and a standard error of

$$\sigma_{\hat{p}} = \sqrt{\frac{p(1-p)}{n}} = \sqrt{\frac{0.38(1-0.38)}{435}} \approx 0.023$$

A sample proportion of 0.59 is about 95 standard deviations above the mean of the sampling distribution, and therefore it would be very unusual to see such a result. The probability is close to 0.

b. Yes, $\hat{p} = 257 / 435 \approx 0.59$. As you saw in part a, it is almost impossible for this many representatives to be Democrats if they were chosen at random from the general population. Possible reasons include the fact that people do not always vote for their party's candidate. In addition, the population that year identified themselves as 28% Republican, 34% Democrat, and 38% Independent, yet there was only one Independent in the House. So one possible explanation is that Independents tended to vote Democrat in this particular election.

c. For example, people don't always vote for the candidate of their party.

P17. First check the guideline: Both $np = 100 \cdot 0.61 = 61$ and $n(1 - p) = 100 \cdot 0.39 = 39$ are greater than 10, so the shape of the distribution will be approximately normal. $\mu_{sum} = np = 61$. $\sigma_{sum} = \sqrt{np(1-p)} = \sqrt{100 \cdot 0.61 \cdot 0.39} \approx 4.88$. We must find z-scores and probabilities for 50 freshmen.

$$z = \frac{50 - 61}{4.88} \approx -2.25, \ P(sum \geq 50) \approx 0.9879$$

Yes, because there is about a 98.8% probability that a sample of 100 freshmen will contain at least 50 freshmen who are attending their first choice college.

P18. a. First check the guideline: Both $np = 1000 \cdot 0.23 = 230$ and $n(1 - p) = 1000 \cdot 0.77 = 770$ are greater than 10, so the shape of the distribution will be approximately normal. $\mu_{sum} = np = 230$. $\sigma_{sum} = \sqrt{np(1-p)} = \sqrt{1000 \cdot 0.23 \cdot 0.77} \approx 13.31$. So, the z-score corresponding to 18% (or 0.18(1000)=180 people) is

$$z = \frac{180 - 230}{13.31} \approx -3.757, \ P(sum \leq 180) = P(z \leq -3.757) \approx 0.0001.$$

So, there is about a 0.01% chance that the sum will not exceed 180 people.

b. Similar to (a), the z-score corresponding to 24% (or 240 people) is

$$z = \frac{240 - 230}{13.31} \approx 0.751, \ P(sum \geq 240) = P(z \geq 0.751) \approx 0.2261.$$

c. The reasonably likely proportions are given by the interval
$$0.23 \pm 1.96(0.013) = 0.23 \pm 0.026.$$

So, there is a 95% chance that it is no farther than 2.6% from 23%.

Exercise Solutions

E1. **a.** $\bar{x} = \dfrac{12.500 + 11.625 + 18.275 + 13.225}{4} = 13.90625$

b. $\bar{x} = \dfrac{6.625 + 10.375 + 13.225 + 8.800}{4} = 9.75625$

c. $P(\bar{x} \geq 13.90625) = \frac{1}{70}$ (because just 1 bar of height 1 occurs to the right of 13.90625).

d. No, the high mean enzyme level probably was not due to chance, but because the shorter days cause higher enzyme levels.

E3. **a.** The distribution is highly skewed toward the larger values. There are two outliers, Alaska and Texas.

b. Number the states from 01 to 50 and then use a table of random digits or the TI-84 entry **randInt(1,50,5)** to select five of them, discarding any duplicates.
Here is a dot plot for one random sample of size five. Although results will vary, it is unlikely that a single sample of size five will include Texas or Alaska, so these values, ranging from 19.9 to 44.6, do appear to capture the essence of where most of the areas lie. The mean of these five values is 30.80.

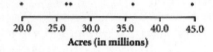

c. The samples will occasionally pick up Texas or Alaska, so the sampling distribution for these small samples is still highly skewed toward the larger values but with a much smaller spread than the population. The mean is still in the low 40's.

d. To get a sample mean between 95 and 105, one of the states must be either Alaska or Texas. The five states could be, for example, Alaska, Minnesota, Florida, Iowa, and Kentucky. If Texas is chosen, the other four states must all be fairly large, say California, Arizona, Idaho, and Montana.

E5. **a.**

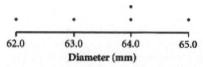

b. 62 and 63; 62 and 64; 62 and 64; 62 and 65; 63 and 64; 63 and 64; 63 and 65; 64 and 64; 64 and 65; 64 and 65.

c. In the same order as the samples listed in part b, their means are 62.5, 63, 63, 63.5, 63.5, 63.5, 64, 64, 64.5, and 64.5. The exact sampling distribution of the sample mean for samples of size 2 is shown in this dot plot.

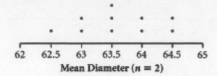

Mean Diameter ($n = 2$)

The sampling distribution has a mean of 63.6 and a standard error of about 0.6245. The population has a mean of 63.6 and a standard deviation of about 1.02. The two means are exactly the same and, as usual, the standard error is smaller than the standard deviation of the population.

Note: *You may have noticed that the standard error is not equal to*
$$\sigma / \sqrt{n} = 1.02 / \sqrt{2} = 0.72.$$
That is because this formula is true only if there is an infinite population or if sampling with replacement is used, which we did not do in this exercise. Because we are taking a sample of size 2 from a population of only five tennis balls, sampling with or without replacement makes quite a difference in the standard error.

If the tennis balls had been sampled with replacement, this formula would give the correct SE. But because there is no replacement a different formula is needed:
$$SE = \frac{\sigma}{\sqrt{n}} \cdot \sqrt{\frac{N-n}{N-1}} = \frac{1.02}{\sqrt{2}} \cdot \sqrt{\frac{3}{4}} = 0.6245$$
This also would be the correct short-cut formula to use in E6, where the sampling distribution for the mean is computed by sampling without replacement in part c.]

d. The sample maximums for the 10 possible samples are 63, 64, 64, 65, 64, 64, 65, 64, 65, 65. A dot plot is shown here.

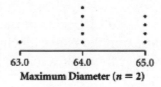

Maximum Diameter ($n = 2$)

e. The sample ranges are 1, 2, 2, 3, 1, 1, 2, 0, 1, and 1. A dot plot is shown here.

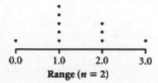

Range ($n = 2$)

E7. a. The sample mean is an unbiased estimator of the population mean because the mean of the sampling distribution of the sample mean is the same as the population mean.

b. The sample maximum is not an unbiased estimator of the population maximum because the mean of the sampling distribution of the sample maximum is less than the population maximum. The sample maximum tends to be too low.

c. The sample range is not an unbiased estimator of the population range because the mean of the sampling distribution of the sample range is less than the population range. The sample range tends to be too low.

E9. a. The population range is $100 - 0 = 100$.

b. The sample range is always less than or equal to 100, so the mean of these values is less than 100. In this case, the mean of the values was 78.725.

c. Joel's estimate is more symmetric, but it is still centered too low. The mean value of doubling the IQR for his 10,000 samples was 78.725. This is still a biased estimator. Additionally, the spread is too large to be able to have much confidence in the result of any one sample.

E11. a. – e.

Sample	Mean	Variance, Dividing by $n = 2$	Variance, Dividing by $n - 1 = 1$
2, 2	2	0	0
2, 4	3	1	2
2, 6	4	4	8
4, 2	3	1	2
4, 4	4	0	0
4, 6	5	1	2
6, 2	4	4	8
6, 4	5	1	2
6, 6	6	0	0
Average	4	$1\frac{1}{3}$	$2\frac{2}{3}$

f. The variance of the population {2, 4, 6} is $\frac{8}{3}$ which is equal to the average of the sample variances when you divide by $n - 1$.

g. Dividing by $n - 1$, because the average of the sample variances is equal to the population variance. Although this phenomenon has been shown for only one example, it is true in all cases.

h. Too small. When you divide by $n - 1$, the sample variance is an unbiased estimator of the population variance. This means that although the sample variance is not always equal to the population variance, the average of the variances of all possible samples is exactly equal to the population variance. If you divide by n, the sample variance tends to be too small, on average. Note that the sample mean, when you divide by n, is an unbiased estimator of the population mean. If you divided by $n - 1$, it would be too big.

E13. a. There would be five approximately equal gaps over a distance of N, so the average gap between each number is (1/5)N.

b. The maximum would be 4/5 of the way to N, so it would be at (4/5)N. Alternatively, in terms of the average gap, the sample maximum would occur after 4 average gaps or at 4 (1/5)N = (4/5)N.

c. From (b), *sample maximum* ≈ (4/5)N, so N ≈ (5/4) · *sample maximum*

d. The sampling distribution of the adjusted maximum from the same batch of 200 random samples of size 5 is shown in the next display. It appears from this simulation, that the adjusted maximum does a better job of estimating the population maximum. However, the mean of this simulated sampling distribution is still a little less than the population maximum of 18. This difference is because the distribution of the areas of the rectangles is not uniform. This sampling distribution has mean 16.8, and standard deviation 4.44.

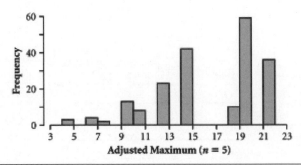

E15. a. The z-score is

$$z = \frac{510 - 500}{100} = 0.10$$

which gives a probability of 0.4602. Alternatively, on a TI-84 the result of **normalcdf (510,9999999,500,100)** is approximately 0.4602.

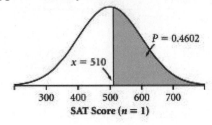

b. The z-score is

$$z = \frac{510 - 500}{100 / \sqrt{4}} = 0.20$$

which gives a probability of 0.4207.

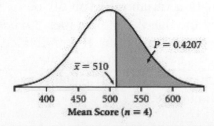

128

c. The z-score is

$$z = \frac{510 - 500}{100 / \sqrt{25}} = 0.50$$

which gives a probability of 0.3085.

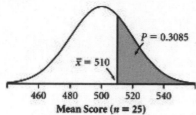

d. The z-score for 490 is

$$z = \frac{490 - 500}{100 / \sqrt{25}} = -0.50.$$

Using this, together with the z-score for 510 given in (c), we find that

$$P(490 < \overline{x} < 510) = P(-0.5 < z < 0.5) = 0.383.$$

E17. a. If the weekends are randomly selected, we would expect to see 2 • 67.4 = 134.8, or a total of about 135 children. The expected number is 134.8 children.

b. Since the numbers of children seen on a summer weekend were approximately normally distributed, the sampling distribution of the mean, even with $n = 2$, will be approximately normally distributed. The z-score is

$$z = \frac{73 - 134.8}{10.4 / \sqrt{2}} = -8.40$$

and the probability is almost 0.

c. We can conclude that the low number of children seen in the emergency room is not due to chance.

E19. a.   I. Histogram B; $n = 25$
II.   Histogram A; $n = 4$
III.   Histogram C; $n = 2$

b. The theoretical standard error is $2.402 / \sqrt{n}$ which turns out to be 1.698, 1.201, and 0.480 for the respective sample sizes of 2, 4, and 25. All of these are fairly close to the observed standard errors.

c. For samples of size 2 and 4, the simulated sampling distributions of the mean reflect the skewness of the population distribution. For samples of size 25, the skewness is essentially eliminated and the simulated sampling distribution looks like a normal distribution.

d. The rule that about 95% of the observations lie within two standard errors of the population mean works well for $n = 25$, and slightly less well for the skewed distributions occurring for $n = 4$ and $n = 2$.

E21. **a.** No. Exactly two accidents happened in about 15% of the days. Two or more accidents happened in about 25% of the days.

b. In the display, the first plot is the one for eight days, and the second plot is the one for four days.

c. For the four-day averages, two accidents are not very likely. It occurs here about 6 times out of 100. For the second part, yes, if the days can be viewed as a random sample of all days. For the eight-day average, an average of two accidents never occurred and, hence, could be deemed very unlikely.

d. The sampling distributions used in parts a–c are based on random samples of four and eight days. A particular period of four (or eight) days may not look like a random sample at all because of the high dependency from day to day. For example, if all days of a sample are taken from the winter season in a region that has ice and snow, the accident rate might be far higher than what is typical for the rest of the year.

E23. **a.** You want the probability that the mean score will be between 490 and 510. For a mean score of 490, the z-score is $z = (490 - 500) / (100/\sqrt{40}) \approx -0.632$ and the probability that the mean is less than or equal to 490 is about 0.2635.

For a mean score of 510, the z-score is $z = (510 - 500) / (100/\sqrt{40}) \approx 0.632$ and the probability that the mean is less than or equal to 510 is about 0.7365. So the probability that the mean score is between 490 and 510 is $0.7365 - 0.2635 = 0.473$.

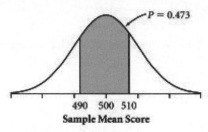

Or, **normalcdf**(490, 510, 500, $100/\sqrt{(40)}$) = 0.473

b. We know that if the sampling distribution is approximately normal, about 95% of all sample means are in the interval $\mu_{\bar{x}} \pm 1.96 \dfrac{\sigma}{\sqrt{n}}$. Thus, to be 95% sure that the sample mean is within a value E of the population mean, we must have $E \geq 1.96 \dfrac{\sigma}{\sqrt{n}}$. Solving for the square root gives: $\sqrt{n} \geq 1.96 \dfrac{\sigma}{E}$. Squaring both sides and simplifying gives:

$$n \geq \frac{3.8416\sigma^2}{E^2} = \frac{3.8416 \cdot 100^2}{10^2} \approx 384.16$$

Rounding up, this gives a sample size of 385.

E25. a. The sampling distribution of the sample total should be approximately normal because this is a very big sample size. It has mean and standard error

$$\mu_{sum} = n\mu = 1000(0.9) = 900$$

$$\sigma_{sum} = \sigma\sqrt{n} = 1.1\sqrt{1000} \approx 34.79$$

The z-score for 1000 families is then

$$z = \frac{sum - n\mu}{\sigma\sqrt{n}} = \frac{1000 - 900}{34.785} \approx 2.875 .$$

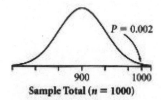

Sample Total ($n = 1000$)

The probability of getting at least 1000 children is about 0.002. Thus, there is almost no chance the network will get 1000 children.

b. Yes, the probability goes from practically 0 to almost certain. The z-score for 1200 families is

$$z = \frac{sum - n\mu}{\sigma\sqrt{n}} = \frac{1000 - 1080}{38.105} \approx -2.099 .$$

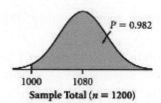

Sample Total ($n = 1200$)

The probability of getting at least 1000 children in a random sample of 1200 families is 0.982.

E27. We will do this problem using the sampling distribution of the sample sum, which has a mean of $4.3n$ and a standard error $1.4 \cdot \sqrt{n}$.

a. The point that cuts off the lower 0.02 (so 98% is above it) on the normal distribution is at $z = -2.054$. If the total number of people is above 100, then we must choose the sample size, n, so that the point 100 lies 2.054 standard errors below $4.3n$, or

$$100 = 4.3n - 2.054 \cdot 1.4\sqrt{n} \quad \text{or} \quad 4.3n - 2.8756\sqrt{n} - 100 = 0$$

This equation is quadratic in form. It can be solved for $\sqrt{n}$ by using the quadratic formula and then squaring the positive solution to get n. The negative solution for $\sqrt{n}$ can be ignored. Alternatively, the solution can be estimated by plotting the equation on a graphing calculator. Because we only need the nearest integer solution, either method works well. The solution is about $n = 27$, so the director should select about 27 names.

b. However, drawing 27 names, he is likely to get more than 100 people. The expected number of people he will get is $4.3 \cdot 27 = 116.1$ people. So 16.1 extra people will cost him $16.1 \cdot \$250 = \$4025$. This was a pretty costly oversight!

E29. a. Observe that

$$\textbf{i. } 0.9535 \quad \textbf{ii. } 0.7107 \quad \textbf{iii. } 0$$

b. For all sample sizes, the *SE* is smaller with the adjustment for non-replacement, not much smaller at first, but decreases to 0 when $n = N$. As the sample size increases, you are sure to have a bigger proportion of the population in the sample and so $\bar{x}$ tends to be closer to μ.

E31. a.

$$\mu_{v+m} = \mu_v + \mu_m = 501 + 515 = 1016$$

$$\sigma_{v+m} = \sqrt{\sigma_v^2 + \sigma_m^2} = \sqrt{112^2 + 116^2} \approx 161.245$$

b. The sampling distribution of the sum will be approximately normal because both distributions are normal. The *z*-score is

$$z = \frac{800 - 1016}{161.245} = -1.340$$

The probability is about 0.0901.

c. $1016 \pm 1.96(161.245)$, or between approximately 700 and 1332.

d. We need $v - m \geq 100$. The sampling distribution of the difference will be approximately normal because both distributions are normal. The mean and standard error of the sampling distribution of the difference are

$$\mu_{v-m} = \mu_v - \mu_m = 501 - 515 = -14$$

$$\sigma_{v-m} = \sqrt{\sigma_v^2 + \sigma_m^2} = \sqrt{112^2 + 116^2} \approx 161.245$$

The *z*-score for a difference of 100 is

$$z = \frac{100 - (-14)}{161.245} \approx 0.707$$

The probability is 0.2969 (or 0.2981 using Table A).

Note: This does not imply that 29.69% of students have an SAT verbal score of at least 100 points higher than their SAT math score. See part e.

e. The mean of the distribution of the sum will still be 1016. The standard error is unpredictable because the two scores are certainly not independent. Students who score high on the verbal portion also tend to score high on the math portion. The shape is also unpredictable.

E33. a. $\mu_{sum} = 3(3.5) = 10.5$; $\sigma^2_{sum.} = 3(2.917) = 8.751$

b. $\mu_{sum} = 7(3.5) = 24.5$; $\sigma^2_{sum.} = 7(2.917) = 20.419$

c. The sampling distribution of the sum is approximately normal because we were told in E31 that the distribution of verbal scores is approximately normal. It has mean and variance

$$\mu_{sum} = 20 \cdot 501 = 10,020$$
$$\sigma^2_{sum} = n\sigma^2 = 20 \cdot 112^2 = 250,880$$

The z-score for a total of 10,000 is

$$z = \frac{10,000 - 10,020}{\sqrt{250,880}} \approx -0.040$$

The probability is about 0.4840.

E35. The sampling distributions for the sample proportion and the number of successes can be considered approximately normal because both $np = 920$ and $n(1 - p) = 80$ are 10 or greater.

a. This distribution is approximately normal with mean and standard error

$$\mu_{\hat{p}} = 0.92$$

$$\sigma_{\hat{p}} = \sqrt{\frac{p(1-p)}{n}} = \sqrt{\frac{0.92(1-0.92)}{1000}} \approx 0.00858$$

The sketch, with scale on the x-axis, is shown in part (c).

b. The z-score is: $z = \dfrac{0.9 - 0.92}{\sqrt{\dfrac{0.92(1-0.92)}{1000}}} \approx -2.33$

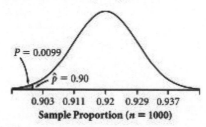

The probability is about 0.0099.

c. The z-score is: $z = \dfrac{925 - 920}{\sqrt{1000(0.92)(1-0.92)}} \approx 0.5828$

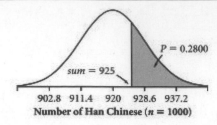

Number of Han Chinese ($n = 1000$)

The probability is about 0.2800.

d. Rare events would be the totals that are outside the interval $920 \pm 1.96(8.58)$; that is, larger than 936.82 or smaller than 903.18. Rare events would be the proportions that are outside the interval $0.92 \pm 1.96(0.00858)$; that is, larger than 0.937 or smaller than 0.903.

E37. a. The shape would be slightly skewed to the left because $np = 100 \cdot 0.92 = 92$ and $n(1 - p) = 100 \cdot 0.08 = 8$ are not both at least 10. The mean of the sampling distribution would still be 0.92 and the spread would be greater because of the smaller sample size.

b. With $n = 100$, the distribution is more spread out and skewed left than the distribution for $n = 1000$. This means a larger percentage of the sample proportions would be further from p and, in this case, in the direction of the skew or in the left tail. Because the sample proportion of 0.9 is in the left portion of the distribution or the part that is with the skew, the probability of getting 90% or fewer in the sample will be greater with a sample size of 100 than with a sample size of 1000.

Note: we cannot accurately calculate this probability using our formula because the distribution is not approximately normal.

E39. a. 0.5

b. The sampling distribution can be considered approximately normal because both np = 25 and $n(1 - p) = 25$ are 10 or greater. The sampling distribution has mean and standard error

$$\mu_{\hat{p}} = 0.5 \text{ and } \sigma_{\hat{p}} = \sqrt{\frac{0.5(1-0.5)}{50}} \approx 0.0707$$

The z-score for a sample proportion of 0.2 is

$$z = \frac{0.2 - 0.5}{0.0707} \approx -4.2433$$

The probability of getting 10 or fewer under the median age is 0.000011.

c. These results are not at all what we would expect from a random sample of 50 people. This group is special in that they are all old enough to have a job. Children are included in computing the 33.1 median age, but they do not hold jobs. Thus, we would expect the median age of job holders to be greater than the median age of all people in the United States, as they are at Westvaco.

E41. You are assuming that the 75 married women were selected randomly from a population in which 60% are employed. Both $np = 75 \cdot 0.6 = 45$ and $n(1-p) = 75 \cdot 0.4 = 30$ are at least 10, so the distribution will be approximately normal. We will use the distribution of the sum which has

$$\mu_{sum} = np = 45 \text{ and } \sigma_{sum} = \sqrt{np(1-p)} = \sqrt{75 \cdot 0.6 \cdot 0.4} \approx 4.24.$$

For 30 employed women in the sample, $z = (30 - 45) / 4.24 = -3.538$.
$P(sum \leq 30) = 0.0002$.

For 40 employed women in the sample, $z = (40 - 45) / 4.24 = -1.179$.
$P(sum \leq 40) = 0.1192$.
$P(30 \leq sum \leq 40) \approx 0.1192 - 0.0002 = 0.1190$.

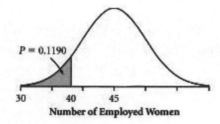

E43. a. The distribution for $n = 100$ is the most normal while the distribution for $n = 10$ is the least normal. Notice that for $n = 10$, $np = 10 \cdot 0.1 = 1 < 10$, and $n(1-p) = 10 \cdot 0.9 = 9 < 10$. The guideline does not hold and the distribution for $n = 10$ is quite skewed. For $n = 20$, $np = 2 < 10$, but $n(1-p) = 18 > 10$. The guideline does not hold and this distribution is less skewed, but the skew is still very noticeable. When $n = 40$, $np = 4 < 10$ but $n(1-p) = 36 > 10$. The guideline does not hold and this distribution is less skewed, but the skew is still noticeable. For $n = 100$, both $np = 10$ and $n(1-p) = 90$ are at least 10. The guideline does hold and the distribution looks very symmetric and fairly smooth. The skew is visible for all but $n = 100$, where both np and $n(1-p)$ are at least 10. It appears that the guideline works quite well.

b. In each of these, the expected proportion of drivers is 10%.

c. The distribution for $n = 10$ has the largest spread, and the distribution for $n = 100$ has the smallest spread.

d. This would be more likely in a sample of size 10 drivers. The area of the bars in the histogram to the right of 0.20 gets smaller as the sample size increases and with larger sample sizes, the proportions tend to cluster closer to 0.1.

E45. a. The means of the sampling distributions definitely depend upon *p,* as the first three center close to $p = 0.2$ and the second three center close to $p = 0.4$. The sampling distributions have centers close to p regardless of the sample size; the centers do not depend upon the sample size.

b. The spreads of the sampling distributions decrease as *n* increases for both values of *p*. The spreads do depend upon the value of *p*, however. For each sample size, the spread for *p* = 0.4 has a larger standard error than the one for *p* = 0.2.

c. For *p* = 0.2, the shape is quite skewed for *n* = 5 and some slight skewness remains at *n* = 25. For *n* = 100, the shape is basically symmetric. For *p* = 0.4, the shape shows a slight skewness at *n* = 5 but is fairly symmetric at *n* = 25 and beyond. The farther *p* is from 0.5, the more skewness in the distribution of the sample proportion.

d. The rule does not work well for samples of size 5 for either value of *p* but works well for samples of size 25 or more for both values of *p*.

E47. a. $\bar{x} = \dfrac{\sum xf}{n} = \dfrac{0 \cdot 14 + 1 \cdot 26}{40} = 0.65$.

This sample mean is the same as $\hat{p}$ because each success adds 1 and each failure adds 0, so it counts the number of successes, then divides by the sample size. This is exactly the procedure for calculating $\hat{p}$.

b. $\mu = \sum x \cdot P(x) = 0 \cdot 0.4 + 1 \cdot 0.6 = 0.6$. The box says $\mu = p = 0.6$, which is the same.

c. The sample proportion is a type of sample mean, if the successes are given the value 1 and failures the value 0.

E49. a. Because the sample size is less than 10% of the population size, it does not really matter that we sampled without replacement, and so we can use the following formulas for the mean and standard error:

$$\mu_{\bar{x}} = \mu = 140.9$$

$$\sigma_{\bar{x}} = \frac{\sigma}{\sqrt{n}} = \frac{119.4}{\sqrt{3}} \approx 68.94$$

(The formula for the mean holds with or without replacement. Although the formula for the standard error holds exactly when sampling is done with replacement, it is only approximately true when sampling is done without replacement. The larger *N* is compared to *n*, the better the approximation.)

b. No. This is a very small sample from a highly skewed population. The sampling distribution of the sample mean will be skewed right.

E51. a. The shape can be considered approximately normal because both *np* = 200(0.68) = 136 and *n*(1 − *p*) = 200(0.32) = 64 are 10 or greater. The mean and standard error are

$$\mu_{\hat{p}} = 0.68$$

$$\sigma_{\hat{p}} = \sqrt{\frac{0.68 \cdot (1-0.68)}{200}} \approx 0.0330$$

b. The z-score for $\hat{p} = 0.75$ is

$$z = \frac{0.75 - 0.68}{0.0330} \approx 2.121$$

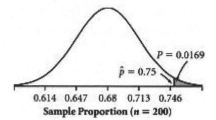

The probability is about 0.0170.

c. Values less than $0.68 - 1.96(0.0330) = 0.615$, or larger than $0.68 + 1.96(0.0330) = 0.745$.

d. The mean and standard error of the sampling distribution of the sum are

$$\mu_{sum} = 200(0.68) = 136 \text{ and } \sigma_{sum} = \sqrt{200(0.68)(1-0.68)} \approx 6.597$$

The z-score for 130 people is

$$z = \frac{130 - 136}{6.597} \approx -0.9095$$

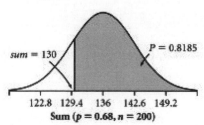

The probability is about 0.8185.

Alternatively, this problem can be done using the mean and standard deviation in part a and with $\hat{p} = \frac{130}{200} = 0.65$.

E53. a. One interesting feature is the regular increase from 40 to 55 followed by a sharp drop after 55 and another sharp drop after 65.

b. Since the distribution of ages has little skew, a sampling distribution of the sample mean with $n = 10$ should be approximately normal. The mean and standard deviation of the distribution are

$$\mu_{\bar{x}} = 56, \quad \sigma_{\bar{x}} = \frac{6.3}{\sqrt{10}} \approx 1.99$$

For a sample mean of 55, $z = \dfrac{55-56}{1.99} \approx -0.503$, $P(\bar{x} < 55) = 0.3085$

For a sample mean of 60, $z = \dfrac{60-56}{1.99} = 2.010$, $P(\bar{x} < 60) = 0.9772$

As such,

$$P(55 < \bar{x} < 60) = 0.9772 - 0.3085 = 0.6687.$$

E55. a. The shape is approximately normal because each of the two distributions is approximately normal. The mean and standard error of the sampling distribution of the difference are

$$\mu_{m-w} = \mu_m - \mu_w = 68 - 64 = 4$$

$$\sigma_{m-w} = \sqrt{\sigma_m^2 + \sigma_w^2} = \sqrt{2.7^2 + 2.5^2} \approx 3.680$$

b. The z-score for a difference of 2 is

$$z = \dfrac{2-4}{3.680} \approx -0.5435$$

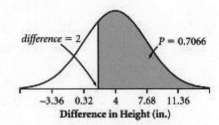

The probability is 0.7066.

c. She is taller than he is if the difference is negative (less than 0). The z-score for a difference of 0 is

$$z = \dfrac{0-4}{3.680} = -1.087$$

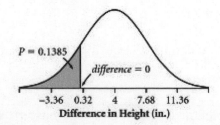

The probability is 0.1385.

E57. Select two integers at random from 0 through 9 and compute their mean. Repeat this. Subtract the second mean from the first. Students should do this three times.

Because the distribution of the results of a single selection from the population {0, 1, 2, 3, 4, 5, 6, 7, 8, 9} has a mean of 4.5 and a standard deviation of 2.87, the sampling distribution of the mean of two selections will have a mean of 4.5 and standard error

$$\frac{\sigma}{\sqrt{n}} = \frac{2.87}{\sqrt{2}} \approx 2.03$$

Then, the mean of the sampling distribution of the difference between the two players' averages is $\mu_{you-opponent} = \mu_{you} - \mu_{opponent} = 4.5 - 4.5 = 0$.

The standard error is

$$\sigma_{you-opponent} = \sqrt{\sigma^2_{you} + \sigma^2_{opponent}} = \sqrt{2.03^2 + 2.03^2} \approx 2.87$$

The shape will still be basically triangular, but rounding out slightly.

E59. The shape becomes more approximately normal, the mean stays fixed at the population mean, and the standard error decreases with (is proportional to) the reciprocal of the square root of n.

E61. Because the distribution from which the samples were selected is highly skewed, the skewness persists in the simulated sampling distribution of the sample mean for samples of size 5 and 10. However, with samples of size 20, most of the skewness disappears and you see a sampling distribution that is nearly normal in shape. The centers of the sampling distributions lie close to the population mean of 42,520 passengers. Observe that the standard deviations of the sampling distributions shrink faster, with increasing sample size, than the theory used earlier for sampling *with replacement* would predict. For $n = 20$, the observed standard deviation of 1,909 is much smaller than $14758/\sqrt{20} = 3299.99$ This is because we are taking a sample of size 20 without replacement from a small population, with only 30 values. However, if we use the adjusted formula from E30, for sampling *without replacement*

$$\sigma_{\bar{x}} = \frac{\sigma}{\sqrt{n}}\sqrt{\frac{N-n}{N-1}} = \frac{14758}{\sqrt{20}}\sqrt{\frac{30-20}{30-1}} = 1937.8$$

we get very close to the SE of the simulated sampling distribution. Recall that the simpler formula is generally 'close enough' if the sample size is less than 10% of the population size.

E63. The cap fits on the bottle if $d_c > d_b$, where d_c is the inside diameter of the cap and d_b is the outside diameter of the bottle. So the problem asks for the probability that $d_c - d_b$ >0. Because both sampling distributions are approximately normal, the sampling distribution of $d_c - d_b$ is approximately normal. This sampling distribution has mean and variance

$$\mu_{difference} = \mu_1 - \mu_2 = 36 - 35 = 1 \text{ mm}$$
$$\sigma^2_{difference} = \sigma^2_1 + \sigma^2_2 = 1^2 + 1.2^2 = 2.44 \text{ mm}^2$$

Taking the square root, the standard deviation is about 1.562 mm. The area to the right of zero is shaded in the sampling distribution of $d_c - d_b$.

The z-score for a difference of zero is

$$z = \frac{\text{sample difference} - \mu_{difference}}{\sigma_{difference}} = \frac{0-1}{1.562} \approx -0.640$$

The area to the right of this z-score is 0.739. The chance the cap fits on the bottle is only 0.739 – not very good for a manufacturing process.

E65. a. $\mu_{morning} = 13.2$ and $\sigma^2_{morning} = 7.76$; $\mu_{afternoon} = 8$ and $\sigma^2_{afternoon} = 4$.

b. $\mu_{total} = 21.2$ and $\sigma^2_{total} = 10.96$.

c. The sum of the means 13.2 + 8 equals 21.2, as we would expect. But the sum of the variances 7.76 + 4 is not equal to 10.96, and we do not expect it to be because the morning and afternoon times were not selected independently when we computed 10.96.

Chapter 8

Practice Problem Solutions

P1. Note that $\hat{p} = \frac{14}{40} = 0.35$. So, the middle 95% of the sampling distribution of $\hat{p}$ is

$$0.35 \pm 1.96\sqrt{\frac{(0.35)(0.65)}{40}}$$

or about 0.202 to 0.498. Hence, we easily classify the given values as plausible or not plausible as follows:

a. Yes   **b.** Yes   **c.** No.

P2. Note that $\hat{p} = 0.08$. So, the middle 95% of the sampling distribution of $\hat{p}$ is

$$0.08 \pm 1.96\sqrt{\frac{(0.08)(0.92)}{1166}}$$

or about 0.064 to 0.096. Hence, we easily classify the given values as plausible or not plausible as follows:

a. No   **b.** No   **c.** Definitely not.

P3. a. Using $n = 40$ and $\hat{p} = \frac{25}{40} = 0.625$, we see that the 95% confidence interval is

$$0.625 \pm 1.96\sqrt{\frac{(0.625)(0.375)}{40}}$$

or about 0.47 to 0.78.

b. p is the proportion of males in the population from which the sample was selected who would give shocks up to the danger level.

P4. a. The population is the collection of all college-bound students at the time of the Survey. The parameter being estimated is the proportion of the population that reported reading about college rankings. And, $\hat{p} = 0.20$.

b. Yes. Note that

$$np = 500(0.20) = 100, \quad n(1-p) = 500(0.80) = 400.$$

The sample is taken from a binomial population (since there are only two responses allowed) and the population size certainly exceeds 5,000.

c. Using $n = 500$ and $\hat{p} = 0.20$, we see that the 90% confidence interval is

$$0.20 \pm 1.645\sqrt{\frac{(0.20)(0.80)}{500}}$$

or about 0.17 to 0.23. This means that among the college-bound students in the population from which this sample was selected, you are 90% confident that the percentage who would report they have read about college rankings is between 17% and 23%.

P5. You would expect $0.95(365) = 346.75$ of the confidence intervals to contain the proportion.

P6. You would expect $0.95(350) = 332.50$ of the confidence intervals to contain the proportion.

P7. a. Using $n = 1144$ and $\hat{p} = 0.71$, observe that $np \geq 10$, $n(1-p) \geq 10$. The sample is taken from a binomial population (since there are only two responses allowed) and the population size certainly exceeds 11,400. So, the conditions are met. But, there might be a non-response bias.

b. The 95% confidence interval is

$$0.71 \pm 1.96\sqrt{\frac{(0.71)(0.29)}{1144}}$$

or about 0.68 to 0.74.

c. Among the population of physicians from which this sample was selected, you are 95% confident that the percentage who would say that they had an obligation to refer the patent is between 68% and 74%.

d. The method used to produce the confidence interval captures the population proportion 95% of the time in repeated usage.

P8. Observe that the confidence interval is given by 0.51 ± 0.031, or about 0.479 to 0.541.
a. Yes
b. Note that

$$E = z^*\sqrt{\frac{(0.51)(0.49)}{1011}} = 0.016z^*.$$

So, if $z^* = 1.96$ (for 95% confidence), then the margin of error is 0.031. So, it is reported correctly.

P9. Observe that the confidence interval is given by 0.46 ± 0.03, or about 0.43 to 0.49.
a. No
b. Note that

$$E = z^*\sqrt{\frac{(0.46)(0.54)}{1400}} = 0.013z^*.$$

So, if $z^* = 1.96$ (for 95% confidence), then the margin of error is $0.026 \approx 0.03$. So, it is reported correctly.

P10. a. $z^* = 1.285$ for 80% confidence level.

b. The formula for margin of error is $E = z^*\sqrt{\frac{p(1-p)}{n}}$. Since the value of z^* increases as the confidence level increases, the margin of error will be larger for 95% confidence level than for 80% confidence level.

P11. a. Decrease because $z^*\sqrt{\frac{p(1-p)}{400}} > z^*\sqrt{\frac{p(1-p)}{800}}$.

b. Decrease because $1.96\sqrt{\frac{p(1-p)}{n}} > 1.645\sqrt{\frac{p(1-p)}{n}}$.

c. Decrease (for same reason as given in (b)).

P12. Using $\hat{p} = 0.36$, we see that the desired value of n is given by

$$n = (1.96)^2\left[\frac{(0.36)(0.64)}{(0.03)^2}\right] \approx 984.$$

P13. Using $\hat{p} = 0.50$ (since a value is not specified), we see that the desired value of n is given by

$$n = (1.96)^2\left[\frac{(0.5)(0.5)}{(0.03)^2}\right] \approx 1067.$$

P14. Because you have no estimate for p, use $p = 0.5$.
a. For $E = 2\%$ with 95% confidence, use
$$n = z^2\left(\frac{p(1-p)}{E^2}\right) = 1.96^2\left(\frac{0.5(1-0.5)}{0.02^2}\right) = 2401.$$

b. For $E = 1\%$ with 99% confidence, use
$$n = z^2\left(\frac{p(1-p)}{E^2}\right) = 2.576^2\left(\frac{0.5(1-0.5)}{0.01^2}\right) = 16,589.44 \approx 16,590.$$

c. For $E = 0.5\%$ with 90% confidence, use
$$n = z^2\left(\frac{p(1-p)}{E^2}\right) = 1.645^2\left(\frac{0.5(1-0.5)}{0.005^2}\right) = 27,060.25 \text{ or } 27,061.$$

P15. a. The standard $p_0 = 0.5$.

b. Since it is asking if there is a change (period), the test should be two-sided. Hence, the null and alternative hypotheses are:

H_o: $p = 0.50$ H_a: $p \neq 0.50$, where p is the proportion of all of this year's juniors who would say they will want extra tickets

c. $\hat{p} = \frac{16}{40} = 0.40$

d. There is not strong evidence of change.

P16. a. The student is guessing, and so the probability the student gets a question right is 0.5. Or, using symbols, $p = 0.5$, where p is the probability a student gets a question right.

b. The test should be one-sided. The null and alternative hypotheses are:
H_o: $p = 0.50$ H_a: $p > 0.50$, where p is the probability that the student gets a question correct

c. The sample proportion is $\hat{p} = \frac{30}{40} = 0.75$.

d. The result is statistically significant. The vertical line leading from a sample proportion of $\hat{p} = 0.75$ does not intersect the middle 95% of the possible values of $\hat{p}$ when $p = 0.5$. Thus 0.75 is not a reasonably likely result if $p = 0.5$. We reject the hypothesis that the student was guessing.

e. This does not prove beyond any doubt that the student was not guessing. The student may have been guessing and been extremely lucky to get a score this high.

P17. <u>For P15:</u> $\dfrac{0.40 - 0.50}{\sqrt{\frac{(0.50)(0.50)}{40}}} = -1.265$ <u>For P16:</u> $\dfrac{0.75 - 0.50}{\sqrt{\frac{(0.50)(0.50)}{40}}} = 3.162$

P18. a. The standard $p_0 = 0.5$.

b. The test should be one-sided. Hence, the null and alternative hypotheses are:

H_o: $p = 0.50$ H_a: $p > 0.50$, where p is the probability that the judge selects the correct dog

c. $\hat{p} = \frac{23}{45} = 0.511$

d. $z = \dfrac{0.511 - 0.50}{\sqrt{\frac{(0.50)(0.50)}{45}}} = 0.1476$

P19. If the 2-year-olds are guessing then $p_0 = 0.5$ and $\hat{p} = \frac{28}{50} = 0.56$. The test statistic is

$$z = \frac{0.56 - 0.5}{\sqrt{\frac{0.5 \cdot 0.5}{50}}} \approx 0.849$$

The *P*-value is 0.198. If the 2-year-olds could not tell a difference between the rakes, the probability of getting the rake with the food 28 times or more is almost 20%. If it is true that 2-year-olds were selecting a rake at random, which would give them a 50% chance of selecting the rake that would get them food, the probability is 0.196 that the 2-year-olds would rake in food as often as they did. This does not provide sufficient evidence that the 2-year-olds were able to deliberately choose the rake that got them the food.

P20. a. It is given that the standard is $p_0 = \frac{2}{3} \approx 0.67$.

b. The test should be two-sided. Hence, the null and alternative hypotheses are:

H_o: $p = 0.67$ H_a: $p \neq 0.67$, where p is the proportion of all U.S. undergraduates who receive financial aid.

c. $\hat{p} = \frac{302}{400} = 0.755$

d. $z = \dfrac{0.755 - 0.67}{\sqrt{\frac{(0.67)(0.33)}{400}}} = 3.615$. The p-value is $2P(Z > 3.615) \approx 0.0003$.

e. There is strong evidence against H_o significant at 0.05.

P21. a. First, we test: H_o: $p = 0.50$ versus H_a: $p < 0.50$, where p is the probability of inattentional blindness.

Observe that the test statistic is
$$z = \frac{0.46 - 0.50}{\sqrt{\frac{(0.50)(0.50)}{192}}} \approx -1.109.$$
The p-value is $P(Z < 1.109) = 0.1561$. As such, there is insufficient evidence to reject the null hypothesis that the inattentional blindness rate is (at least) 50%.

b. The plausible proportions are given by the confidence interval centered around the value of p in H_0. That is,
$$0.5 \pm 1.96\sqrt{\frac{(0.50)(0.50)}{192}} = 0.5 \pm 0.0707,$$
or about 0.4293 to 0.5707.

P22. a. Note that $np = 266(.02) = 5.3$, which is less than 10, and the sampling distribution is skewed toward the larger values. So, the conditions are not met for this test. As such, a normal approximation might not be the best approximation to Display 8.10.

b. No, because the skewness is not drastic.

P23. a. The alternative hypothesis is:
H_a: $p < 0.05$, where p is the defective rate for the new supplier.

b. Using $n = 300$, $\hat{p} = 0.03$, and $p_0 = 0.05$, we see that the test statistic is
$$z = \frac{0.03 - 0.05}{\sqrt{\frac{(0.05)(0.95)}{300}}} \approx -1.589.$$
The p-value is $P(Z < -1.589) = 0.056$.

c. No, there is not sufficient evidence at the 1% level since the p-value is greater than 0.01.

P24. a. The test should be one-sided since you want to make certain the proportion is at least a certain value.

b. The null and alternative hypotheses are:
H_o: $p = 0.87$ H_a: $p > 0.87$, where p is the proportion of Down syndrome detected by the new test

c. Using $n = 200$, $\hat{p} = \frac{182}{200}$, and $p_0 = 0.87$, we see that the test statistic is

$$z = \frac{\frac{182}{200} - 0.87}{\sqrt{\frac{(0.87)(0.13)}{200}}} \approx 1.68 .$$

The p-value is P(Z>1.68) =0.0465.

d. Yes, there is evidence to support the alternative that the detection rate exceeds 87% since the p-value is less than 0.05.

P25. a. The test should be one-sided.

b. The null and alternative hypotheses are:
H_o: $p = 0.55$ H_a: $p < 0.55$, where p is the proportion of all likely voters who support the bond issue

c. Using $n = 700$, $\hat{p} = 0.53$, and $p_0 = 0.55$, we see that the test statistic is

$$z = \frac{0.53 - 0.55}{\sqrt{\frac{(0.55)(0.45)}{700}}} \approx -1.064 .$$

The p-value is P(Z<-1.064) =0.1437.

d. No, there isn't statistically significant evidence to support the alternative that the support has dropped below 55% since the p-value exceeds 0.01.

P26. a. The student is guessing, and so the probability that he or she gets any one question correct is 0.2.
Or, $p = 0.2$, where p is the probability that the student gets any one question correct.

b. $p = 0.104$, where p is the proportion of veterans in your county.

c. Of the people who wash their car once a week, $\frac{1}{7}$ wash them on Saturday.

Or, $p = \frac{1}{7}$, where p is the proportion of people who wash their cars once a week who wash them on Saturday.

P27. B

P28. The null and alternative hypotheses are:

H_o: $p = (18/38)$ H_a: $p \neq (18/32)$, where p is the probability of a red

Using $n = 500$, $\hat{p} = \frac{194}{500}$, and $p_0 = \frac{18}{38}$, we see that the test statistic is

$$z = \frac{\frac{194}{500} - \frac{18}{38}}{\sqrt{\frac{\left(\frac{18}{38}\right)\left(\frac{20}{38}\right)}{500}}} \approx -3.837 .$$

The p-value is 2P(Z<-3.837) nearly 0.0001.

Hence, we reject the null hypothesis that the probability of red is (18/38) for this wheel.

P29. The null and alternative hypotheses are:

H_o: $p = 0.07$, H_a: $p > 0.07$, where p is the probability that a bear will select a mini-van to break into.

Using $n = 412$, $\hat{p} = \frac{120}{412}$, and $p_0 = 0.07$, we see that the test statistic is

$$z = \frac{\frac{120}{412} - 0.07}{\sqrt{\frac{(0.07)(0.93)}{412}}} \approx 17.60 .$$

The p-value is P(Z >17.60), which is nearly 0.

Hence, we reject the null hypothesis that the probability that a bear will select a mini-van to break into is 0.07 and conclude that bears break into mini-vans more often than would be predicted by knowing that 7% of all vehicles parked overnight are mini-vans.

P30. *Check conditions.* The conditions are met for doing a test of significance for a proportion because you have a random sample, both $np_0 = 169(0.51) = 86.19$ and $n(1 - p_0) = 169(1 - 0.51) = 82.81$ are at least 10, and the number of teens in your community (probably) is at least 10(169) = 1690. If this latter condition doesn't hold, the test will be conservative.

State your hypotheses. The null hypothesis is that 51% (or even fewer) of teens in your community know this fact. The alternate hypothesis is that the percentage is greater than 51%.

Compute the test statistic and draw a sketch. The test statistic is

$$z = \frac{\hat{p} - p_0}{\sqrt{\dfrac{p_0(1 - p_0)}{n}}} = \frac{0.55 - 0.51}{\sqrt{\dfrac{0.51(1 - 0.51)}{169}}} \approx 1.04$$

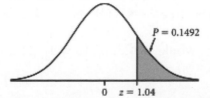

The *P*-value from Table A is about 0.1492. Thus, 0.55 is much the sort of sample

147

proportion you might get when taking a random sample of size 169 from a population with $p = 0.51$. Doing the test on the calculator, $z = 1.0479$ and the P-value is about 0.1473.

Write a conclusion in context. You should not reject the null hypothesis. If it is true that the percentage of teens in your community who know epilepsy is not contagious is only 51%, then it is reasonably likely to get 55% who know this in a random sample of 169 teens.

P31. a. Because Hila is using a significance level of 0.05, she should reject the null hypothesis if her test statistic is larger than 1.96 or smaller than −1.96. The value of her test statistic is

$$z = \frac{\hat{p} - p_0}{\sqrt{\dfrac{p_0(1 - p_0)}{n}}} = \frac{0.25 - (1/6)}{\sqrt{\dfrac{0.167(1 - 0.167)}{100}}} \approx 2.234$$

Hila rejects the null hypothesis.

b. She has made a Type I error. The null hypothesis is true, and she rejected it.

P32. a. The two-sided p-value is about 0.088. Hence, he should not reject the null hypothesis that the lottery is fair with regard to the number 1.

b. Jack did not make an error, at the 5% level of significance.

P33. a. The choices are random; since there are five such choices, we have $p = 0.2$, where p is the probability of a correct choice. The researchers failed to reject the null hypothesis.

b. The researchers could have made a Type II error because the null hypothesis was not rejected.

c. The researchers would have failed to discover that the subjects could choose the dog food more often than chance would predict.

P34. a. A Type-I error would be made when concluding that the proportion of children living near freeways in Los Angeles have a higher rate of asthma than the general population of children when they don't.

A serious consequence of this would be scaring parents into moving away from freeways.

b. A Type-II error would be made when concluding that it's plausible that the rates are equal when children living near freeways really have a higher rate.

A serious consequence of this would be failing to warn parents about a serious health problem for their children.

P35. The test with the higher significance level ($\alpha = 0.10$) has greater power because there is a greater chance of rejecting the null hypothesis. In the graphs below, the hypothesized value p_0 is 0.6 and the unknown population proportion p is 0.55. The solid shaded region shows the reasonably likely values of the normal approximation for the binomial distribution if the null hypotheses were true. The top graph has the middle 95% shaded ($\alpha = 0.05$) and the bottom graph has the middle 90% shaded ($\alpha = 0.10$). The normal distribution to the left in each graph is the normal approximation for the actual sampling distribution of the sample proportion for $p = 0.55$. The region indicated by the striped shading shows the portion of that normal distribution that lies in the rejection region for the hypothesized value p_0. This region is larger in the bottom graph indicating a greater likelihood of rejecting the null hypothesis and thus, greater power.

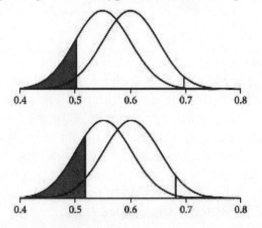

P36. The test with the larger sample size (200) has greater power. In the graphs below, the hypothesized value p_0 is 0.6 and the unknown population proportion p is 0.55. The solid shaded region shows the reasonably likely values of the normal approximation for the binomial distribution if the null hypotheses were true. Both graphs have the middle 95% shaded ($\alpha = 0.05$). The normal distribution to the left in each graph is the normal approximation for the actual sampling distribution of the sample proportion for $p = 0.55$.

The larger sample size results in a smaller standard error for the sampling distribution of the sample proportion and the distribution is clustered more closely to the mean, as shown in the second graph. There is less overlap of the two distributions in the second graph. The region indicated by the striped shading shows the portion of that normal distribution that lies in the rejection region for the hypothesized value p_0. This region is larger in the bottom graph indicating a greater likelihood of rejecting the null hypothesis and thus, greater power.

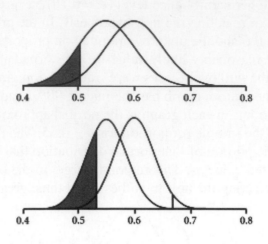

P37. The power is greater when the true population proportion is farther from p_0. In the graphs below, the hypothesized value p_0 is 0.06 and the unknown population proportion p is 0.08 in the first graph and 0.10 in the second graph. The solid shaded region shows the reasonably likely values of the normal approximation for the binomial distribution if the null hypotheses were true. Both graphs have the middle 95% shaded ($\alpha = 0.05$). The normal distribution to the right in each graph is the normal approximation for the actual sampling distribution of the sample proportion for $p = 0.08$ in the first graph and $p = 0.10$ in the second. Notice that there is less overlap of the two normal distributions in the second graph. The region indicated by the striped shading shows the portion of that normal distribution that lies in the rejection region for the hypothesized value p_0. This region is larger in the bottom graph indicating a greater likelihood of rejecting the null hypothesis and thus, greater power.

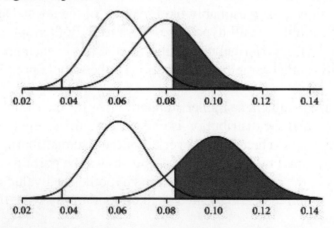

Notice that there is less overlap of the two normal distributions in the lower graph.

150

Exercise Solutions

E1. a. Using $n = 1100$, $\hat{p} = 0.19$, $z^* = 1.96$, we see that the 95% confidence level is

$$0.19 \pm 1.96\sqrt{\frac{(0.19)(0.81)}{1100}}$$

or about 0.167 to 0.213.

b. The margin of error is $E = 1.96\sqrt{\frac{(0.19)(0.81)}{1100}} \approx 0.023184$.

E3. a. The population is all college students graduating from 4-year colleges; the parameter being estimated is the proportion who are proficient in quantitative literacy; $\hat{p} = 0.34$ since it is measured from the sample.

b. Using $n = 1000$ and $\hat{p} = 0.34$, observe that $np \geq 10$, $n(1-p) \geq 10$. The sample is taken from a binomial population (since there are only two responses allowed) and the population size certainly exceeds 10,000. So, the conditions are met.

c. Using $n = 1000$, $\hat{p} = 0.34$, $z^* = 1.96$, we see that the 95% confidence level is

$$0.34 \pm 1.96\sqrt{\frac{(0.34)(0.66)}{1000}}$$

or about 0.311 to 0.369. This means that among the college students graduating from 4-year colleges from which this sample was selected, you are 95% confident that the percentage who would test proficient in quantitative literacy is between 31% and 37%.

d. Using $n = 800$, $\hat{p} = 0.18$, $z^* = 1.96$, we see that the 95% confidence level is

$$0.18 \pm 1.96\sqrt{\frac{(0.18)(0.82)}{800}}$$

or about 0.153 to 0.207. This means that among the college students graduating from two-year colleges from which this sample was selected, you are 95% confident that the percentage who would test proficient in quantitative literacy is between 15% and 21%.

e. Yes, because the confidence intervals do not overlap.

E5. a. The population is men of the Oxford cohort of patients; the parameter being estimated is proportion of men classified as overweight.

b. Using $n = 279$ and $\hat{p} = 0.409$, observe that $np \geq 10$, $n(1-p) \geq 10$. The sample is taken from a binomial population (since there are only two responses allowed) and the population size certainly exceeds 2,790. So, the conditions are met.

c. Using $n = 279$, $\hat{p} = 0.409$, $z^* = 1.96$, we see that the 95% confidence level is

$$0.409 \pm 1.96\sqrt{\frac{(0.409)(0.591)}{279}}$$

or about 0.35 to 0.47.

d. Using $n = 279$, $\hat{p} = 0.409$, $z^* = 1.645$, we see that the 90% confidence level is

$$0.409 \pm 1.645\sqrt{\tfrac{(0.409)(0.591)}{279}}$$

or about 0.36 to 0.46.

e. The 95% interval because to have a greater chance of capturing the true parameter value, the interval must be wider.

E7. a. Using $n = 600$ and $\hat{p} = 0.65$, observe that $np \geq 10$, $n(1-p) \geq 10$. The sample is taken from a binomial population (since there are only two responses allowed) and the population size certainly exceeds 6,000. So, the conditions are met.

b. Using $n = 600$, $\hat{p} = 0.65$, $z^* = 1.96$, we see that the 95% confidence level is

$$0.65 \pm 1.96\sqrt{\tfrac{(0.65)(0.35)}{600}}$$

or about 0.61 to 0.69. Note that the margin of error is $1.96\sqrt{\tfrac{(0.65)(0.35)}{600}}$, or about 0.04.

c. Using $n = 600$, $\hat{p} = 0.65$, $z^* = 1.645$, we see that the 90% confidence level is

$$0.65 \pm 1.645\sqrt{\tfrac{(0.65)(0.35)}{600}}$$

or about 0.62 to 0.68. Note that the margin of error is $1.645\sqrt{\tfrac{(0.65)(0.35)}{600}}$, or about 0.03.

d. The 95% interval because to have a greater chance of capturing the true parameter value, the interval must be wider.

E9. a. Using $n = 4884$, $\hat{p} = 0.0108$, $z^* = 1.96$, we see that the 95% confidence level is

$$0.0108 \pm 1.96\sqrt{\tfrac{(0.0108)(0.9892)}{4884}}$$

or about 0.008 to 0.014.

b. Using $n = 4884$, $\hat{p} = 0.0613$, $z^* = 1.96$, we see that the 95% confidence level is

$$0.0613 \pm 1.96\sqrt{\tfrac{(0.0613)(0.9387)}{4884}}$$

or about 0.054 to 0.068.

c. Yes, because the intervals do not overlap.

E11. a. The 95% confidence intervals are as follows:

<u>Black</u>: Using $n = 634$, $\hat{p} = 0.74$, $z^* = 1.96$, we see that the 95% confidence level is

$$0.74 \pm 1.96\sqrt{\tfrac{(0.74)(0.26)}{634}}$$

or about 0.71 to 0.77.

<u>White:</u> Using $n = 567$, $\hat{p} = 0.70$, $z^* = 1.96$, we see that the 95% confidence level is

$$0.70 \pm 1.96 \sqrt{\frac{(0.70)(0.30)}{567}}$$

or about 0.66 to 0.737.

<u>Hispanic:</u> Using $n = 314$, $\hat{p} = 0.66$, $z^* = 1.96$, we see that the 95% confidence level is

$$0.66 \pm 1.96 \sqrt{\frac{(0.66)(0.34)}{314}}$$

or about 0.61 to 0.71.

b. Black and Hispanic may differ because their intervals barely overlap.

c. For the population of black youth from which this sample was selected, you are 95% confident that the percentage who believes they have the skills to participate in politics is between 71% and 77%.

d. Only students with telephones could have been selected; payment for services may have attracted students from lower income brackets.

E13. a. The text doesn't say whether or not this is a random sample of adults, but, given that it was done by *U.S. News & World Report*, it is likely some randomization was involved. Therefore, our formulas give a reasonable approximation to the margin of error. Both $n\hat{p} = 0.81(1000) = 810$ and $n(1 - \hat{p}) = 0.19(1000) = 190$ are at least 10. Finally, the number of adults in the United States is greater than 10(1000).

We are 95% confident that if we were to ask *all* adults from the general public whether they thought TV contributed to a decline in family values, the percentage would be between 78.6% and 83.4%. The computations follow:

$$\hat{p} \pm z^* \cdot \sqrt{\frac{\hat{p}(1-\hat{p})}{n}} = 0.81 \pm 1.96 \sqrt{\frac{0.81(1-0.81)}{1000}} \approx 0.81 \pm 0.243$$

b. No, the low response rate of 9.4% could seriously bias the results.

E15. D, F, and H are the correct interpretations.

E17. You should have used a sample size 9 times as big, or $9n$. The margin of error is given by the formula

$$z^* \cdot \sqrt{\frac{p(1-p)}{n}}$$

If you want this to be $\frac{1}{3}$ as large as it was before, you must solve for the new sample size, *m,* in the equation

$$\frac{1}{3} \cdot z^* \cdot \sqrt{\frac{p(1-p)}{n}} = z^* \cdot \sqrt{\frac{p(1-p)}{m}}$$

which will give the result $m = 9n$.

Hence, using $n = 200$, we would use a sample of size 1800.

E19. You would expect $0.90(80) = 72$ of these intervals to include the proportion 0.60.

E21. The symbol p is used for the proportion of successes in the population from which we are drawing a sample. This is the unknown parameter—the value that we are trying to estimate. The symbol $\hat{p}$ is used for the proportion of successes in a sample drawn from the population with proportion of successes p. The value of $\hat{p}$ varies from sample to sample. When constructing a confidence interval, the value of $\hat{p}$ from the sample is at the center of the confidence interval and so is always in it. The value of p may or may not be in the confidence interval.

E23. a. The graph is the graph of the parabola $y = x - x^2$. This parabola opens down as shown here:

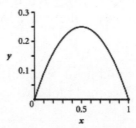

b. Because $x(1 - x)$ is of the form of $p(1 - p)$, which we are trying to maximize. Using y and x allows you to graph the function on a graphing calculator as you are asked to do in part b. The domain of x is restricted because a probability can be at most 1 and must be at least 0.

c. The maximum y-value occurs at the vertex. The vertex for a parabola $y = ax^2 + bx + c$ occurs at $x = \frac{-b}{(2a)}$. Here, $a = -1$ and $b = 1$, so the vertex is $x = \frac{-1}{2(-1)} = \frac{1}{2}$. The value of y at $x = \frac{1}{2}$ is $y = \frac{1}{2}(1 - \frac{1}{2}) = \frac{1}{4}$. If you graph this function on a graphing calculator you could also locate the coordinates of the vertex using the **maximum** function.

d. The standard deviation of $\hat{p}$ is maximized at $p = 0.5$.

E25. a. No, because the sample size 20 is greater than $0.10(135)$.

b. Using $n = 20$, $\hat{p} = 0.5$, $z^* = 1.96$, we see that the 95% confidence level is

$$0.5 \pm 1.96 \sqrt{\frac{(0.5)(0.5)}{20}}$$

or about 0.281 to 0.719.

c. Using $N = 135$, $n = 20$, $\hat{p} = 0.5$, $z^* = 1.96$, we see that the 95% confidence level is

$$0.5 \pm 1.96 \sqrt{\frac{(0.5)(0.5)}{20}} \sqrt{\frac{135-20}{135-1}}$$

or about 0.297 to 0.703.

d. The interval with the finite population correction is shorter; you should be able to estimate more accurately if the sample consists of a high percentage of the population values.

e. In such case, $\frac{N-n}{N-1} \sim 1$, and so $\sqrt{\frac{N-n}{N-1}} \sim 1$.

E27. a. $\frac{2}{3}$. You can explain this in two ways. The first way is to have the first student sit down anywhere. Then the probability the second student sits corner-to-corner is $\frac{2}{3}$ because, of the three seats that are left, two are adjacent seats. The other explanation is to list all 12 of the ways two students, A and B, can sit in chairs 1, 2, 3, and 4.

b. The null hypothesis is that the proportion of pairs of students who are seated on adjacent sides of a table is $\frac{2}{3}$.

c. The test should be one-sided. So, the alternate hypothesis is that the proportion of pairs is greater than $\frac{2}{3}$. (This indicates a preference for sitting on adjacent sides.)

d. The test statistic is

$$z = \frac{\hat{p} - p_0}{\sqrt{\dfrac{p_0(1-p_0)}{n}}} = \frac{0.70 - 2/3}{\sqrt{\dfrac{(2/3)(1-2/3)}{50}}} \approx 0.5$$

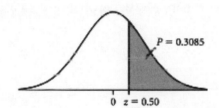

$P = 0.3085$

$0 \quad z = 0.50$

e. The p-value for this one-sided test is about 0.3085. We don't reject the null hypothesis.

f. The probability that two students will sit on adjacent sides of a table just by chance is $\frac{2}{3}$. In the psychologist's sample of size 50, 70% of the students were sitting on adjacent sides. This is just about what you would expect from chance variation alone. There is no evidence that students prefer sitting on adjacent sides. To establish that, the psychologist would need a larger sample size or a sample proportion quite a bit higher than 0.7.

E29. a. H_0: $p = 0.69$, where p is the proportion of houses in your community that are occupied by their owners.
H_A: $p \neq 0.69$

155

b.

The z-value from the Minitab output is -1.38.

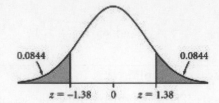

c. The *P*-value is 0.169. If the proportion of houses in your community that are occupied by their owners is indeed 69%, then the probability of getting 30 or fewer owner-occupied homes or 39 or more owner-occupied homes in a random sample of 50 homes is 16.9%.

d. Because this *P*-value is relatively high (17%), you do not have sufficient evidence to reject the null hypothesis. There is insufficient evidence that the proportion of houses occupied by their owners in your community is different from that of the country as a whole.

E31. a. $p_0 = 0.50$

b. The test should be one-sided. The null and alternative hypotheses are:
H_o: $p = 0.50$ H_a: $p < 0.50$, where p is the probability an astrologer can chose the correct CPI test result.

c. $\hat{p} = (40/116) = 0.345$

d. The test statistic is

$$z = \frac{0.345 - 0.50}{\sqrt{\frac{(0.50)(0.50)}{116}}} = -3.339 \, .$$

The p-value is P(Z < -3.339) = 0.0004.

e. There is strong evidence against the null hypothesis in favor of the claim that an astrologer chooses correct CPI test results less than 50% of the time.

E33. We test:
H_o: $p = 0.50$ H_a: $p < 0.50$, where p is the probability that a U.S. adult says he believes in ghosts.

The test statistic is

$$z = \frac{0.345 - 0.50}{\sqrt{\frac{(0.50)(0.50)}{1013}}} = -10.185 \, .$$

The p-value is P(Z < -10.185), which is nearly zero. So, there is very strong evidence to reject the hypothesis that 50% (or more) believe in ghosts in favor of the alternative that the proportion is less.

E35. a. $p_o = 0.50$

b. The test should be one-sided. The null and alternative hypotheses are:
H_o: $p = 0.50$ H_a: $p < 0.50$, where p is the probability that a spun penny lands heads up.

c. $\hat{p} = (10/40) = 0.25$

d. The test statistic is
$$z = \frac{0.25 - 0.50}{\sqrt{\frac{(0.50)(0.50)}{40}}} = -3.162.$$
The p-value is $P(Z < -3.162) = 0.0008$.

e. There is strong evidence against the null hypothesis in favor of the claim that a spun penny lands heads-up less than 50% of the time.

E37. a. The null hypothesis is:
H_o: $p = 0.60$, where p is the proportion of all students on campus who carry a backpack to class; that is, 60% of the students on campus carry backpacks to class.

b. Note that $\hat{p} = (28/50) = 0.56$. So, the test statistic is
$$z = \frac{0.56 - 0.60}{\sqrt{\frac{(0.60)(0.40)}{50}}} = -0.577.$$
So, the p-value is $2P(Z < -0.577) = 0.5639$. This means that under the hypothesis that 60% of students carry backpacks, in a sample of 50 the chance of at most 28, or at least 32, carrying backpacks is 0.5639.

c. There is no evidence against the null hypothesis that 60% of the students carry backpacks.

E39. B

E41. We test:
H_o: $p = 0.5$ H_a: $p > 0.5$, where p is the proportion of Americans who worry a great deal about the pollution of drinking water

The test statistic is
$$z = \frac{0.59 - 0.50}{\sqrt{\frac{(0.50)(0.50)}{1012}}} = 5.726.$$
The p-value is $P(Z > 5.726)$, which is nearly zero. Hence, there is strong evidence to reject the null hypothesis in favor of the alternative that more than half of Americans worry a great deal about the pollution of drinking water.

E43. The confidence interval does not contain 0.5, so 0.5 is a not plausible value for the proportion of all Americans who worry a great deal about the pollution of drinking water. As such, we conclude that $p > 0.5$.

E45. a. The null and alternative hypotheses are:

H_0: The proportion p of all adults in the United States who would say they are satisfied with the quality of K–12 education in the country is 0.5.

H_a: $p < 0.5$.

The question asks whether the poll results imply that less than half of adults are satisfied with the quality of education. This implies a one-sided alternative hypothesis.

b. The conditions are met for doing a test of significance for a proportion because you have a random sample, both $np_0 = 1000(0.5) = 500$ and $n(1 - p_0) = 1000(1 - 0.5) = 500$ are at least 10, and the number of adults in the United States is at least $10(1000) = 10,000$.

c. The test statistic is: $z = \dfrac{\hat{p} - p_0}{\sqrt{\frac{p_0(1-p_0)}{n}}} = \dfrac{0.49 - 0.5}{\sqrt{\frac{0.5 \cdot 0.5}{1000}}} \approx -0.632$. The p-value is about 0.2635.

d. There is insufficient evidence to reject the null hypothesis that at least half of the adult residents are satisfied with the quality of K-12 education.

E47. We test:

H_o: $p = 0.5$ versus H_a: $p < 0.5$, where p is the percentage of all Americans approving.

The test statistic is

$$z = \frac{0.45 - 0.50}{\sqrt{\frac{(0.50)(0.50)}{1004}}} = -3.169.$$

The p-value is $P(Z < -3.169) = 0.001$. Yes, the statement is fair. So, it's not plausible that half (or more) of all Americans give him positive marks.

E49. Neither C nor D is true. For C, this is an incorrect interpretation of a confidence interval and D is an incorrect interpretation of margin of error.

E51. a. The test is one-sided.

b. The expected proportion of deaths is $p = (11.3/196) = 0.058$, so the null and alternative hypotheses are:

H_o: $p = 0.058$ and H_a: $p > 0.058$

The test statistic is

$$z = \frac{\frac{39}{196} - 0.058}{\sqrt{\frac{(0.058)(0.942)}{196}}} = 8.44.$$

The p-value is near zero.

c. No, the result cannot reasonably be attributed to chance since the p-value is so small.

E53. a. You would expect $0.05(240) = 12$ of these people to make a Type-I error.

b. You cannot tell from the given information.

E55. a. Yes, it's possible for both to make a Type I error if they reject this true null hypothesis. They are equally likely to do so because they have the same value of α.

b. No, it isn't possible for them to fail to reject a false null hypothesis because the null hypothesis isn't false.

E57. a. No, it isn't possible for them to reject a true null hypothesis because the null hypothesis isn't true.

b. Yes, it's possible for both to make the Type II error of failing to reject this false null hypothesis. Jeffrey is more likely to do so because he has the smaller sample size.

E59. a. $\hat{p} = 0.668$, $p_0 = 0.5$.

b. The test statistic is

$$z = \frac{\hat{p} - p_0}{\sqrt{\frac{p_0(1-p_0)}{n}}} = \frac{0.668 - 0.5}{\sqrt{\frac{0.5 \cdot 0.5}{600}}} \approx 8.23.$$

The difference is due to rounding error. To see this note that the reported $\hat{p}$ is 66.8% and the sample size is 600. 66.8% of 600 is 400.8 which means the number of correct guesses in the sample was 401. $\frac{401}{600} = 0.668\overline{3}$, and the test statistic calculated using this value for $\hat{p}$ gives a result of 8.2466, rounding to 8.25.

c. Yes. A P-value of 0.001 corresponds to a z-score of around 3. A z-score of 8 is *much* more extreme than this, in fact it is $8.01 \cdot 10^{-17}$.

d. Not really. While the sample size always affects power, a smaller sample would have been sufficient in this case because 66.8% is so far from 50%. Even if the sample size had been only 60, with 40 successes, the result would have been statistically significant with a P-value of 0.01.

e. Since you are rejecting the null hypothesis, you could be making a Type-I error. The error may be that the null hypothesis is actually true. Specifically, the consequence would be believing that the winning candidate is more likely to be perceived as competent when this is not the case among the population from which the subjects were taken.

E61. a. The test statistic is

$$z = \frac{\frac{23}{45} - 0.50}{\sqrt{\frac{(0.50)(0.50)}{45}}} = 0.149.$$

The p-value for the one-sided test is $P(Z > 0.149) = 0.4404$. So, we do not reject the null hypothesis (in a big way!).

b. We could be making a Type-II error. The consequence is believing that the judge cannot select the correct dog at better than chance level when in fact he or she can.

c. Use a larger sample size.

E63. a. Since 82% of studies had a Type-II error, they failed to reject the null hypothesis.

b. C

E65. a.

x, the number of heads	p(x)
0	$\binom{20}{0} \cdot 0.5^{20} \approx 9.54 \cdot 10^{-7}$
1	$\binom{20}{1} \cdot 0.5^{20} \approx 1.91 \cdot 10^{-5}$
2	$\binom{20}{2} \cdot 0.5^{20} \approx 1.81 \cdot 10^{-4}$
3	$\binom{20}{3} \cdot 0.5^{20} \approx 0.00109$
4	$\binom{20}{4} \cdot 0.5^{20} \approx 0.00462$
5	$\binom{20}{5} \cdot 0.5^{20} \approx 0.0148$
6	$\binom{20}{6} \cdot 0.5^{20} \approx 0.0370$
7	$\binom{20}{7} \cdot 0.5^{20} \approx 0.0739$
8	$\binom{20}{8} \cdot 0.5^{20} \approx 0.1201$
9	$\binom{20}{9} \cdot 0.5^{20} \approx 0.1602$
10	$\binom{20}{10} \cdot 0.5^{20} \approx 0.1762$
11	$\binom{20}{11} \cdot 0.5^{20} \approx 0.1602$
12	$\binom{20}{12} \cdot 0.5^{20} \approx 0.1201$

13	$\binom{20}{13} \cdot 0.5^{20} \approx 0.0739$
14	$\binom{20}{14} \cdot 0.5^{20} \approx 0.0370$
15	$\binom{20}{15} \cdot 0.5^{20} \approx 0.0148$
16	$\binom{20}{16} \cdot 0.5^{20} \approx 0.00462$
17	$\binom{20}{17} \cdot 0.5^{20} \approx 0.00109$
18	$\binom{20}{18} \cdot 0.5^{20} \approx 1.81 \cdot 10^{-4}$
19	$\binom{20}{19} \cdot 0.5^{20} \approx 1.91 \cdot 10^{-5}$
20	$\binom{20}{20} \cdot 0.5^{20} \approx 9.54 \cdot 10^{-7}$

b. The outer 95% can be found by adding the probabilities, starting with the probability of 0 successes and working your way up, until you get a sum of 0.025. Since the distribution is symmetric, take the same number of outcomes starting from 20 successes and working your way down. The closest you can come to the outer 5% is to add up the probabilities of 0 through 5 successes and 15 to 20 successes. This sum is approximately 0.04. You would reject the null hypothesis if you have 5 or fewer heads or 15 or more heads.

c. Here we need the part of the probability distribution for 40% successes that represents 0 to 5 outcomes and 15 to 20 outcomes.

X, the number of successes	*p(x)*
0	$\binom{20}{0} 0.4^0 \cdot 0.6^{20} \approx 3.656 \cdot 10^{-5}$
1	$\binom{20}{1} 0.4^1 \cdot 0.6^{19} \approx 4.875 \cdot 10^{-4}$
2	$\binom{20}{2} 0.4^2 \cdot 0.6^{18} \approx 3.087 \cdot 10^{-3}$
3	$\binom{20}{3} 0.4^3 \cdot 0.6^{17} \approx 0.0123$

4	$\binom{20}{4}0.4^4 \cdot 0.6^{16} \approx 0.0350$
5	$\binom{20}{5}0.4^5 \cdot 0.6^{15} \approx 0.0746$
15	$\binom{20}{15}0.4^{15} \cdot 0.6^5 \approx 0.00129$
16	$\binom{20}{16}0.4^{16} \cdot 0.6^4 \approx 2.697 \cdot 10^{-4}$
17	$\binom{20}{17}0.4^{17} \cdot 0.6^3 \approx 4.230 \cdot 10^{-5}$
18	$\binom{20}{18}0.4^{18} \cdot 0.6^2 \approx 4.700 \cdot 10^{-6}$
19	$\binom{20}{19}0.4^{19} \cdot 0.6^1 \approx 3.299 \cdot 10^{-7}$
20	$\binom{20}{20}0.4^{20} \cdot 0.6^0 \approx 1.0995 \cdot 10^{-8}$
Total	0.1271 or 12.7%

There is a 12.7% chance that the null hypothesis will be rejected if the true proportion of getting heads when spinning a penny is 40%.

d. To increase the power of the test your friend should increase the sample size.

E67. a. *Blinded* means that a TT practitioner could not tell whether the investigator's hand was placed above their left hand or above their right hand. The way it was done was to have the TT practitioner rest their hands, palms up, on a flat surface. A tall screen with cutouts on its base was placed over the TT practitioner's arms so that they couldn't see the investigator's hand on the other side of the screen. Double-blinding would mean that the person who placed his or her hand above the TT practitioner's hands would not hear what the person's response was. Although this might have made the experiment a bit better, there was no judgment on the part of the investigator in evaluating the response from the TT practitioner. (The response was either right or wrong.)

b. H_0: The TT practitioners did no better than chance in identifying the correct hand. That is, $p = 0.5$, where p is the proportion of times the TT practitioners identified the correct hand.
H_a: The proportion of hands correctly identified was greater than 0.5.

c. No. The alternative hypothesis was that the TT practitioners should be able to identify the correct hand more often than random guessing. Because they actually performed more poorly than random guessing, the P-value will be more than 0.5 and the null hypothesis would be rejected.

d. This means that the sample size was large enough so that if TT practitioners could identify the correct hand with any consistency, the null hypothesis would have been rejected. Specifically, to reject the null hypothesis that the probability that they select the correct hand is 0.5, the value of z for a one-sided test would have to be 1.645. With a sample size of 280, the practitioners would only have to get 54.9% correct:

$$z = \frac{0.549 - 0.5}{\sqrt{\frac{0.5(1-0.5)}{280}}} \approx 1.645$$

E69. a. $(0.05)(200) = 10.$ **b.** $(0.95)^{200} = 0.000035$

E71. C

E73. You should include some explanation of a margin of error and how it is relatively small (around 3%) even with a sample size as small as 1000. The part that people tend to have the hardest time understanding is that, for a fixed $\hat{p}$, the margin of error depends almost entirely on the sample size n and not on how large the population is. That is, a random sample of size 1000 from the residents of Seattle has about the same margin of error as a random sample of size 1000 from the residents of the United States. If the sample size n is large relative to the size N of the population (more than about 10% of the population size), then you should use a correction factor for the formula for the margin of error that makes it smaller. Specifically, an approximate confidence interval is

$$\hat{p} - z^* \sqrt{\frac{\hat{p}(1-\hat{p})}{n-1}} \sqrt{\frac{N-n}{N}}$$

where N is the size of the population.
(See also the explanation from the Gallup Organization in E72.)

E75. a. Using $n = 54,461$, $\hat{p} = 0.76$, $z^* = 1.96$, we see that the 95% confidence level is

$$0.76 \pm 1.96 \sqrt{\frac{(0.76)(0.24)}{54,461}}$$

or about 0.756 to 0.764.

b. The margin of error is $1.96\sqrt{\frac{(0.76)(0.24)}{54,461}} \approx 0.004$.

c. Observe that $1.96\sqrt{\frac{(0.695)(0.305)}{1000}} \approx 0.028$.

E77. a. False. There should be a theoretical basis for the formulation of hypotheses. Using the data to form them would be "stacking the deck."

b. False. A two-tailed test p-value is 2 times the p-value of a corresponding one-sided test.

c. False. The p-value is the likelihood of rejecting a true null hypothesis.

d. True.

Chapter 9

Practice Problem Solutions

P1. a. As before, the samples can be considered random samples, and the samples were selected independently of each other. Each of

$$n_1 \hat{p}_1 = 100 \cdot 0.56 = 56 \qquad n_1(1-\hat{p}_1) = 100 \cdot 0.44 = 44$$
$$n_2 \hat{p}_2 = 100 \cdot 0.63 = 63 \qquad n_2(1-\hat{p}_2) = 100 \cdot 0.37 = 37$$

are at least 5, where n_1 and n_2 are the numbers of households sampled in 1994 and this year, respectively, and $\hat{p}_1$ and $\hat{p}_2$ are the proportions of households in 1994 and this year, respectively, that had a pet. The number of U.S. households in each year is larger than 10 times 100 or 1000.

b. The confidence interval is

$$(\hat{p}_1 - \hat{p}_2) \pm z * \sqrt{\frac{\hat{p}_1(1-\hat{p}_1)}{n_1} + \frac{\hat{p}_2(1-\hat{p}_2)}{n_2}} = (0.56 - 0.63) \pm 1.96 \sqrt{\frac{0.56(1-0.56)}{100} + \frac{0.63(1-0.63)}{100}}$$

$$\approx -0.07 \pm 0.136$$

or about –0.206 to 0.066. You are 95% confident that the difference in the two rates of pet ownership is between –0.206 to 0.066. This means that it is plausible that *20.6% less* households owned pets in 1994 than own pets now, and it is also plausible that 6.6% more households owned a pet in 1994 than own a pet now.

c. Yes. A difference of 0 does lie within the confidence interval. This means that if the difference in the proportion of pet owners now and in 1994 is actually 0, getting a difference of –0.07 in the samples is reasonably likely. Thus, it is plausible that there is no difference between the proportion of all households that owned a pet in 1994 and the proportion of all households that own a pet now. There is insufficient evidence to support a claim that there was a change in the percentage of households that own a pet between 1994 and now.

P2. a. The Gallup poll uses what can be considered a simple random sample. The populations are binomial (answering "yes" or "no"), and the samples would be independent of each other. Each of $n_1 \hat{p}_1$ = 325 • 0.5 = 162.5, $n_1(1 - \hat{p}_1)$ = 325 • 0.5 = 162.5, $n_2 \hat{p}_2$ = 224 • 0.28 = 62.72, $n_2(1 - \hat{p}_2)$ = 224 • 0.72 = 161.28 are at least five. There are more than 325 • 10 = 3,250 13 to 15-year olds and more than 224 • 10 = 2,240 16 to 17-year olds in the U.S. The conditions for a confidence interval for the difference of two proportions are met.

b.

$$(\hat{p}_1 - \hat{p}_2) \pm z * \sqrt{\frac{\hat{p}_1(1-\hat{p}_1)}{n_1} + \frac{\hat{p}_2(1-\hat{p}_2)}{n_2}} = (0.50 - 0.28) \pm 1.96 \sqrt{\frac{0.50(1-0.50)}{325} + \frac{0.28(1-0.28)}{224}}$$

$$\approx 0.22 \pm 0.08$$

or about 0.14 to 0.30.

c. You are 95% confident that the difference between the proportion of all 13 to 15-year olds who respond, "yes" and the proportion of all 16 to 17-year olds who would respond, "yes" to the question that it was appropriate for parents to install a special device on the car to allow parents to monitor teenagers' driving speeds is between 14% and 30%.

d. 0 is not in the confidence interval, which implies that it is *not* plausible that there is no difference between these proportions. This means that if the difference in the proportion of 13 to 15-year olds who respond, "yes" and the proportion of all 16 to 17-year olds who would respond, "yes" is actually 0, getting a difference of 0.14 in the samples is not at all likely. Thus, you are convinced that there is a difference in opinion between these two ages groups on this question.

P3. As before, the samples can be considered random samples, and the samples were selected independently of each other. Here,

$$n_1 = 1000, \ \hat{p}_1 = 0.34, \ n_2 = 800, \ \hat{p}_2 = 0.18$$

and certainly each of the products $n_1\hat{p}_1$, $n_1(1-\hat{p}_1)$, $n_2\hat{p}_2$, and $n_2(1-\hat{p}_2)$ is at least 5.

The 95% confidence interval is

$$(\hat{p}_1 - \hat{p}_2) \pm z^* \sqrt{\frac{\hat{p}_1(1-\hat{p}_1)}{n_1} + \frac{\hat{p}_2(1-\hat{p}_2)}{n_2}} = (0.34 - 0.18) \pm 1.96 \sqrt{\frac{0.34(1-0.34)}{1000} + \frac{0.18(1-0.82)}{800}}$$

or about (0.121, 0.199).

This means that you are 95% confident that the difference between the proficiency percentages for 4-year and 2-year colleges is between 12% and 20%.

P4. As before, the samples can be considered random samples, and the samples were selected independently of each other. Here,

$$n_1 = 273, \ \hat{p}_1 = \tfrac{48}{273} = 0.176, \ n_2 = 442, \ \hat{p}_2 = \tfrac{28}{442} = 0.063$$

and certainly each of the products $n_1\hat{p}_1$, $n_1(1-\hat{p}_1)$, $n_2\hat{p}_2$, and $n_2(1-\hat{p}_2)$ is at least 5.

The 95% confidence interval is

$$(\hat{p}_1 - \hat{p}_2) \pm z^* \sqrt{\frac{\hat{p}_1(1-\hat{p}_1)}{n_1} + \frac{\hat{p}_2(1-\hat{p}_2)}{n_2}} = (0.176 - 0.063) \pm 1.96 \sqrt{\frac{0.176(1-0.176)}{273} + \frac{0.063(1-0.063)}{442}}$$

or about (0.062, 0.164).

This means that you are 95% confident that the difference between the percentages playing video games online for DC versus DV types is between 6.2% and 16.4%.

P5. a. As before, the samples can be considered random samples, and the samples were selected independently of each other. Here,

$$n_1 = 663, \ \hat{p}_1 = 0.35, \ n_2 = 1591, \ \hat{p}_2 = 0.29$$

and certainly each of the products $n_1\hat{p}_1$, $n_1(1-\hat{p}_1)$, $n_2\hat{p}_2$, and $n_2(1-\hat{p}_2)$ is at least 5.

b. The 95% confidence interval is

$$(\hat{p}_1 - \hat{p}_2) \pm z^* \sqrt{\frac{\hat{p}_1(1-\hat{p}_1)}{n_1} + \frac{\hat{p}_2(1-\hat{p}_2)}{n_2}} = (0.35 - 0.29) \pm 1.96 \sqrt{\frac{0.35(0.65)}{663} + \frac{0.29(0.71)}{1591}}$$

or about (0.017, 0.103).

c. You are 95% confident that the difference between the proportion of those who Twitter who live in urban areas the proportion of Internet users who live in urban areas is between 0.017 and 0.103.

d. No; yes, because only positive differences are in the CI

e. If you were to repeat the process of taking two random samples and constructing a confidence interval for the difference over and over, in the long run, you expect that 95% of them contain the true difference in the proportion of people who live in urban areas.

P6. a. The expected value would be $p_1 - p_2 = 0.12 - 0.12 = 0$.

b. The *SE* is $\sqrt{\dfrac{p_1(1-p_1)}{n_1} + \dfrac{p_2(1-p_2)}{n_2}} = \sqrt{\dfrac{0.12 \cdot 0.88}{1000} + \dfrac{0.12 \cdot 0.88}{800}} \approx 0.0154$

c. Since $n_1 p_1$, $n_1(1 - p_1)$, $n_2 p_2$, and $n_2(1 - p_2)$ are all at least 5 you can approximate the sampling distribution of the difference with a normal model with mean 0 and *SD* 0.514.

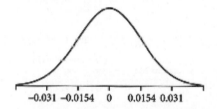

-0.031 -0.0154   0   0.0154 0.031

d. You can use your calculator. **normalcdf(0.05,1E99,0,0.0154)** will give approximately 0.00058. Alternatively, you can use the *z*-score.

$$z = \frac{(\hat{p}_1 - \hat{p}_2) - (p_1 - p_2)}{\sqrt{\dfrac{p_1(1-p_1)}{n_1} + \dfrac{p_2(1-p_2)}{n_2}}} = \frac{0.05 - 0}{\sqrt{\dfrac{0.12 \cdot 0.88}{1000} + \dfrac{0.12 \cdot 0.88}{800}}} \approx 3.244 .$$

According to Table A, the probability of a *z*-score greater than 3.24 is approximately 0.0006.

P7. a. False. The values of $\hat{p}_1$ and $\hat{p}_2$ vary from sample to sample.
b. True. We are told that the proportion of successes in the two populations are equal.
c. True.
d. True. We have $\mu_{\hat{p}_1 - \hat{p}_2} = p_1 - p_2 = 0$
e. True. As you can see from the dot plots in the display, the sample differences $\hat{p}_1 - \hat{p}_2$ have less variability in the samples of size 100 than in the samples of size 30. They

cluster more closely to 0, so we can see there is more of a chance of having the sample difference $\hat{p}_1 - \hat{p}_2$ nearer 0 with a larger sample size.

P8. This situation calls for a one-sided significance test for the difference of two proportions because we are asked whether the data support the conclusion that there was a decrease in voter support for the candidate.

Check conditions. You are told that you have two random samples from a large population (potential voters in some city). It's reasonable to assume that the samples are independent. For the first survey $n_1 = 600$ and $\hat{p}_1 = \frac{321}{600} = 0.535$. For the second survey, $n_2 = 750$ and $\hat{p}_1 = \frac{382}{750} \approx 0.509$. Each of $n_1 \hat{p}_1 = 321$, $n_1(1 - \hat{p}_1) = 279$, $n_2 \hat{p}_2 = 382$, and $n_2(1 - \hat{p}_2) = 368$ is at least 5. The number of potential voters at both times is much larger than 10 times the sample size for both samples.

State your hypotheses.
H_0: The proportion, p_1, of potential voters who favored the candidate in the first survey is equal to the proportion, p_2, of potential voters who favored the candidate one week before the election, or $p_1 = p_2$.
H_a: The proportion, p_1, of potential voters who favored the candidate in the first survey is greater than the proportion, p_2, of potential voters who favored the candidate one week before the election, or $p_1 > p_2$.

Compute the test statistic and draw a sketch. The test statistic is

$$z = \frac{(\hat{p}_1 - \hat{p}_2) - (p_1 - p_2)}{\sqrt{\hat{p}(1 - \hat{p})\left(\frac{1}{n_1} + \frac{1}{n_2}\right)}} \approx \frac{(0.535 - 0.5093) - 0}{\sqrt{0.521(1 - 0.521)\left(\frac{1}{600} + \frac{1}{750}\right)}} \approx 0.939$$

where

$$\hat{p} = \frac{\textit{total number of successes in both samples}}{n_1 + n_2} = \frac{321 + 382}{600 + 750} \approx 0.521$$

Using the table, the P-value for this one-sided test is 0.1736. From the TI-84+, the test statistic is $z = 0.938$ and the P-value is 0.1741. In this case, the **2-PropZTest** gives us the most accurate answer because there is less rounding.

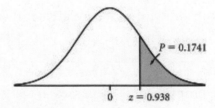

Write a conclusion in context. If there is no difference between the proportion of potential voters who favored the candidate at three weeks and the proportion of potential voters who favored the candidate at one week, then there is a 0.1741 chance of getting a

168

difference of 0.0257 or larger with samples of these sizes. This difference is not statistically significant—it can reasonably be attributed to chance variation. We do not reject the null hypothesis and can not conclude that there has been a drop in support for the new candidate.

P9. **a.** H_o: $p_1 - p_2 = 0$, H_a: $p_1 - p_2 \neq 0$, where p_1 is the proportion of conforming pellets from Method A and p_2 is the proportion of conforming pellets from Method B.

b. Here,
$$n_1 = n_2 = 100, \quad \hat{p}_1 = 0.38 \text{ (Method A)}, \quad \hat{p}_2 = 0.29.$$
and certainly each of the products $n_1\hat{p}_1$, $n_1(1-\hat{p}_1)$, $n_2\hat{p}_2$, and $n_2(1-\hat{p}_2)$ is at least 5. Also, the pooled estimate is
$$\hat{p} = \frac{\text{total number of successes from both treatments}}{n_1 + n_2} = \frac{67}{200} \approx 0.335.$$
So, the test statistic is
$$z = \frac{(\hat{p}_1 - \hat{p}_2) - (p_1 - p_2)}{\sqrt{\hat{p}(1-\hat{p})\left(\dfrac{1}{n_1} + \dfrac{1}{n_2}\right)}} = \frac{(0.38 - 0.29) - 0}{\sqrt{0.335(1-0.335)\left(\dfrac{1}{100} + \dfrac{1}{100}\right)}} \approx 1.348.$$

c. Since this is a two-tailed test, the p-value is $2P(Z > 1.348) = 2(0.0889) = 0.1778$.

d. There is insufficient evidence to conclude that the proportion of conforming pellets from Method A differs from the proportion of conforming pellets from Method B.

P10. **a.** Construct 95% confidence intervals for each of the 5 behaviors. If the confidence interval contains 0, then that behavior does not yield a statistically significant difference.

Registered to vote:
$$(\hat{p}_1 - \hat{p}_2) \pm z^* \cdot \sqrt{\frac{\hat{p}_1(1-\hat{p}_1)}{n_1} + \frac{\hat{p}_2(1-\hat{p}_2)}{n_2}} = (0.89 - 0.78) \pm 1.96\sqrt{\frac{(0.89)(0.11)}{3011} + \frac{(0.78)(0.22)}{1055}}$$
$$= 0.11 \pm 0.027, \text{ or } 0.083 \text{ to } 0.137$$

Active in community:
$$(\hat{p}_1 - \hat{p}_2) \pm z^* \cdot \sqrt{\frac{\hat{p}_1(1-\hat{p}_1)}{n_1} + \frac{\hat{p}_2(1-\hat{p}_2)}{n_2}} = (0.68 - 0.41) \pm 1.96\sqrt{\frac{(0.68)(0.32)}{3011} + \frac{(0.41)(0.59)}{1055}}$$
$$= 0.27 \pm 0.034, \text{ or } 0.236 \text{ to } 0.304$$

Suffer from personal addiction:
$$(\hat{p}_1 - \hat{p}_2) \pm z^* \cdot \sqrt{\frac{\hat{p}_1(1-\hat{p}_1)}{n_1} + \frac{\hat{p}_2(1-\hat{p}_2)}{n_2}} = (0.12 - 0.13) \pm 1.96\sqrt{\frac{(0.12)(0.88)}{3011} + \frac{(0.13)(0.87)}{1055}}$$
$$= -0.01 \pm 0.023, \text{ or } -0.033 \text{ to } 0.013$$

Overweight:
$$(\hat{p}_1 - \hat{p}_2) \pm z^* \cdot \sqrt{\frac{\hat{p}_1(1-\hat{p}_1)}{n_1} + \frac{\hat{p}_2(1-\hat{p}_2)}{n_2}} = (0.41 - 0.26) \pm 1.96\sqrt{\frac{(0.41)(0.59)}{3011} + \frac{(0.26)(0.74)}{1055}}$$
$$= 0.15 \pm 0.032, \text{ or } 0.118 \text{ to } 0.182$$

Stressed out:

$$(\hat{p}_1 - \hat{p}_2) \pm z^* \cdot \sqrt{\frac{\hat{p}_1(1-\hat{p}_1)}{n_1} + \frac{\hat{p}_2(1-\hat{p}_2)}{n_2}} = (0.26 - 0.37) \pm 1.96 \sqrt{\frac{(0.26)(0.74)}{3011} + \frac{(0.37)(0.63)}{1055}}$$

$$= -0.11 \pm 0.033, \text{ or } -0.143 \text{ to } -0.077$$

So, all are significant except "suffer from personal addiction."

b. The strongest evidence is given by "active in community" since it is the furthest away from 0.

c. Activist Christians may have been more willing to participate than non-Christian activists, especially those from higher economic strata.

P11. a. These are not independent because men and women are paired.
b.　These are independent because the sampling is done with replacement.
c.　These are nearly independent because the men and women are not paired, and can be considered independent for this large population.

P12. Proceed as follows:

Check conditions. First, the conditions for an experiment are met, which allows the computing of a confidence interval for the difference of two proportions: Treatments were randomly assigned to subjects. Each of $n_1\hat{p}_1 = 169$, $n_1(1 - \hat{p}_1) = 10{,}868$, $n_2\hat{p}_2 = 138$, and $n_2(1 - \hat{p}_2) = 10{,}896$ is at least 5.

Do computations. The 95% confidence interval is:

$$(\hat{p}_1 - \hat{p}_2) \pm z^* \cdot \sqrt{\frac{\hat{p}_1(1-\hat{p}_1)}{n_1} + \frac{\hat{p}_2(1-\hat{p}_2)}{n_2}} = (0.0153 - 0.0125) \pm 1.96 \sqrt{\frac{(0.0153)(0.9847)}{11{,}037} + \frac{(0.0125)(0.9875)}{11{,}034}}$$

$$= 0.0028 \pm 0.0031, \text{ or } -0.0003 \text{ to } 0.0059$$

Write a conclusion in context. Suppose all of the subjects could have been given the aspirin treatment and all of the subjects could have been given the placebo treatment. Then you are 95% confident that the difference in the proportion who would get ulcers is in the interval (-0.0003, 0.0059). Because 0 is in this interval, it is plausible that there is no difference in the proportions who would get ulcers. The term "95% confident" means that this method of constructing confidence intervals results in $p_1 - p_2$ falling in an average of 95 out of every 100 confidence intervals you construct.

P13. The hypotheses are:
H_0: $p_1 - p_2 = 0$, H_a: $p_1 - p_2 \neq 0$, where p_1 is the proportion of successes if all subjects could have been asked for a quarter and p_2 is the proportion of successes if all subjects could have been asked for 17 cents.

Here,

$$n_1 = 72, \ \hat{p}_1 = 0.306, \ n_2 = 72, \ \hat{p}_2 = 0.431$$

and certainly each of the products $n_1\hat{p}_1$, $n_1(1-\hat{p}_1)$, $n_2\hat{p}_2$, and $n_2(1-\hat{p}_2)$ is at least 5. The randomness conditions are also met.

Also, the pooled estimate is

$$\hat{p} = \frac{\text{total number of successes from both treatments}}{n_1+n_2} = \frac{22.032+31.032}{72+72} \approx 0.369 .$$

So, the test statistic is

$$z = \frac{(\hat{p}_1-\hat{p}_2)-(p_1-p_2)}{\sqrt{\hat{p}(1-\hat{p})\left(\dfrac{1}{n_1}+\dfrac{1}{n_2}\right)}} = \frac{(0.431-0.306)-0}{\sqrt{0.369(1-0.369)\left(\dfrac{1}{72}+\dfrac{1}{72}\right)}} \approx 1.555$$

Since this is a two-sided test, the p-value is $2P(Z>1.555)=0.1212$. Hence, there is insufficient evidence, at the 5% level, to say that asking for 17 cents will increase the percentage of success over asking for 25 cents.

P14. a. Because we are simply looking for a difference we will use a two-sided significance test for a difference in proportions.

Check conditions. The problem does not state whether treatments were randomly assigned. The other condition is met, however. Each of $n_1\ddot{p}_1 = 411$, $n_1(1-\ddot{p}_1) = 4009$, $n_2\ddot{p}_2 = 463$, and $n_2(1-\hat{p}_2) = 3989$ is at least 5.

State your hypotheses.
H_0: If all patients could have been given Lipitor, the proportion p_1 of them that had heart attacks would be the same as the proportion p_2 that would have had heart attacks had they all been given Zocor.

H_a: $p_1 \neq p_2$

Calculate the test statistic and draw a sketch.

$$z = \frac{(\hat{p}_1-\hat{p}_2)-(p_1-p_2)}{\sqrt{\hat{p}(1-\hat{p})\left(\dfrac{1}{n_1}+\dfrac{1}{n_2}\right)}} = \frac{(0.093-0.104)-0}{\sqrt{0.0985(1-0.0985)\left(\dfrac{1}{4420}+\dfrac{1}{4452}\right)}} \approx -1.738$$

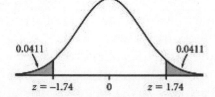

Here the pooled estimate is

$$\hat{p} = \frac{\text{total number of successes from both treatments}}{n_1 + n_2} = \frac{874}{8872} \approx 0.0985$$

The z-score of -1.738 corresponds to a P-value of $2 \cdot 0.0411 = 0.0822$.
A 2-PropZTest on the TI-84+ gives a z-score of -1.7402 and a P-value of 0.0818.

State your conclusion in context. Because the P-value is greater than 0.05 you would not reject the null hypothesis. There is not sufficient evidence to conclude that, if all experimental units would have been treated with Lipitor, the proportion who had heart attacks would have been different than if all patients had been treated with Zocor.

b. The conditions and test statistic will be the same as in part a. You need to restate your hypotheses, calculate the new P-value, and state your conclusions.

State your hypotheses.
H_0: If all patients could have been given Lipitor, the proportion p_1 of them that had heart attacks would be the same as the proportion p_2 that would have had heart attacks had they all been given Zocor.
H_a: $p_1 < p_2$ (If Lipitor is more effective, you would expect the proportion of patients having heart attacks to be lower.)

P-value. The test statistic is still -1.738. The P-value is now half what it was for a two-sided test. A 2-PropZTest for alternative hypothesis $p_1 < p_2$ now shows a P-value of 0.0409. The sketch in part (a) would be shaded only in the left tail.

State your conclusion in context. The P-value of 0.0409 is less than 0.05. You would reject the null hypothesis that if all patients could have been given Lipitor, the proportion of them that had heart attacks would be the same as the proportion that would have had heart attacks had they all been given Zocor. If there would have been no difference in the proportion of patients who had heart attacks if they had all taken Lipitor and the proportion of patients who had heart attacks if they had all taken Zocor, then there is a 0.0409 chance of getting a difference of -1.74 or smaller in the proportions from random assignment of these treatments to the subjects. This difference can not be reasonably attributed to chance variation. There is evidence that Lipitor is more effective than Zoloc.

A one-sided test makes it easier to reject the null hypothesis if the difference is in the direction your alternative hypothesis states. Mathematically, this happens because the entire 5% rejection region is on that side, meaning a less extreme z-score will allow rejection. Philosophically, the fact that you suspect one direction may be due to evidence in favor of that alternative hypothesis. Less additional evidence is needed to verify this.

P15. Because the null hypothesis was not rejected in Part A, a Type II error could have been made. In part B the null hypothesis was rejected, so a Type I error could have occurred. A Type I error would mean the patient receives a different drug even though there is no actual difference in their effectiveness. A Type II error means a patient is not given a new drug that would actually have a better chance of success. Both could be serious errors as both mean the patient is not receiving the most effective drug.

P16. In all cases below, the hypotheses (in symbols) are:
H_0: $p_1 - p_2 = 0$, H_a: $p_1 - p_2 \neq 0$.

TV in bedroom: Here,
$$n_1 = 92, \ \hat{p}_1 = 0.435, \ n_2 = 100, \ \hat{p}_2 = 0.43$$
and certainly each of the products $n_1\hat{p}_1$, $n_1(1-\hat{p}_1)$, $n_2\hat{p}_2$, and $n_2(1-\hat{p}_2)$ is at least 5. The randomness conditions are also met.

Also, the pooled estimate is
$$\hat{p} = \frac{\text{total number of successes from both treatments}}{n_1 + n_2} = \frac{40.02 + 43}{100 + 92} \approx 0.432 .$$
So, the test statistic is
$$z = \frac{(\hat{p}_1 - \hat{p}_2) - (p_1 - p_2)}{\sqrt{\hat{p}(1-\hat{p})\left(\dfrac{1}{n_1} + \dfrac{1}{n_2}\right)}} = \frac{(0.435 - 0.43) - 0}{\sqrt{0.432(1-0.432)\left(\dfrac{1}{92} + \dfrac{1}{100}\right)}} \approx 0.070$$

Since this is a two-sided test, the p-value is $2P(Z > 0.070) = 0.9442$. Hence, there is insufficient evidence, at the 5% level, to say these proportions are different.

College grads: Here,
$$n_1 = 92, \ \hat{p}_1 = 0.45, \ n_2 = 100, \ \hat{p}_2 = 0.21$$
and certainly each of the products $n_1\hat{p}_1$, $n_1(1-\hat{p}_1)$, $n_2\hat{p}_2$, and $n_2(1-\hat{p}_2)$ is at least 5. The randomness conditions are also met.

Also, the pooled estimate is
$$\hat{p} = \frac{\text{total number of successes from both treatments}}{n_1 + n_2} = \frac{21 + 41.4}{100 + 92} \approx 0.325 .$$
So, the test statistic is
$$z = \frac{(\hat{p}_1 - \hat{p}_2) - (p_1 - p_2)}{\sqrt{\hat{p}(1-\hat{p})\left(\dfrac{1}{n_1} + \dfrac{1}{n_2}\right)}} = \frac{(0.45 - 0.21) - 0}{\sqrt{0.325(1-0.325)\left(\dfrac{1}{92} + \dfrac{1}{100}\right)}} \approx 3.547$$

Since this is a two-sided test, the p-value is $2P(Z > 3.547) = 0.0004$. Hence, there is sufficient evidence, at the 5% level, to the difference in these proportions is not attributed to chance.

Female participant: Here,
$$n_1 = 92, \ \hat{p}_1 = 0.45, \ n_2 = 100, \ \hat{p}_2 = 0.485$$
and certainly each of the products $n_1\hat{p}_1$, $n_1(1-\hat{p}_1)$, $n_2\hat{p}_2$, and $n_2(1-\hat{p}_2)$ is at least 5. The randomness conditions are also met.

Also, the pooled estimate is

$$\hat{p} = \frac{\text{total number of successes from both treatments}}{n_1 + n_2} = \frac{41.4 + 48.5}{100 + 92} \approx 0.468 \,.$$

So, the test statistic is

$$z = \frac{(\hat{p}_1 - \hat{p}_2) - (p_1 - p_2)}{\sqrt{\hat{p}(1-\hat{p})\left(\dfrac{1}{n_1} + \dfrac{1}{n_2}\right)}} = \frac{(0.45 - 0.485) - 0}{\sqrt{0.468(1-0.468)\left(\dfrac{1}{92} + \dfrac{1}{100}\right)}} \approx -0.4855$$

Since this is a two-sided test, the p-value is $2P(Z < -0.4855) = 0.6273$. Hence, there is insufficient evidence, at the 5% level, to say these proportions are different.

Thus, we see that only the "college grads" issue shows a difference that could not be easily relegated to chance alone.

b. Randomization to the larger units (schools) rather than to the smaller units (students) generally is not a good idea because it reduces the number of randomly assigned units and that reduces the effective sample size. In the extreme, if all the students in one school of 500 students acted alike, the result would be one new piece of information for the school rather than 500 pieces of information that could have been obtained if the 500 students had been randomly selected from a large population of students.

P17. a. This is an observational study. There was no random sampling done, and no random assignment of treatments.

b. *Check conditions.* We already know there was no random assignment of treatments. Each of $n_1 \cdot \hat{p}_1 = 103$, $n_1(1 - \hat{p}_1) = 805$, $n_2 \hat{p}_2 = 53$, and $n_2(1 - \hat{p}_2) = 614$ is at least 5, where $\hat{p}_1$ is the proportion of the n_1 people observed who had been abused as children who later went on to commit violent crime, and $\hat{p}_2$ is the proportion of the n_2 people observed who had not been abused as children who later went on to commit violent crime. This second condition is met.

Calculate the interval.

$$(\hat{p}_1 - \hat{p}_2) \pm z^* \sqrt{\frac{\hat{p}_1(1-\hat{p}_1)}{n_1} + \frac{\hat{p}_2(1-\hat{p}_2)}{n_2}} = (0.113 - 0.079) \pm 1.645 \sqrt{\frac{0.113 \cdot 0.887}{908} + \frac{0.079 \cdot 0.921}{667}}$$

$$\approx 0.034 \pm 0.024$$

or about from about 0.01 to 0.058.

c. We can conclude that the difference in proportions of the people in this study that were abused as children who later committed crimes and the people in this study who were not abused as children who later committed crimes cannot be reasonably attributed to chance. There may be, and probably are, many other factors that contributed to this difference, so we cannot conclude from this study alone that abuse of children causes them to be more likely to commit violent crime later in life.

P18. The riders on greenways show the strongest association between helmet use and the law. Note that this still does not imply causation.

Exercise Solutions

E1. B and C

E3. a. You do not know that this is a random sample of all purebred dog owners or all mutt owners in the San Diego area. However, you could consider guessing to be a random event and you want to compare the probability of guessing correctly with purebred dog owners and mutt owners. Each of

$$n_1\hat{p}_1 = 16 \qquad n_1(1-\hat{p}_1) = 9$$
$$n_2\hat{p}_2 = 7 \qquad n_2(1-\hat{p}_2) = 13$$

are at least 5, where n_1 and n_2 are the numbers of guesses made with pure-bred dog owners and with mutt owners, respectively, and $\hat{p}_1$ and $\hat{p}_2$ are the proportions of correct guesses made with purebred dog owners and with mutt owners, respectively. There are probably more than 25 • 10 = 250 purebred dog owners and more than 20 • 10 = 200 mutt owners in the San Diego area.

b.

$$(\hat{p}_1-\hat{p}_2)\pm z*\sqrt{\frac{\hat{p}_1(1-\hat{p}_1)}{n_1}+\frac{\hat{p}_2(1-\hat{p}_2)}{n_2}} = (0.64-0.35)\pm 1.96\sqrt{\frac{0.64(1-0.64)}{25}+\frac{0.35(1-0.35)}{20}},$$

$$\approx 0.29\pm 0.281$$

or about 0.009 to 0.571.

c. You are 95% confident that if the judges had been given a choice of two dogs for each owner, the difference in the proportion of correct guesses for all purebred owners and the proportion of correct guesses for all mutt owners in the San Diego area would be between 0.009 and 0.571.

d. No. This implies that the difference in the proportions in the study were probably not due to chance. You have sufficient evidence that there is a higher probability of judges guessing correctly with purebred owners than with mutt owners.

e. The researchers need to enlarge their sample sizes and to select dog-owners and judges randomly from their location of interest. As it is, we can't know whether these results mean anything in terms of guessing correctly or whether there is something distinctive about either San Diego dogs or the judges from San Diego that led to these results. Or perhaps there was something distinctive about the dogs and owners that were chosen that influenced the judges' guesses.

f. 95% of all possible samples would yield a difference in proportions that is between 0.009 and 0.571.

E5. a. Here,

$$n_1 = 4775, \quad \hat{p}_1 = 0.52, \quad n_2 = 2685, \quad \hat{p}_2 = 0.61$$

and certainly each of the products $n_1\hat{p}_1$, $n_1(1-\hat{p}_1)$, $n_2\hat{p}_2$, and $n_2(1-\hat{p}_2)$ is at least 5. The 90% confidence interval is

$$(\hat{p}_1 - \hat{p}_2) \pm z^* \sqrt{\frac{\hat{p}_1(1-\hat{p}_1)}{n_1} + \frac{\hat{p}_2(1-\hat{p}_2)}{n_2}} = (0.61-0.52) \pm 1.645\sqrt{\frac{0.61(1-0.61)}{2685} + \frac{0.52(1-0.52)}{4775}}$$

or about 0.070 to 0.110. This means that you are 90% confident that the true difference is in the interval (0.070, 0.110).

b. More of those who have a negative view of hazing may have responded, biasing the reported percentages toward the high side. If both male and female samples are similarly biased, the difference in sample percentages may be a valid estimate.

E7. a. The conditions are met for constructing a confidence interval for the difference of two proportions:

- You were told that you may assume that the samples are equivalent to simple random samples.
- The number of men and number of women in the United States are more than $425 \cdot 10 = 4250$.
- Each of

$$n_1\hat{p}_1 = 425(0.23) = 98$$
$$n_1(1 - \hat{p}_1) = 425(1 - 0.23) = 327$$
$$n_2\hat{p}_2 = 425(0.34) = 145$$
$$n_2(1 - \hat{p}_2) = 425(1 - 0.34) = 281$$

is at least 5, where n_1 and n_2 are the sample sizes for men and women respectively, and $\hat{p}_1$ and $\hat{p}_2$ are the proportions of men and women, respectively, in the sample who said they would prefer to be addressed by their last name.

b.

$$(\hat{p}_1 - \hat{p}_2) \pm z^* \cdot \sqrt{\frac{\hat{p}_1(1-\hat{p}_1)}{n_1} + \frac{\hat{p}_2(1-\hat{p}_2)}{n_2}} = (0.23-0.34) \pm 2.576\sqrt{\frac{(0.23)(1-0.23)}{425} + \frac{(0.34)(1-0.34)}{425}}$$

$$= -0.11 \pm 0.079$$

or about (-0.189, -0.031).

c. You are 99% confident that the difference in the percentage of all men and the percentage of all women who prefer to be addressed by their last name is in the interval −0.189 to −0.031. (Alternatively, you are 99% confident that the difference in the percentage of all women and the percentage of all men who prefer to be addressed by their last name is in the interval 0.031 to 0.189.)

d. 0 is not in the confidence interval. This means that the statement, "There is no difference in the proportions of all men who would prefer to have their last name used and the proportion of all women who would prefer to have their last name used" is not plausible. If the difference in the proportion of men who prefer being addressed by their last name and the proportion of women who prefer being addressed by their last name is actually 0, getting a difference of 11% in the samples is not at all likely. Thus, you are convinced that there is a difference between the percentage of women and percentage of men who prefer to be addressed by their last name.

E9. *Check conditions.* The conditions are met for constructing a confidence interval for the difference of two proportions:

• You were told that you may assume that the samples are equivalent to simple random samples.
• There are more than 76,000 male students and more than 76,000 female students in the United States.
• Each of

$$n_1 \hat{p}_1 = 7{,}600 \bullet 0.59 = 4484$$
$$n_1(1 - \hat{p}_1) \approx 7{,}600 \bullet 0.41 = 3116$$
$$n_2 \hat{p}_2 = 7{,}600 \bullet 0.48 = 3648$$
$$n_2(1 - \hat{p}_2) = 7{,}600 \bullet 0.52 = 3952$$

is at least 5, where n_1 and n_2 are the numbers of male and female high school seniors sampled, respectively, and $\hat{p}_1$ and $\hat{p}_2$ are the proportions of male and female high school seniors sampled, respectively, who have played on sports teams run by their school during the 12 months preceding the survey

Do computations. The 95% confidence interval for the difference in the proportions of male and female seniors who have played on sports teams run by their school during the 12 months preceding the survey is

$$(\hat{p}_1 - \hat{p}_2) \pm z^* \sqrt{\frac{\hat{p}_1(1-\hat{p}_1)}{n_1} + \frac{\hat{p}_2(1-\hat{p}_2)}{n_2}} = (0.59 - 0.48) \pm 1.96 \sqrt{\frac{0.59(1-0.59)}{7{,}600} + \frac{0.48(1-0.48)}{7{,}600}}$$
$$\approx 0.11 \pm 0.016$$

or between 0.094 and 0.126.

Write a conclusion in context. Because 0 isn't included in this confidence interval, it is acceptable to say that senior boys are "significantly more likely" than senior girls to have played on sports teams run by their school in the previous 12 months.

E11. In general, as sample sizes get larger, the length of the confidence interval gets smaller. (If the sample size for only one sample gets larger, that part of the formula for the standard error goes to zero. This alone won't make the standard error itself go to zero unless the other sample size also gets larger.)

E13. You can use z in this way only because the sampling distribution of the estimate $\hat{p}_1 - \hat{p}_2$ is approximately normal. How do you know that the distribution of $\hat{p}_1 - \hat{p}_2$ is approximately normal? A theorem in mathematical statistics given in text says that the sampling distribution of the difference of two normally distributed random variables is normal. So the sampling distribution of $\hat{p}_1 - \hat{p}_2$ will be approximately normal if the separate sampling distributions of $\hat{p}_1$ and $\hat{p}_2$ are normal. They are approximately normal if each of $n_1\hat{p}_1, n_1(1-\hat{p}_1), n_2\hat{p}_2,$ and $n_2(1-\hat{p}_2)$ is at least 10. However, this condition is stronger than necessary in the case of a difference—the sampling distribution of the difference will be approximately normal as long as each one of these is at least 5.

E15. The method is not correct because the respondents weren't selected independently from two different populations. These people were all from the same population and are differentiated only by their answer to the question. The appropriate method to use is a confidence interval for a proportion from a single population.

$$\hat{p} \pm 1.96\sqrt{\frac{\hat{p}(1-\hat{p})}{n}} = 0.75 \pm 1.96\sqrt{\frac{0.75(1-0.75)}{1008}} \approx 0.75 \pm 0.027 \, .$$

You are 95% confident that the proportion of online respondents who would favor the legal drinking age as 21 is in the interval 0.723 to 0.777.

E17. a. $\mu_{\hat{p}_1-\hat{p}_2} = 0.24 - 0.20 = 0.04.$

b. Don't use the pooled variance here because the two populations do not have a common variance.

$$\sigma_{\hat{p}_1-\hat{p}_2} = \sqrt{\frac{p_1(1-p_1)}{n_1} + \frac{p_2(1-p_2)}{n_2}} = \sqrt{\frac{0.24 \cdot 0.76}{100} + \frac{0.20 \cdot 0.80}{100}} \approx 0.0585$$

c.

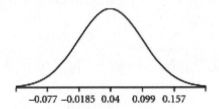

-0.077 -0.0185 0.04   0.099   0.157

d. You can use your calculator. **normalcdf(0.05,1E99,.04,.0585)** will give approximately 0.432. Alternatively, you can use the z-score.

$$z = \frac{(\hat{p}_1 - \hat{p}_2) - (p_1 - p_2)}{\sqrt{\frac{p_1(1-p_1)}{n_1} + \frac{p_2(1-p_2)}{n_2}}} = \frac{0.05 - 0.04}{\sqrt{\frac{0.24 \cdot 0.76}{100} + \frac{0.20 \cdot 0.80}{100}}} \approx 0.171 \, .$$

According to Table A, the probability of a z-score greater than 0.17 is approximately 0.4325.

E19. *Check conditions.* Although the situation probably is actually more complicated, you can assume that you have two independent random samples. All of $n_1\hat{p}_1 = 177(0.30) = 53.1$, $n_1(1 - \hat{p}_1) = 177(1 - 0.30) = 123.9$, $n_2\hat{p}_2 = 616(0.24) = 147.84$, and $n_2(1 - \hat{p}_2) = 616(1 - 0.24) = 468.16$ are at least 5. The number of people in each age group is much larger than 10 times the sample size.

State your hypotheses.
H_0:   The proportion, p_1, of all people aged 18 to 29 who sleep eight hours or more on a weekday is equal to the proportion, p_2, of all people aged 30 to 49 who sleep eight hours or more on a weekday.
H_a: $p_1 \neq p_2$

Compute the test statistic and draw a sketch. The test statistic is

$$z = \frac{(\hat{p}_1 - \hat{p}_2) - (p_1 - p_2)}{\sqrt{\hat{p}(1-\hat{p})\left(\dfrac{1}{n_1} + \dfrac{1}{n_2}\right)}} = \frac{(0.30 - 0.24) - 0}{\sqrt{0.253 \cdot 0.747 \left(\dfrac{1}{177} + \dfrac{1}{616}\right)}} \approx 1.62$$

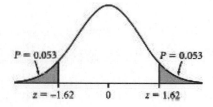

Here the pooled estimate, $\hat{p}$, is

$$\hat{p} = \frac{\text{total number of successes in both samples}}{n_1 + n_2} = \frac{53 + 148}{177 + 616} \approx 0.253.$$

The *P*-value from the calculator for a two-sided test is
$2 \cdot$ **normalcdf(–1E99,–1.62)** ≈ 0.105.

Using the table with $z = -1.62$ gives a *P*-value of $2(0.0526) = 0.1052$. Using the **2-PropZTest** command on your calculator with $x_1 = 53$ and $x_2 = 148$ gives a test statistic $z = 1.5951$ and a *P*-value of 0.1107.

Write a conclusion in context. No significance level was given, so you can assume 0.05. The *P*-value is larger than 0.05 so this difference is not statistically significant and you do not reject the null hypothesis. If the proportion of all Americans aged 18 to 29 who sleep eight hours or more on a workday is equal to the proportion of all Americans aged 30 to 49 who do so, the probability of getting a difference in sample proportions of 6% or larger from samples of these sizes is 0.11. Because this *P*-value is larger than 0.05, you can reasonably attribute the difference to chance variation. You have no evidence that the proportions would be different if you were to ask everyone in each of these two age groups whether they sleep more than eight hours on a workday.

E21. The question asks "Was NASA being looked upon *more favorably* by the American public in 2007 than in 1999?" This suggests you should do a one-sided significance test for the difference of two proportions.

This was a Gallup poll so we can assume these samples are equivalent to simple random samples. Each of $n_1\hat{p}_1 = 1000 \cdot 0.46 = 460$, $n_1(1 - \hat{p}_1) = 1000(1 - 0.46) = 640$, $n_2\hat{p}_2 = 1010 \cdot 0.56 = 565.6$, and $n_2(1 - \hat{p}_2) = 1010(1 - 0.56) = 444.4$ is at least 5. There are more than $10 \cdot 1010 = 10{,}100$ adult Americans. The conditions for inference are met.

The hypotheses are:

H_0: The proportion, p_1, of all adult Americans who gave NASA a favorable rating in 1999 is equal to the proportion, p_2, of all adult Americans who gave NASA a favorable rating in 2007.
H_a: $p_1 < p_2$

Note that the pooled estimate is

$$\hat{p} = \frac{\text{total number of successes from both treatments}}{n_1 + n_2} = \frac{460 + 565.6}{1000 + 1010} \approx 0.510.$$

So, the test statistic is

$$z = \frac{(\hat{p}_1 - \hat{p}_2) - (p_1 - p_2)}{\sqrt{\hat{p}(1 - \hat{p})\left(\dfrac{1}{n_1} + \dfrac{1}{n_2}\right)}} = \frac{(0.56 - 0.46) - 0}{\sqrt{0.510(0.490)\left(\dfrac{1}{1{,}000} + \dfrac{1}{1{,}010}\right)}} \approx 4.484$$

The *p*-value is < 0.0001.

Hence, with a *p*-value as low as 0.0001, which is well below 0.05, you would reject the null hypothesis. If the proportion of all adult Americans who gave NASA a favorable rating in 2007 is equal to the proportion of all adult Americans who gave NASA a favorable rating in 1999, then there is at most a 1 out of 10,100 chance of getting a difference in sample proportions of 10% or larger. There is strong evidence that NASA was being looked upon more favorably in 2007 than it was in 1999.

E23. a. Here,

$$n_1 = 663, \quad \hat{p}_1 = 0.35, \quad n_2 = 1591, \quad \hat{p}_2 = 0.29$$

and certainly each of the products $n_1\hat{p}_1$, $n_1(1 - \hat{p}_1)$, $n_2\hat{p}_2$, and $n_2(1 - \hat{p}_2)$ is at least 5. The randomness conditions are also met.

The hypotheses, in symbols, are:
$$H_0: p_1 - p_2 = 0, \quad H_a: p_1 - p_2 > 0.$$

Note that the pooled estimate is

$$\hat{p} = \frac{\text{total number of successes from both treatments}}{n_1 + n_2} = \frac{232.05 + 461.39}{663 + 1591} \approx 0.308.$$

So, the test statistic is

$$z = \frac{(\hat{p}_1 - \hat{p}_2) - (p_1 - p_2)}{\sqrt{\hat{p}(1-\hat{p})\left(\dfrac{1}{n_1} + \dfrac{1}{n_2}\right)}} = \frac{(0.35 - 0.29) - 0}{\sqrt{0.308(1 - 0.308)\left(\dfrac{1}{663} + \dfrac{1}{1591}\right)}} \approx 2.811$$

Since this is a one-sided test, the p-value is $P(Z > 2.811) = 0.0024$. Hence, there is sufficient evidence to say that a larger proportion of those who twitter live in urban areas.

b. Here,

$$n_1 = 663, \ \hat{p}_1 = 0.76, \ n_2 = 1591, \ \hat{p}_2 = 0.60$$

and certainly each of the products $n_1\hat{p}_1$, $n_1(1-\hat{p}_1)$, $n_2\hat{p}_2$, and $n_2(1-\hat{p}_2)$ is at least 5. The randomness conditions are also met.

The hypotheses, in symbols, are:

$$H_0: p_1 - p_2 = 0, \ H_a: p_1 - p_2 > 0.$$

Note that the pooled estimate is

$$\hat{p} = \frac{\text{total number of successes from both treatments}}{n_1 + n_2} = \frac{503.88 + 954.6}{663 + 1591} \approx 0.647.$$

So, the test statistic is

$$z = \frac{(\hat{p}_1 - \hat{p}_2) - (p_1 - p_2)}{\sqrt{\hat{p}(1-\hat{p})\left(\dfrac{1}{n_1} + \dfrac{1}{n_2}\right)}} = \frac{(0.76 - 0.60) - 0}{\sqrt{0.647(1 - 0.647)\left(\dfrac{1}{663} + \dfrac{1}{1591}\right)}} \approx 7.241$$

Since this is a one-sided test, the p-value is $P(Z > 7.241) < 0.0001$. Hence, there is strong evidence to say that those who twitter read newspapers online at a higher percentage than those who do not twitter.

c. No; you need the number of people sampled in each age group.

E25. Observe that $z = 7.51$ and the p-value is near zero. So, there is strong evidence that the male and female populations differ with respect to the percentages that have experienced hazing.

E27. a. We wish to test:

$H_0: p_1 - p_2 = 0$, $H_a: p_1 - p_2 \neq 0$, where p_1 is the proportion of domestic fruit showing no residue and p_2 is the proportion of imported fruit showing no residue.

Here,

$$n_1 = 344, \ \hat{p}_1 = 0.442, \ n_2 = 1136, \ \hat{p}_2 = 0.704$$

and certainly each of the products $n_1\hat{p}_1$, $n_1(1-\hat{p}_1)$, $n_2\hat{p}_2$, and $n_2(1-\hat{p}_2)$ is at least 5. The randomness conditions are also met.

Note that the pooled estimate is

$$\hat{p} = \frac{\text{total number of successes from both treatments}}{n_1 + n_2} = \frac{152.048 + 799.744}{344 + 1136} \approx 0.643.$$

So, the test statistic is

$$z = \frac{(\hat{p}_1 - \hat{p}_2) - (p_1 - p_2)}{\sqrt{\hat{p}(1-\hat{p})\left(\dfrac{1}{n_1} + \dfrac{1}{n_2}\right)}} = \frac{(0.704 - 0.442) - 0}{\sqrt{0.643(1-0.643)\left(\dfrac{1}{344} + \dfrac{1}{1136}\right)}} \approx 8.89$$

The p-value is near zero. Hence, there is strong evidence of a difference between the domestic and imported fruits with regard to the proportions showing no residue.

b. We wish to test:

H_0: $p_1 - p_2 = 0$, H_a: $p_1 - p_2 \neq 0$, where p_1 is the proportion of domestic vegetables showing no residue and p_2 is that proportion of imported vegetables showing no residue.

Here,

$$n_1 = 672, \ \hat{p}_1 = 0.738, \ n_2 = 2447, \ \hat{p}_2 = 0.604$$

and certainly each of the products $n_1\hat{p}_1$, $n_1(1-\hat{p}_1)$, $n_2\hat{p}_2$, and $n_2(1-\hat{p}_2)$ is at least 5. The randomness conditions are also met.

Note that the pooled estimate is

$$\hat{p} = \frac{\text{total number of successes from both treatments}}{n_1 + n_2} = \frac{495.936 + 1477.988}{672 + 2447} \approx 0.633.$$

So, the test statistic is

$$z = \frac{(\hat{p}_1 - \hat{p}_2) - (p_1 - p_2)}{\sqrt{\hat{p}(1-\hat{p})\left(\dfrac{1}{n_1} + \dfrac{1}{n_2}\right)}} = \frac{(0.738 - 0.604) - 0}{\sqrt{0.633(1-0.633)\left(\dfrac{1}{672} + \dfrac{1}{2447}\right)}} \approx 6.384$$

The p-value is near zero. Hence, there is strong evidence of a difference between the domestic and imported vegetables with regard to the proportions showing no residue (but in a different direction for the result for fruit).

c. No, the differences may be valid estimates even though the proportions are biased toward the higher values.

E29. a. The question "Is this a significant increase?" not, "Is this significantly different?" You would do a one-sided test.

The polls were conducted by Gallup, who uses what can be considered a simple random sample. Each of

$$n_1\hat{p}_1 = 1{,}000 \bullet 0.48 = 480, \ n_1(1 - \hat{p}_1) = 1{,}000(1 - 0.48) = 520,$$

$$n_2\hat{p}_2 = 1{,}000 \cdot 0.43 = 430, \text{ and } n_2(1 - \hat{p}_2) = 1{,}000(1 - 0.43) = 570$$

is at least 5, where n_1 and n_2 are the numbers of adults polled in 2009 and 2008, respectively, and $\hat{p}_1$ and $\hat{p}_2$ are the proportions of polled adults in 2009 and 2008, respectively, that logged onto the Internet for an hour or more daily. There were more than $10 \cdot 1{,}000 = 10{,}000$ adults both years in the United States. Hence, the conditions for inference are met.

We wish to test the hypotheses:

H_0: The proportion, p_1, of adults in 2009 who logged onto the Internet for at least an hour daily is equal to the proportion, p_2, of adults in 2008 who logged onto the Internet for at least an hour daily.
H_a: $p_1 < p_2$

Note that the pooled estimate is
$$\hat{p} = \frac{\text{total number of successes from both treatments}}{n_1 + n_2} = \frac{480 + 430}{1000 + 1000} \approx 0.455 .$$

So, the test statistic is
$$z = \frac{(\hat{p}_1 - \hat{p}_2) - (p_1 - p_2)}{\sqrt{\hat{p}(1-\hat{p})\left(\dfrac{1}{n_1} + \dfrac{1}{n_2}\right)}} = \frac{(0.48 - 0.43) - 0}{\sqrt{0.455(1-0.545)\left(\dfrac{1}{1000} + \dfrac{1}{1000}\right)}} \approx 2.245$$

The p-value is $P(Z > 2.245) = 0.0123$. As such, you would reject the null hypothesis that the proportion of adults in 2009 who logged onto the Internet for at least an hour daily is equal to the proportion of adults in 2008 who did so. You have sufficient evidence that this proportion has increased over the year.

b. The one-sided test of H_0: $p = 0.5$, H_a: $p < 0.5$ has $z \approx -1.265$ and a P-value of about 0.1030. You cannot conclude that less than a majority use the Internet more than an hour per day in 2009.

E31. B

E33. a. The question asks you to determine if you have statistically significant evidence that more males are left-handed than females, so use a one-sided test.

Check conditions. You saw that the conditions were met in the example in the text.

State your hypotheses.
H_0: $p_1 - p_2 = 0.02$, where p_1 is the proportion of all males who are left-handed and p_2 is the proportion of all females who are left-handed.
H_a: $p_1 - p_2 > 0.02$.

*Calculate the test statistic and **P**-value.*

$$z = \frac{(\hat{p}_1 - \hat{p}_2) - (p_1 - p_2)}{\sqrt{\dfrac{\hat{p}_1(1-\hat{p}_1)}{n_1} + \dfrac{\hat{p}_2(1-\hat{p}_2)}{n_2}}} = \frac{(0.106 - 0.079) - 0.02}{\sqrt{\dfrac{0.106 \cdot 0.894}{1,067} + \dfrac{0.079 \cdot 0.921}{1,170}}} \approx 0.570$$

According to Table A, the z-score for 0.570 corresponds to a 1-sided P-value of 0.2843.

State your conclusion in context. Since the P-value of 0.2843 is greater than 0.05 you would not reject the null hypothesis that the proportion of all males who are left-handed is 2% more than the proportion of all females who are left-handed. If the proportion of all males who are left-handed is equal to 2% more than the proportion of all females who are left-handed, then, the chance of seeing a difference in sample proportions greater than the observed 2.7% is 0.2843. Because this probability is so large, the observed difference can be attributed to chance alone and there is insufficient evidence to support the claim that the proportion of males who are left-handed is at least 2% greater than the proportion of females who are left-handed.

E35. **a.** We wish to test:

H_0: $p_1 - p_2 = 0$, H_a: $p_1 - p_2 \neq 0$, where p_1 is the proportion of subjects getting colds if all subjects could have been given vitamin C and p_2 is the proportion of subjects getting colds if all subjects could have been given the placebo.

Here,

$$n_1 = 139 \text{ (vitamin C)}, \ \hat{p}_1 = 0.122, \ n_2 = 140, \ \hat{p}_2 = 0.221$$

and certainly each of the products $n_1\hat{p}_1$, $n_1(1-\hat{p}_1)$, $n_2\hat{p}_2$, and $n_2(1-\hat{p}_2)$ is at least 5. The randomness conditions are also met.

Note that the pooled estimate is

$$\hat{p} = \frac{\text{total number of successes from both treatments}}{n_1 + n_2} = \frac{17 + 31}{139 + 140} \approx 0.172.$$

So, the test statistic is

$$z = \frac{(\hat{p}_1 - \hat{p}_2) - (p_1 - p_2)}{\sqrt{\hat{p}(1-\hat{p})\left(\dfrac{1}{n_1} + \dfrac{1}{n_2}\right)}} = \frac{(0.221 - 0.122) - 0}{\sqrt{0.172(1 - 0.172)\left(\dfrac{1}{139} + \dfrac{1}{140}\right)}} \approx 2.191$$

The p-value is $2P(Z > 2.191) = 0.0285$. So, there is sufficient evidence, at the 5% level, to conclude that the proportions getting colds differs for the two treatments, and vitamin C seems to have a positive effect.

b. There is insufficient evidence of a difference at the 1% level because the p-value is larger than 0.01.

E37. We wish to test:

H_0: $p_1 - p_2 = 0$, H_a: $p_1 - p_2 \neq 0$, where p_1 is the proportion of subjects getting polio if all subjects could have been given the Salk vaccine and p_2 is the proportion of subjects getting polio if all subjects could have been given the placebo.

Here,

$$n_1 = 200,745 \text{ (Salk vaccine)}, \quad \hat{p}_1 = 0.0004, \quad n_2 = 201,229 \text{ (placebo)}, \quad \hat{p}_2 = 0.0008$$

and certainly each of the products $n_1\hat{p}_1$, $n_1(1-\hat{p}_1)$, $n_2\hat{p}_2$, and $n_2(1-\hat{p}_2)$ is at least 5. The randomness conditions are also met.

Note that the pooled estimate is

$$\hat{p} = \frac{\text{total number of successes from both treatments}}{n_1 + n_2} = \frac{82 + 162}{200,745 + 201,229} \approx 0.0006.$$

So, the test statistic is

$$z = \frac{(\hat{p}_1 - \hat{p}_2) - (p_1 - p_2)}{\sqrt{\hat{p}(1-\hat{p})\left(\dfrac{1}{n_1} + \dfrac{1}{n_2}\right)}} = \frac{(0.0008 - 0.0004) - 0}{\sqrt{0.0006(1 - 0.0006)\left(\dfrac{1}{200,745} + \dfrac{1}{201,229}\right)}} \approx 5.178$$

The p-value is $2P(Z > 5.178)$ is near 0; there is strong evidence to conclude that the proportion getting polio is smaller among those getting the vaccine. (The difference in proportions may seem small, but the vaccine cut the incidence of polio about in half.)

E39. Because it was hypothesized before the experiment began that aspirin was beneficial, you should conduct a one-sided significance test for the difference of two proportions. (If you do a two-sided test, the computations and conclusion will be the same in this case.)

Check conditions. The subjects weren't selected randomly from a larger population, but the treatments were randomly assigned to the subjects, so you can use this significance test for the difference of two proportions. For the group taking aspirin, $n_1 = 11,037$ and $\hat{p}_1 = \frac{139}{11,037} \approx 0.0126$. For the group taking a placebo, $n_2 = 11,034$ and $\hat{p}_2 = \frac{239}{11,034} \approx 0.0217$.

Therefore, each of $n_1\hat{p}_1 = 139$, $n_1(1 - \hat{p}_1) = 10,898$, $n_2\hat{p}_2 = 239$, and $n_2(1 - \hat{p}_2) = 10,795$ is at least 5.

State your hypotheses.

H_0: If all of the men could have been given aspirin, the proportion, p_1, who had a heart attack would have been equal to the proportion, p_2, of the men who had a heart attack if all could have been given the placebo.

H_a: $p_1 < p_2$

Write a conclusion in context. Using the output, we conclude that if there is no difference in the proportion of men who would have had a heart attack if they had all taken aspirin and the proportion who would have had a heart attack if they had all taken

185

the placebo, then there is almost no chance of getting a difference of 0.0091 or smaller in the two proportions from a random assignment of these treatments to the subjects. This difference can not reasonably be attributed to chance variation. You reject the null hypothesis.

Note that although the difference in proportions is very small, only 0.0091, this difference is statistically significant because of the large sample sizes. Further, men who take low-dose aspirin cut their chance of a heart attack almost in half.

E41. a. This is an observational study.

b. *Check Conditions.* There was no randomization, so this is an observational study. Each of $n_1 \hat{p}_1 \approx 912$, $n_1(1 - \hat{p}_1) \approx 273$, $n_2 \hat{p}_2 = 106$, and $n_2(1 - \hat{p}_2) = 52$ is at least 5.

State hypotheses.
H_0: The difference between the proportion p_1 of dementia-free people who exercise three or more times a week, and the proportion p_2 of those with signs of dementia who exercise three or more times a week can be reasonably attributed to chance variation.

H_a: The difference cannot be reasonably attributed to chance variation.

Calculate the test statistic and draw a sketch.

$$z = \frac{(\hat{p}_1 - \hat{p}_2) - (p_1 - p_2)}{\sqrt{\hat{p}(1 - \hat{p})\left(\frac{1}{n_1} + \frac{1}{n_2}\right)}} = \frac{(0.77 - 0.67) - 0}{\sqrt{0.758 \cdot 0.242\left(\frac{1}{1185} + \frac{1}{158}\right)}} \approx 2.757$$

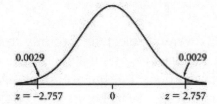

0.0029   0.0029

$z = -2.757$ 0 $z = 2.757$

Here, the pooled estimate $\hat{p}$ is

$$\frac{\text{the total number of exercisers in both groups}}{n_1 + n_2} = \frac{912 + 106}{1185 + 158} \approx 0.758$$

A z-score of 2.757 corresponds to a P-value of $2(0.0029) = 0.0058$.

State conclusion in context. Because this P-value is so low, much less than 0.05, you would reject the null hypothesis that the difference in proportions could be reasonably attributed to chance. There is evidence of an association between exercise and a delay of dementia for this group of persons in this study.

This study cannot demonstrate any causal relationship due to the lack of randomization, but the issue may warrant more study. This is an example of a newspaper reporting a result that is not indicated by the study, and demonstrates the importance of clearly stating what your study shows and what it does not show.

E43. a. H_0: $p_1 - p_2 = 0$, H_a: $p_1 - p_2 > 0$, where p_1 is the proportion of subjects eating goldfish if all subjects could have seen the host eating goldfish and p_2 is the proportion of subjects eating goldfish if all subjects could have seen the host eating animal crackers.

b. Here,
$$n_1 = 29, \ \hat{p}_1 = 0.724, \ n_2 = 26, \ \hat{p}_2 = 0.462$$
and certainly each of the products $n_1\hat{p}_1$, $n_1(1-\hat{p}_1)$, $n_2\hat{p}_2$, and $n_2(1-\hat{p}_2)$ is at least 5. The randomness conditions are also met.

Note that the pooled estimate is
$$\hat{p} = \frac{\text{total number of successes from both treatments}}{n_1 + n_2} = \frac{21+12}{29+26} \approx 0.60.$$
So, the test statistic is
$$z = \frac{(\hat{p}_1 - \hat{p}_2) - (p_1 - p_2)}{\sqrt{\hat{p}(1-\hat{p})\left(\frac{1}{n_1} + \frac{1}{n_2}\right)}} = \frac{(0.724 - 0.462) - 0}{\sqrt{0.6(1-0.6)\left(\frac{1}{29} + \frac{1}{26}\right)}} \approx 1.980$$

The p-value is $P(Z > 1.980) = 0.0238$. So, there is sufficient evidence, at the 5% level, to conclude that students watching the host eat goldfish have a higher proportion of goldfish eaters than if they watch the host eat animal crackers.

E45. a. It seems reasonable that larger tumors would be more likely to spread than smaller tumors.

b. No. There is not a random sample of patients with tumors of either size. Instead there is a group of patients enrolled in a particular program. It is true that the other conditions have been met. Each of
$$n_1\hat{p}_1 = 234, \ n_1(1-\hat{p}_1) = 24$$
$$n_2\hat{p}_2 = 98, \ n_2(1-\hat{p}_2) = 20$$
is at least 5. The number of cancer patients with tumors of each given size is more than ten times the respective sample size given in the problem.

Note: You could do a test to see whether the observed difference in proportions can be reasonably attributed to chance. The conditions *are* met for such a test.

c. $z = \dfrac{(\hat{p}_1 - \hat{p}_2) - (p_1 - p_2)}{\sqrt{\hat{p} \cdot (1-\hat{p})\left(\dfrac{1}{n_1} + \dfrac{1}{n_2}\right)}} = \dfrac{\left(\frac{234}{258} - \frac{98}{118}\right) - 0}{\sqrt{\dfrac{332}{376} \cdot \dfrac{44}{376}\left(\dfrac{1}{258} + \dfrac{1}{118}\right)}} \approx 2.14.$

Table A gives a one-sided P-value of 0.0162. 0.02 would be a correct conservatively rounded approximation. Here, $\hat{p} = \dfrac{234+98}{258+118} = \dfrac{332}{376}$.

d. If tumors measuring 15 mm or less and tumors measuring 16-25 mm are equally likely to metastasize, then there is about a 2% probability of seeing a difference in proportions of metastases at least as large as that seen in this study.

E47. *Is a significance test legal in this case?* Purists would say that we should not use a test of significance in this situation. They have two reasons. The first is that the numbers given are not a random sample from any population—in fact, they are the population of Reggie Jackson's "at bats." (He is retired, so there will be no further at bats.) We know all of his at bats, and we can see that, in fact, he did have a higher batting average in the World Series than in regular season play.

The second reason is that this is a classic example of "data snooping." There are hundreds of baseball players. Even if some underlying batting average is the same in regular season play as in the World Series for all players, by definition some players are certain to be rare events and do better in the World Series than in regular season play. Reggie is simply the player that stands out as the rarest of the predictable rare events.
Note that the question asks whether Reggie's better average in the World Series can reasonably be attributed to chance. This is the first question we should ask before assigning him the nickname "Mr. October." If it turns out that we can't reasonably attribute this to chance, then we have to look for some other explanation. That explanation might in fact be that we did some data snooping and ended up with a Type I error. On the other hand, the explanation might be that he came through in the World Series. At any rate, the data must pass the test that the results can't reasonably be attributed to chance before we take any further steps in comparing the performance of Reggie Jackson in the World Series to regular season play.

Check conditions. The two samples aren't random; they are the entire populations. Thus, a significance test will tell us only whether such a difference can reasonably be attributed to chance. Each of $n_1\hat{p}_1 \approx 2584$, $n_1(1 - \hat{p}_1) \approx 7280$, $n_2\hat{p}_2 \approx 35$, and $n_2(1 - \hat{p}_2) \approx 63$ is at least 5.

State your hypotheses.

H_0: The difference between the proportion of hits in regular season play and the proportion of hits in the World Series can reasonably be attributed to chance variation.
H_a: The difference between the proportion of hits in regular season play and the proportion of hits in the World Series is too large to be attributed to chance variation.

Compute the test statistic and draw a sketch.
The test statistic is

$$z = \frac{(\hat{p}_1 - \hat{p}_2) - (p_1 - p_2)}{\sqrt{\hat{p}(1-\hat{p})\left(\frac{1}{n_1} + \frac{1}{n_2}\right)}} = \frac{(0.262 - 0.357) - 0}{\sqrt{0.263(1 - 0.263)\left(\frac{1}{9864} + \frac{1}{98}\right)}} \approx -2.126,$$

or, using the TI-84+'s **2-PropZTest**, $z = 2.1299$. Here,

$$\hat{p} = \frac{\textit{total number of successes in both samples}}{n_1 + n_2} = \frac{2584 + 35}{9864 + 98} \approx 0.263$$

The one-sided *P*-value is about 0.017.

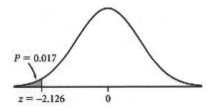

Write a conclusion in context. A difference as large as Reggie's between regular season and World Series play would happen by chance to fewer than 17 players in 1000. Therefore, Reggie's record is indeed unusual. (It is interesting that Reggie hit only 0.227 in 163 bats in crucial League Championship Series play.)

E49. *Check conditions.* The two samples may be considered random samples. They were taken independently from the population of U.S. adults in 2008 and in 1974 The number of adults in each year is larger than ten times 1702. Finally, each of $n_1\hat{p}_1 = 1702(0.48) = 817$, $n_1(1 - \hat{p}_1) = 1702(1 - 0.48) = 885$, $n_2\hat{p}_2 = 1002(0.46) = 461$, and $n_2(1 - \hat{p}_2) = 1002(1 - 0.46) = 541$ is at least 5.

Do computations. The 90% confidence interval for the difference of the two population proportions p_1 and p_2 is

$$(\hat{p}_1 - \hat{p}_2) \pm z^* \cdot \sqrt{\frac{\hat{p}_1(1-\hat{p}_1)}{n_1} + \frac{\hat{p}_2(1-\hat{p}_2)}{n_2}} = (0.48 - 0.46) \pm 1.65\sqrt{\frac{(0.48)(1-0.48)}{1702} + \frac{(0.46)(1-0.46)}{1002}}$$

$$= 0.02 \pm 0.033$$

Alternatively, you can write this confidence interval as (-0.013, 0.053).

Write a conclusion in context. You are 90% confident that the difference between the proportion of all adults who would assign a grade of A or B in 2008 and in 1974 is between -0.013 and 0.053. Because 0 is in this confidence interval, the increase is insignificant.

E51. a. *Check conditions.* The treatments were assigned randomly to the subjects, so you can use this significance test for the difference of two proportions. For the group taking the medication, $n_1 = 25$ and $\hat{p}_1 = 13/25 = 0.52$. For the group taking a placebo, $n_2 = 26$ and $\hat{p}_2 = 10/26 = 0.385$. Each of

$$n_1\hat{p}_1 = 13, \; n_1(1 - \hat{p}_1) = 12, \; n_2\hat{p}_2 = 10, \text{ and } n_2(1 - \hat{p}_2) = 16$$

is at least 5.

State your hypotheses.

H_0: The proportion, p_1, of people who would have responded if everyone had been given medication is equal to the proportion, p_2, of people who would have responded if everyone had been given the placebo.

H_a: $p_1 \neq p_2$

Compute the test statistic and draw a sketch. The test statistic is

$$z = \frac{(\hat{p}_1 - \hat{p}_2) - (p_1 - p_2)}{\sqrt{\hat{p}(1-\hat{p})\left(\dfrac{1}{n_1} + \dfrac{1}{n_2}\right)}} = \frac{(0.52 - 0.385) - 0}{\sqrt{0.451(1-0.451)\left(\dfrac{1}{25} + \dfrac{1}{26}\right)}} \approx 0.9686$$

or, using **2-PropZTest**, $z = 0.9713$.

Here,

$$\hat{p} = \frac{total\ number\ who\ responded}{n_1 + n_2} = \frac{13 + 10}{25 + 26} \approx 0.451$$

The two-sided P-value is 0.3314.

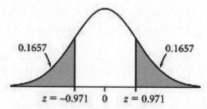

$z = -0.971$ 0 $z = 0.971$

Write a conclusion in context. If there had been no difference between the proportion of people who would have responded if they had all taken the medication and the proportion who would have responded if they had all taken the placebo, then there is a 0.331 chance of getting a difference of 0.135 or larger in the two proportions from a random assignment of subjects to treatment groups. This difference can reasonably be attributed to chance variation. You do not reject the null hypothesis.

Note: Although the difference in the proportion who responded isn't statistically significant, the main point of the study was that "Brain physiology in placebo responders was altered in a different manner than in the medication responders."

b. Neither the subject nor the examining physician knew which treatment they were getting. This can be done by making sure the antidepressant and placebo look alike, and having the random assignment made by a third party.

E53. a. We wish to test:

H_0: $p_1 - p_2 = 0$,

H_a: $p_1 - p_2 < 0$, where p_1 is the population proportion favoring stricter gun control laws in 2009 and p_2 is that proportion in 1990.

Here,

$$n_1 = 1023, \ \hat{p}_1 = 0.39 \ (2009), \ n_2 = 1023, \ \hat{p}_2 = 0.78 \ (1990)$$

and certainly each of the products $n_1\hat{p}_1$, $n_1(1-\hat{p}_1)$, $n_2\hat{p}_2$, and $n_2(1-\hat{p}_2)$ is at least 5. The randomness conditions are also met.

Note that the pooled estimate is
$$\hat{p} = \frac{\text{total number of successes from both treatments}}{n_1 + n_2} = \frac{398.97 + 797.94}{1023 + 1023} \approx 0.585 .$$

So, the test statistic is
$$z = \frac{(\hat{p}_1 - \hat{p}_2) - (p_1 - p_2)}{\sqrt{\hat{p}(1-\hat{p})\left(\dfrac{1}{n_1} + \dfrac{1}{n_2}\right)}} = \frac{(0.39 - 0.78) - 0}{\sqrt{0.585(1-0.585)\left(\dfrac{1}{1023} + \dfrac{1}{1023}\right)}} \approx -17.90$$

The p-value $P(Z < -17.90)$ is near zero. So, there is strong evidence to conclude that the proportion favoring stricter gun control laws has decreased between 1990 and 2009.

b. Here,
$$n_1 = 1023, \ \hat{p}_1 = 0.39 \ (2009), \ n_2 = 100, \ \hat{p}_2 = 0.78 \ (1990)$$

and certainly each of the products $n_1\hat{p}_1$, $n_1(1-\hat{p}_1)$, $n_2\hat{p}_2$, and $n_2(1-\hat{p}_2)$ is at least 5. The randomness conditions are also met.

Note that the pooled estimate is
$$\hat{p} = \frac{\text{total number of successes from both treatments}}{n_1 + n_2} = \frac{78 + 398.97}{100 + 1023} \approx 0.425 .$$

So, the test statistic is
$$z = \frac{(\hat{p}_1 - \hat{p}_2) - (p_1 - p_2)}{\sqrt{\hat{p}(1-\hat{p})\left(\dfrac{1}{n_1} + \dfrac{1}{n_2}\right)}} = \frac{(0.39 - 0.78) - 0}{\sqrt{0.425(1-0.425)\left(\dfrac{1}{100} + \dfrac{1}{1023}\right)}} \approx -7.53$$

The p-value $P(Z < -7.53)$ is near zero. So, there is still strong evidence of a decrease.

c. We construct two 95% confidence intervals, one for the samples from (a) and one for the samples from (b):

Samples from (a):
$$(\hat{p}_1 - \hat{p}_2) \pm z * \cdot \sqrt{\frac{\hat{p}_1(1-\hat{p}_1)}{n_1} + \frac{\hat{p}_2(1-\hat{p}_2)}{n_2}} = (0.78 - 0.39) \pm 1.96 \sqrt{\frac{(0.39)(1-0.39)}{1023} + \frac{(0.78)(1-0.78)}{1023}}$$
or about (.351, 429).

Samples from (b):
$$(\hat{p}_1 - \hat{p}_2) \pm z * \cdot \sqrt{\frac{\hat{p}_1(1-\hat{p}_1)}{n_1} + \frac{\hat{p}_2(1-\hat{p}_2)}{n_2}} = (0.78 - 0.39) \pm 1.96 \sqrt{\frac{(0.39)(1-0.39)}{1023} + \frac{(0.78)(1-0.78)}{100}}$$
or about (0.304, 0.476).
The interval formed using the smaller sample size is about twice as wide as the other.

E55. First, we wish to test:

H_0: $p_1 - p_2 = 0$, H_a: $p_1 - p_2 \neq 0$, where p_1 is the proportion of deaths if all subjects could have been given the intensive treatment and p_2 is the proportion of deaths if all subjects could have been given the conventional treatment.

Here,

$n_1 = 3054$ (intensive treatment), $\hat{p}_1 = 0.271$, $n_2 = 3050$ (conventional treatment), $\hat{p}_2 = 0.246$

and certainly each of the products $n_1\hat{p}_1$, $n_1(1-\hat{p}_1)$, $n_2\hat{p}_2$, and $n_2(1-\hat{p}_2)$ is at least 5. The randomness conditions are also met.

Note that the pooled estimate is

$$\hat{p} = \frac{\text{total number of successes from both treatments}}{n_1 + n_2} = \frac{829 + 751}{3054 + 3050} \approx 0.259 .$$

So, the test statistic is

$$z = \frac{(\hat{p}_1 - \hat{p}_2) - (p_1 - p_2)}{\sqrt{\hat{p}(1-\hat{p})\left(\dfrac{1}{n_1} + \dfrac{1}{n_2}\right)}} = \frac{(0.271 - 0.246) - 0}{\sqrt{0.259(1-0.259)\left(\dfrac{1}{3054} + \dfrac{1}{3050}\right)}} \approx 2.229$$

The p-value is $2P(Z > 2.229) = 0.0258$. So, there is sufficient evidence to conclude that the death proportions differ for the two treatments.

Next, assume that

$n_1 = 300$ (intensive treatment), $\hat{p}_1 = 0.271$, $n_2 = 300$ (conventional treatment), $\hat{p}_2 = 0.246$

and certainly each of the products $n_1\hat{p}_1$, $n_1(1-\hat{p}_1)$, $n_2\hat{p}_2$, and $n_2(1-\hat{p}_2)$ is at least 5. The randomness conditions are also met.

Note that the pooled estimate is

$$\hat{p} = \frac{\text{total number of successes from both treatments}}{n_1 + n_2} = \frac{81.3 + 73.8}{300 + 300} \approx 0.259 .$$

So, the test statistic is

$$z = \frac{(\hat{p}_1 - \hat{p}_2) - (p_1 - p_2)}{\sqrt{\hat{p}(1-\hat{p})\left(\dfrac{1}{n_1} + \dfrac{1}{n_2}\right)}} = \frac{(0.271 - 0.246) - 0}{\sqrt{0.259(1-0.259)\left(\dfrac{1}{300} + \dfrac{1}{300}\right)}} \approx 0.699$$

The p-value is $2P(Z > 0.699) = 0.485$. So, there is not insufficient evidence to conclude that the death rates for the treatments differ.

E57. No, because the two sample proportions making up the difference are dependent. If one is very large, the other has to be small.

Chapter 10

Practice Problem Solutions

P1. a. $\bar{x} = 5.019$, $s \approx 1.104$. The interval is $5.019 \pm 1.96 \dfrac{1.104}{\sqrt{10}}$, or about (4.335, 5.703) or (4.335, 5.703) with no rounding.

b. When σ is estimated by s, the centers of the intervals, which are determined only by the sample mean, are the same as if you were using σ. However, the width will vary directly with s rather than being constant. Because s is smaller than σ more often than it is larger, the confidence interval will be too narrow more often than it is too wide. This makes the capture rate smaller than the advertised value.

P2. a. 2.262
b. 2.447
c. 3.106
d. 2.920
e. 2.704 (Use $df = 40$)
f. 2.639 (Use $df = 80$)

Using **invT(** on the TI-84+ will also give these values. **invT(** requires the tail probability and the degrees of freedom, so for a 99% confidence interval, you would use the tail probability 0.005 or 0.995. One will give you the positive value of t^*, the other will give the negative value. For part e, **invT(0.005,43)** gives –2.695, which you would report as t^* = 2.695. For part f, **invT(0.005,81)** results in a t^* of 2.638. Table B gives values that are close enough for practical purposes.

P3. a. $n = 4$, $\bar{x} = 27$, $s = 12$: $t^* = 3.182$, $t^* \cdot s / \sqrt{n} = 19.092$
The interval is 27 ± 19.092, or 7.908 to 46.092. The TInterval on the calculator gives (7.905, 46.095). You are 95% confident that the unknown mean is in the interval 7.908 to 46.092.

b. $n = 9$, $\bar{x} = 6$, $s = 3$: $t^* = 2.306$, $t^* \cdot s / \sqrt{n} = 2.306$ The interval is 6 ± 2.306, or 3.694 to 8.306. The TInterval on the calculator gives the same values. You are 95% confident that the unknown mean is in the interval 3.694 to 8.306.

c. $n = 16$, $\bar{x} = 9$, $s = 48$: $t^* = 2.131$, $t^* \cdot s / \sqrt{n} = 25.572$
The interval is 9 ± 25.572, or –16.572 to 34.572. The TInterval on the calculator gives (–16.58, 34.577). You are 95% confident that the unknown mean is in the interval –16.572 to 34.572.

P4. The confidence interval is

$$\bar{x} \pm t^* \cdot \frac{s}{\sqrt{n}} \text{ or } 5.019 \pm 2.262 \cdot \frac{1.104}{\sqrt{10}}$$

which yields an interval of (4.23, 5.81). The TInterval on the calculator gives

(4.229, 5.809). This is just a little wider than the *z*-interval constructed in P1, which was (4.335, 5.703). You are 95% confident that the mean level of aldrin in the Wolf River downstream from the toxic waste site is in the interval 4.23 to 5.81.

P5. a. $df = 121$. Using Table B the closest value given is $df = 100$.

$$\bar{x} \pm t^* \frac{s}{\sqrt{n}} = 98.1 \pm 1.984 \frac{0.73}{\sqrt{122}} \approx 98.1 \pm 0.131,$$

or about (97.969, 98.231). The **TInterval** on the TI-84+ gives the same values.

b. You are 95% confident that the true mean body temperature of all men is between 97.969°F and 98.231°F.

c. No. The mean body temperature of men is less than 98.6°F.

P6. a. No. This is not a random sample of people but a group of volunteers. Also, the distribution is skewed a little right, making it questionable that the population is normally distributed, although with a sample size of 30, the sampling distribution of the mean would probably be normally distributed. The only condition that is met is the guideline that the population be at least 10 times the sample size. There are certainly more than 30 · 10, or 300, people in the U.S., assuming that is the population of interest. The main concern is the lack of a random sample.

b. $df = 29$, $t^* = 2.756$

$$\bar{x} \pm t^* \frac{s}{\sqrt{n}} = 58.37 \pm 2.756 \frac{15.26}{\sqrt{30}} \approx 58.37 \pm 7.678, \text{ or } (50.692, 66.048)$$

With no rounding, the interval is (50.687, 66.046).

c. If this were a random sample of people, you would be 99% confident that if you blindfolded all people and had them attempt to walk the length of a football field, the mean distance walked before crossing a sideline would be between 50.692 and 66.048 yards.

P7. a. The data are skewed right, so it is unlikely that the sample comes from a normal distribution. However, a sample size of 61 is large enough that this should not matter. If many samples of size 61 were taken, the sampling distribution of the mean would still be approximately normal.

b. t^* for 60 degrees of freedom and 90% confidence is 1.671.

$$\bar{x} \pm t^* \frac{s}{\sqrt{n}} = 10.26 \pm 1.671 \frac{6.22}{\sqrt{61}} \approx 10.26 \pm 1.331,$$

or (8.929, 11.591). **TInterval** on the TI-84+ with these values gives (8.9295, 11.59). Using the data in the stemplot and with no rounding, the TInterval is (8.932, 11,592).

c. You are 90% confident that the mean number of hours of study per week for *all* students taking this course is between 8.929 and 11.591.

By 90% confidence you mean that if you were to take 100 random samples of size 61 from all students who take this course and construct a confidence interval from each, you expect that 90 intervals would contain the mean number of hours of study for all students.

P8. D

P9. a. False. A confidence interval is a statement about plausible values for a population mean, not about the individual values within a population.

b. False. A confidence interval is a statement about plausible values for a population mean, not about the values in the sample.

c. False. The method has a 95% chance of success, but after this particular interval is calculated, it either is successful or it isn't.

d. False. The sample mean is the center of the confidence interval.

e. True. The capture rate equals the confidence level provided the population is nearly normal and the sample is random.

P10. a. Men: 0.397, Women: 0.377

b. The margin of error is larger for men because the standard deviation of the sample of men is larger than for the sample of women.

P11. $\bar{x} \approx 70.143$, $s \approx 1.676$. So,
$$t = \frac{\bar{x} - \mu_0}{s/\sqrt{n}} = \frac{70.143 - 72}{1.676/\sqrt{7}} \approx -2.9315 \text{ (or } -2.9314 \text{ without rounding).}$$

P12. We have
$$\mu = 301{,}050{,}000/83{,}246 \approx 3616.390$$
$$t = \frac{\bar{x} - \mu_0}{s/\sqrt{n}} = \frac{3862.14 - 3616.39}{754/\sqrt{35}} \approx 1.928.$$

P13. The lighter graph is the standard normal distribution. One way you can tell is that the standard normal distribution has less area in the tails. The t-distribution is more spread out with heavier tails.

P14. a. Since you want 6 degrees of freedom you will need samples of size 7. To select from a population that is normally distributed you will need to use a calculator or computer. One way to do this on the TI-84+ is **randNorm(100,15,7)->L$_1$,** which selects seven members of the population and stores the values in L$_1$.

To calculate $t = \dfrac{\bar{x} - \mu_0}{s \big/ \sqrt{n}}$, you can do **(mean(L1)–100)/(stdDev(L1)/$\sqrt{}$ (7))**.

b. On the TI-84+, on one line enter **randNorm(100,15,7)->L₁: (mean(L1)– 100)/(stdDev(L1)/√(7))**. The colon allows you to execute two commands as one step. Then each time you press ENTER you generate a new value of t.

P15. The test statistic computed in P10 was –2.9315 (or –2.9314 without rounding). Table B lists only a few critical tail probabilities, and only gives the positive values of t. For 6 degrees of freedom you can see that $t = 2.9315$ corresponds to a P-value between 0.01 and 0.02. You might estimate it at 0.015. Since the question was posed as a two-sided situation you must double this, so the P-value is approximately 0.03. A calculator will give you a more precise value. **2*tcdf(–1E99,–2.9315,6)** gives a P-value of about 0.0262. If the mean temperature at your desk is actually 72°, the probability that temperatures taken on seven randomly selected days would give a value of t greater than 2.9315 or less than –2.9315 is about 0.0262. (The P-value is the same if you use 2.9314).

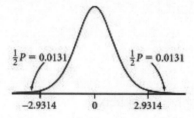

$\frac{1}{2}P = 0.0131$ $\frac{1}{2}P = 0.0131$

–2.9314   0   2.9314

P16. The test statistic computed in P12 was 1.928 for 34 degrees of freedom. The closest you can get using Table B is for 30 degrees of freedom and a one-sided P-value between 0.05 and 0.025. Double this to get a two-sided P-value between 0.05 and 0.10.

If the mean Pell Grant for Minnesota college students is still $3616.39, the probability that a random sample of 35 students would have a t-statistic greater in absolute value than 1.928 is about 0.0622.

P17. a. This situation calls for a one-sided (right-sided) alternate because the claim (research or alternate hypothesis) is that the mean price increased.

b. $\bar{x}$ = mean selling price for a sample of houses sold this month.
μ = true mean selling price of all houses sold in the city for this month.

c. The null hypothesis is that μ is the same as the mean for last month. The alternate is that it is greater.

P18. *Check conditions:* The temperatures were taken on randomly selected days. The dotplot of the data is fairly symmetric with no outliers, so it is reasonable to assume that they were taken from a normal population. The population of days on which the measurements could be taken is more than ten times the sample size.

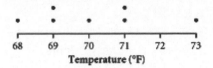

Temperature (°F)

State your hypotheses:

H_0: The mean temperature μ at your seat is 72°F.

H_a: $\mu < 72$°F

Compute the test statistic, find the P-value, and draw a sketch.

$$t = \frac{\bar{x} - \mu_0}{s/\sqrt{n}} = \frac{70.143 - 72}{1.676/\sqrt{7}} \approx -2.931$$

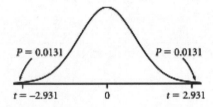

$P = 0.0131$ $P = 0.0131$

$t = -2.931$ 0 $t = 2.931$

Write your conclusion in context, linked to your computations: at the 5% level, the critical *t*-value for a one-sided test with 6 degrees of freedom is $t* = -1.943$. Because the observed *t*-value exceeds the critical value, you have sufficient evidence to reject the null hypothesis. The data suggest that the mean temperature at your desk is less than 72°F. Alternately, if the mean temperature at your desk is indeed 72", there is only a 1.3% chance of getting a result from a sample of size 7 as extreme as or more extreme than the one from your sample ($t = -2.431$). So there is sufficient evidence to support the claim that the mean temperature at your desk is less than 72°.

P19. *State your hypotheses:*

H_0: $\mu = 290$

H_a: $\mu > 290$,

where μ = mean QL score for 2-year college students.

Compute the test statistic and find the P-value:

Note that $\bar{x} \approx 310$, $s \approx 79$. So,

$$t = \frac{\bar{x} - \mu_0}{s/\sqrt{n}} = \frac{310 - 290}{79/\sqrt{800}} \approx 7.161.$$

With df = $800 - 1 = 799$, we see that $P(T > 7.161) < 0.0005$.

State you conclusion: The data provide strong evidence that the mean quantitative literacy score exceeds 290.

P20. *State your hypotheses:*

H_0: $\mu = 350$

H_a: $\mu < 350$,

where μ = mean QL score for 2-year college students.

Compute the test statistic and find the P-value:
Note that $\bar{x} \approx 330$, $s \approx 110$. So,

$$t = \frac{\bar{x} - \mu_0}{s/\sqrt{n}} = \frac{330 - 350}{111/\sqrt{1000}} \approx -5.698.$$

With df = 1000 − 1 = 999, we see that $P(T < -5.698) < 0.0005$.

State you conclusion: The data provide strong evidence that the mean quantitative literacy score is less than 350.

P21. a. H_0: The mean temperature at your desk μ is 72°.

H_a: $\mu \neq 72°$

b. Because the *P*-value is less than both 10% and 5%, you would reject the null hypothesis at the 10% and 5% levels. However, the *P*-value is greater than 1% so you would not reject the null hypothesis at the 1% level.

P22. a. H_0: The mean Pell Grant amount in Minnesota, μ, is \$3,616.39 per student

H_a: $\mu \neq \$3,616.39$.

b. Since the t-statistic is 1.928 and the *P*-value is about 0.062, which is less than 10%, you would reject the null hypothesis at this level, but not at the 5% or 1% levels.

P23. a. H_0: The mean SAT score of State University students μ is 1700.

H_a: $\mu \neq 1700$

b. The test statistic is

$$t = \frac{\bar{x} - \mu_0}{s/\sqrt{n}} = \frac{1535 - 1700}{250/\sqrt{9}} = -1.98.$$

With 8 degrees of freedom this corresponds to a two-sided *P*-value of 0.0830. This is less than 10% so you would reject the null hypothesis at this level. However, you would not reject the null hypothesis at the 5% or 1% level.

P24. *Check conditions:* There is no indication given of how this sample was taken, so you cannot tell whether it can be considered a simple random sample. The plot of the sample data below is not highly skewed and has no extreme outliers, so it is reasonable to assume that the underlying population of all possible measurements is normally distributed. The population of all possible Aldrin level measurements from the Wolf

River is more than ten times the sample size.

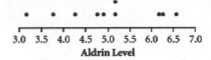

Aldrin Level

State your hypotheses:

H_0: $\mu = 4$

H_a: $\mu \neq 4$

where μ is the mean Aldrin level downstream.

Compute the test statistic, find the P-value, and draw a sketch:
With the sample mean of 5.019 and standard deviation of about 1.104, the test statistic is

$$t = \frac{\bar{x} - \mu_0}{s / \sqrt{n}} \approx \frac{5.019 - 4}{1.1044 / \sqrt{10}} \approx 2.919 .$$

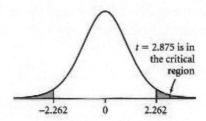

$t = 2.875$ is in the critical region

-2.262 0   2.262

Write your conclusion in context, linked to your computations: At the 5% level, the critical t-value for 9 degrees of freedom is $t^* = 2.262$. Because the observed t-value exceeds the critical value, you have sufficient evidence to reject the null hypothesis. The data suggest that the mean aldrin level differs from 4 nanograms.

P25. *Check conditions:* This was done in the text.

State your hypotheses: The research claim here is that the mean may differ from the advertised standard of 98.6. This claim forms the alternative (or research) hypothesis. The standard against which the sample mean is compared forms the null hypothesis.

   H_0: $\mu = 98.6$

   H_a: $\mu \neq 98.6$,

where μ is the mean body temperature of all males in the population under study.

Compute the test statistic, find the P-value, and draw a sketch:

$$t = \frac{\bar{x} - \mu_0}{s / \sqrt{n}} = \frac{97.88 - 98.6}{0.555 / \sqrt{10}} \approx -4.10$$

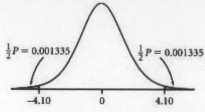

The *P*-value from a calculator is about 0.00267 (or 0.00268 using the data with no rounding). If you must use the table to find a *P*-value all you can say is that the *P*-value is between 2(0.001) and 2(0.0025), or 0.002 < *P*-value < 0.005. The partial *t*-table below shows these values.

Tail probability	0.0025	0.001
$df = 9$	3.690	4.297

Write a conclusion in context, linked to the computations: The small *P*-value of 0.00267 implies that the chance of seeing a value of *t* this unusual if the null hypothesis were true is extremely rare. This is very strong evidence against the null hypothesis. Thus, 98.6 is not a plausible value for the mean body temperature of all males in this population.

To test the same null hypothesis for the females, everything is the same, except that the observed *t* statistic is

$$t = \frac{98.52 - 98.6}{0.527 / \sqrt{10}} \approx -0.48$$

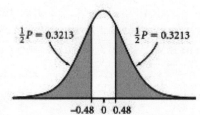

The resulting *P*-value is 0.6426 (or 0.6424 using the data and no rounding). A value of *t* this size is not at all unusual under the null hypothesis, so you don't have convincing evidence against the null hypothesis. Therefore, the claim that the mean body temperature for females is 98.6 is plausible. (But so are many other values.)

P26. a. Yes, the distribution of the underlying population is symmetric and roughly normal, and certainly the population size is at least 310. The randomness of the sample seems likely based on the periodic nature in which the windows were chosen to be sampled.

b. We wish to test:
$H_o: \mu = 32$, $H_a: \mu < 32$, where μ denotes the mean glass strength.

From the given display, $t = -0.91$ with *p*-value of p=0.184. Hence, the data provide insufficient evidence to declare that the standard of 32 is not being met.

c. We wish to test:
H_o: $\mu = 35$, H_a: $\mu < 35$, where μ denotes the mean glass strength.

From the given display, $t = -3.22$ with p-value of p=0.002. Hence, the data provide strong evidence to declare that the standard of 32 is not being met.

P27. a. The null hypothesis was rejected. If the null hypothesis was, in fact, true a Type I error was made.

b. The null hypothesis was rejected. If the null hypothesis was, in fact, true a Type I error was made.

c. The null hypothesis was not rejected. If the null hypothesis was, in fact, false a Type II error was made.

P28. a. The test with $n = 16$. A larger sample size makes it easier to detect a difference from the null hypothesis, which means greater power.

b. The test with $\alpha = 0.05$. A larger significance level makes it easier to reject the null hypothesis which means the power is greater.

Exercise Solutions

E1. a. Here, $\bar{x} = 330$, $\sigma \approx 111$, $n = 1000$. So, the 95% confidence interval is

$$\bar{x} \pm 1.96 \frac{\sigma}{\sqrt{1000}} = 330 \pm 1.96 \frac{111}{\sqrt{1000}},$$

or about (323.1, 336.9).

b. Here, $\bar{x} = 310$, $\sigma \approx 79$, $n = 800$. So, the 95% confidence interval is

$$\bar{x} \pm 1.96 \frac{\sigma}{\sqrt{800}} = 310 \pm 1.96 \frac{79}{\sqrt{800}},$$

or about (304.5, 315.5).

c. Yes because the two interval estimates do not overlap.

E3. The sample mean is $\bar{x} = 15.974$, with $s \approx 0.08$. You could use a confidence interval to estimate the true mean of the weight of the water.

Check conditions: The bottles are a random sample from the day's production. The distribution of the data is fairly symmetric, with a possible slight skew to the left. The skew seems slight enough that a sample size of 10 would be adequate for the sampling distribution of the mean to be approximately normal. Also, you are told that the distribution of the number of ounces in the bottle is approximately normal.

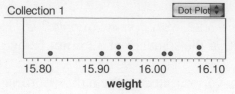

The day's production is most likely more than 100 bottles, which is 10 times the sample size.

Do computations: For 95% confidence and 9 degrees of freedom, $t^* = 2.262$, which results in

$$\bar{x} \pm t^* \frac{s}{\sqrt{n}} = 15.974 \pm 2.262 \frac{0.08}{\sqrt{10}} \approx 15.974 \pm 0.057$$

or an interval of (15.917, 16.031). With no rounding, it is (15.9167, 16.0312).

Give interpretation in context: You are 95% confident that the mean weight of the bottles produced that day is between 15.917 oz. and 16.031 oz. Because 16 oz. is one of the plausible values for the population mean, there is no need to adjust the machine. (This is an example of an important principle in quality improvement; don't make adjustments until you are pretty sure things need adjusting.)

E5. a. Normality cannot be determined from such a small sample; the nature of the population would not suggest extreme skewness.

b. Here, $n = 3$ and

$$\bar{x} = \frac{3 + 3.5 + 3}{3} = 3.17$$

$$s = \sqrt{\frac{(3 - 3.17)^2 + (3.5 - 3.17)^2 + (3 - 3.17)^2}{3 - 1}} = 0.289$$

Also, the df $= 3 - 1 = 2$ and $t^* = 4.303$. Hence, the confidence interval is

$$\bar{x} \pm t^* \left(\frac{s}{\sqrt{n}} \right) = 3.17 \pm 4.303 \left(\frac{0.289}{\sqrt{3}} \right)$$

or about (2.45, 3.89).

c. No. The extreme value is not in the confidence interval; this may be due to the non-normality of the population, with an outlier, or simply an unlucky draw, but three small values would not be an unusual sample from this population.
d. Yes, they would work satisfactorily, if not better given the increased sample size.

E7. a. This is not a random sample of bags of fries. The fact that the thirty-two bags of fries came from two time periods indicates that their masses are not necessarily independent as it may have been the same person preparing many of them. Also, because the bags all came from this one restaurant, even if you had a random sample, it would only allow you to infer about bags of fries from this store only. The distribution is slightly skewed right with one possible outlier on the right, which indicates that it is highly questionable that the population is normally distributed. However, since the skew is not great, a sample size of 32 would be large enough that the sampling distribution of the mean would be approximately normal. It is probably also true that the number of bags of fries sold by this McDonald's restaurant is more than 320, which would be more than ten times the sample size. So the random sample condition is not met but the other conditions are.

b. $\bar{x} \approx 73.72$, $s \approx 10.52$, $df = 31$, $t^* = 2.042$ (for 30 degrees of freedom from Table B)

$$\bar{x} \pm t^* \frac{s}{\sqrt{n}} \approx 73.72 \pm 2.042 \frac{10.52}{\sqrt{32}} \approx 73.72 \pm 3.80 \text{, or } (69.92, 77.52).$$ **TInterval** on the TI-84+ gives (69.927, 77.513).

If this were a random sample, then you would be 95% confident that the mean mass of a bag of fries at this McDonald's restaurant is between 69.92 and 77.52 g.

c. No. The confidence interval contains 74 g, so 74 g is a plausible value for the mean mass of a small bag of fries.

E9. a. No, as there is no indication this is a random sample of brochures. (Note: If these are all of the cancer brochures published by these two organizations, this is essentially a stratified random sample with a sample size of $n = 1$ page in each stratum (brochure). If these are a random sample of the cancer brochures published by these two organizations, inference can proceed if you consider the page tested to be typical of each brochure.)

b. $\bar{x} = 9.8$, $s \approx 2.92$. For $df = 29$, $t^* = 1.699$

$$\bar{x} \pm t^* \frac{s}{\sqrt{n}} \approx 9.8 \pm 1.699 \frac{2.92}{\sqrt{30}} \approx 9.8 \pm 0.91,$$

or (8.89, 10.71) or (8.894, 10.706) with no rounding.

c. If these 30 brochures could be considered a random sample of all brochures and if the readability found on the sampled page for each is typical of that brochure, you are 95% confident that if you tested the readability of all cancer brochures, the average reading level would be between 8.894 and 10.706.

E11. a. <u>4-year college:</u> Here, $s \approx 111$, $n = 1000$. Also, the df = 999 and $t^* = 1.962$. So, the 95% margin of error is

$$E = t^* \frac{s}{\sqrt{n}} = 1.962 \left(\frac{111}{\sqrt{1000}} \right) = 6.9.$$

<u>2-year college:</u> Here, $s \approx 79$, $n = 800$. Also, the df = 799 and $t^* = 1.962$. So, the 95% margin of error is

$$E = t * \frac{s}{\sqrt{n}} = 1.962 \left(\frac{79}{\sqrt{800}} \right) = 5.5$$

b. The 4-year is larger because of the much larger sample standard deviation for that group.

E13. D

E15. B. The 99% confidence interval will have a better chance of capturing the true mean. It will be wider than the 90% interval. The capture rate does not depend on the sample size; it depends only on the confidence level.

E17. a. At first glance, the shape looks to be approximately normal, but closer inspection will reveal slight skewness toward the large values. Both the mean and the median of the distribution are slightly below the population standard deviation of 112, so s tends to underestimate σ. In fact in this simulation s is less than 112 about 64% of the time.

b. Because the median is smaller than 112, you know that more than half of the values of the sample standard deviation must be below 112. A rough count on the histogram shows that at least 61 of the 100 sample standard deviations fall below the population standard deviation of 112. Therefore, s is more often smaller than σ. (The sixth bar from the left on the histogram straddles 112 so it is impossible to tell exactly how many cases in that bar are less than 112.)

c. Because s tends to underestimate σ, the advertised 95% confidence intervals will be too narrow when s is used as an estimate of σ. This means the actual capture rate will be less than 95%.

E19. *Check Conditions:* The bottles were randomly selected from the day's production. The data shown in the dotplot below show no strong skewness and no outliers, so it is reasonable to assume the amounts in the bottles are normally distributed. (The problem states that the distribution is approximately normal). The population, which is the day's total production, is probably more than 100 bottles, which is ten times the sample size.

```
                    • •         •
    •        • • •      • •      •
 ┌──┬────┬────┬────┬────┬────┬────┐
 15.80 15.85 15.90 15.95 16.00 16.05 16.10
              Bottle Weight (oz)
```

State your hypotheses:

H_0: The mean amount in the bottles μ is 16 oz.

H_a: $\mu \neq 16$ oz.

Compute the test statistic, find the P-value, and draw a sketch.

Note that $\overline{x} = 15.974$, $s = 0.0804$. So,

$$t = \frac{\overline{x} - \mu_0}{s / \sqrt{n}} = \frac{15.974 - 16}{0.0804 / \sqrt{10}} \approx -1.023 \text{ (or } -1.022 \text{ without rounding)}.$$

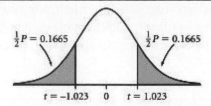

Write your conclusion in context, linked to your computations: *Write your conclusion in context, linked to your computations:* If the mean bottle weight were 16 oz., a *t*-statistic at least as large in absolute value as 1.0221 would occur about 1/3 of the time. This means you do not have sufficient evidence to reject the null hypothesis. The data suggest that it is plausible that the mean amount of water in the bottles is 16 oz.

E21. *Check Conditions:* This is probably not a random sample of students. It is possible that this group of students is not special in their ability (or lack of it) to draw the midpoint of a segment, but you cannot assume that. However, the dotplot of the data is fairly symmetric with no outliers, so it is reasonable to conclude that the average errors come from a normal distribution, and the population of 'students' is more than 150, which is ten times the sample size.

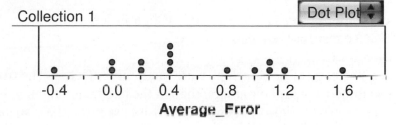

State your hypotheses:

H_0: The mean μ of the average errors is 0.

H_a: $\mu \neq 0$

Compute the test statistic, find the P-value, and draw a sketch:

Note that $\bar{x} = 0.56$, $s \approx 0.550$. So,

$$t = \frac{\bar{x} - \mu_0}{s/\sqrt{n}} = \frac{0.56 - 0}{0.550/\sqrt{15}} \approx 3.943.$$

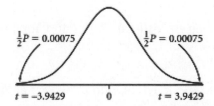

Write your conclusion in context, linked to your computations: Since a *P*-value of 0.0015 is quite small, you have sufficient evidence to reject the null hypothesis. If it were true that the mean of the average errors is 0, a *t*-statistic as extreme as or more extreme than 3.943 in absolute value, would occur only 0.15% of the time. It is not plausible that the mean of the average errors when students draw the midpoint of a vertical segment is 0. Because this is not a random sample of students you cannot generalize to a larger population. Judging from the data, this group appears to guess too high a value.

E23. a. A one-sided (left-sided) test is appropriate because the research hypothesis is that the freezing point will be lower for saltwater.

b. $\bar{x}$ = mean freezing point for a sample of 10 bowls of saltwater.
μ = the true freezing point for saltwater with this degree of salinity.

c. The null hypothesis is that the true freezing point of saltwater μ is the same as for pure water. Or, using symbols:
H_0: $\mu = 32°F$ or H_0: $\mu = 0°C$

The alternate hypothesis is that the true freezing point of saltwater is lower than it is for pure water. Symbolically:
H_a: $\mu < 32°F$ or H_a: $\mu < 0°C$

E25. Since the consumer group says the average weight is less than 10 oz. this will be a one-sided significance test for the mean.

Check conditions: The bags of potato chips were randomly selected. The distribution of weights is fairly symmetric with no outliers, though the bag that weighs 9.3 oz. is far enough from the rest of the weights to suggest a possible skew to the distribution. Because the skewness is slight, a sample size of 15 should be large enough that the sampling distribution of the mean will be approximately normal. Also the company claims that the weight of all bags is normally distributed. The population of bags of potato chips would be more than 150 bags, which is ten times the sample size.

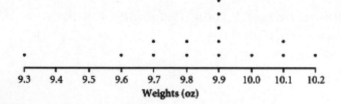

State your hypotheses:
H_0: The mean weight μ of a bag of Munchie's Potato Chips is 10 oz.
H_a: $\mu < 10$

Compute the test statistic, find the P-value, and draw a sketch:
Note that $\bar{x} \approx 9.853$, $s \approx 0.2232$. So,

$$t = \frac{\bar{x} - \mu_0}{s/\sqrt{n}} = \frac{9.853 - 10}{0.2232/\sqrt{15}} \approx -2.55 \ (t = -2.545 \text{ with no rounding with a } P\text{-value of } 0.012).$$

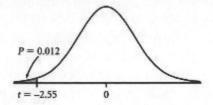

Write your conclusion in context, linked to your compuations:
If it is true that the mean weight of Munchie's Potato Chips is 10 oz., there is a probability of only 0.012 that you would get a *t*-statistic at most –2.55. You should reject the null hypothesis because there is sufficient evidence that the mean weight of a bag of Munchie's Potato Chips is less than 10 oz.

E27. This situation calls for a one-sided significance test for the mean.

Check conditions: You do not have a simple random sample. Instead, you have the entire population of this basketball team. You need to determine how likely it would be for heights from a random sample of U.S. women to have a distribution like that of the Chicago Sky basketball team. The distribution of heights is not symmetric, and this alone makes it seem unlikely that it is a simple random sample from this population you have been told is normally distributed.

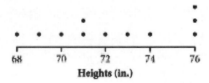

Continue with the test anyway.

State your hypotheses:
H$_0$: The players' heights are equivalent to a simple random sample from the normally distributed population of U.S. young women's heights with mean $\mu = 64.8$ inches and any difference can be reasonably attributed to chance variation.
H$_a$: The players' heights are not equivalent to a simple random sample from this population and any difference can not be reasonably attributed to chance variation.

Compute the test statistic, find the P-*value, and draw a sketch.*
Note that $\bar{x} \approx 72.36$, $s \approx 2.87$. So,

$$t = \frac{\bar{x} - \mu}{s / \sqrt{n}} = \frac{72.36 - 64.8}{2.87 / \sqrt{11}} \approx 8.74.$$

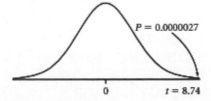

Write your conclusion in context, linked to your computations: If this were a formal one-sided test of significance, the *P*-value would be $2.7 \cdot 10^{-6}$. This indicates that, if the players were randomly selected from the population of young women in the United States there is very little chance of seeing a sample with such a large *t*-statistic. You would reject the null hypothesis that this is a simple random sample from the given population and that the difference in mean height from the normal can be attributed to chance variation.

E29. Choice D is the correct answer. Many students will want to say that the *P*-value measures the chance that the null hypothesis is true because you reject the null hypothesis when the *P*-value is small. You can never calculate this probability based on sample data alone; all you can do is calculate the conditional probability of getting particular results from a random sample under the condition that the null hypothesis is true.

E31. Because $P = 0.333$ is greater than 0.10, you do not have sufficient evidence to reject the null hypothesis. The data suggest that it is plausible that the mean amount of water in the bottles is 16 oz.

E33. a. Because the *P*-value of 0.012 is less than 0.05, you would reject the null hypothesis. There is sufficient evidence that the mean weight of a bag of Munchie's Potato Chips is less than 10 oz.

b. The null hypothesis was that $\mu = 10$ and you rejected it. Because $\mu = 9.9$, the null hypothesis is indeed false and you made the correct decision. No error was made.

c. If you had constructed a 95% confidence interval, you would have gotten an interval that did not contain 10 oz. So you could reject that 10 oz was a plausible value for the mean weight. The actual 95% confidence interval for the mean weight is (9.73, 9.98) which does suggest that the mean weight is lower than 10 oz.

E35. By definition of Type I error, the choices that correctly describe the probability of Type I error are A, C, and D.

B is not correct because Type I error occurs when you reject a true null hypothesis.

E37. a. We wish to test:
H$_o$: $\mu = 68$, H$_a$: $\mu < 68$, where μ denotes the mean battery weight.

The randomness condition is met and the size of the population of all batteries certainly exceeds 300. The distribution is slightly skewed toward smaller values, but is still symmetric, so it is reasonable to assume the underlying population is essentially normal.

Note that using technology yields $\bar{x} \approx 64.327$, $s \approx 1.428$. So,
$$t = \frac{\bar{x} - 68}{s / \sqrt{30}} = \frac{64.327 - 68}{1.428 / \sqrt{30}} \approx -14.09 .$$

Since df = 30 − 1 = 29, we see that $P(T < -14.09)$ is near zero. Hence, the data provide strong evidence to declare that the standard of 68 pounds is not being met.

b. We wish to test:
H_o: $\mu = 65$, H_a: $\mu < 65$, where μ denotes the mean battery weight.

The conditions are still met (as in (a)). This time, the test statistic is given by

$$t = \frac{\bar{x} - 65}{s / \sqrt{30}} \approx -2.581.$$

Since df = 30 − 1 = 29, we see that $P(T < -2.581) = 0.0075$. Hence, the data provide sufficient evidence to declare that the standard of 65 pounds is not being met either.

E39. The two probabilities are the same: both are equal to 0.05.

E41. a. a test with $\alpha = 0.10$.
b. a test with $n = 45$.
c. a one-sided test (assuming the alternative hypothesis is in the correct direction).

E43. a. They are both mound-shaped and centered at about zero although histogram B has a bit of a skew to the left. The spread of A is less than that of B.

b. B is the simulated t-distribution because it has the larger spread and heavier tails. The respective formulas are

$$z = \frac{\bar{x} - 100}{20 / \sqrt{4}} \text{ and } t = \frac{\bar{x} - 100}{s / \sqrt{4}}$$

Because s tends to be smaller than σ more often than it is larger, t will tend to be larger in absolute value than z.

E45. a. The athletes were selected randomly. The stemplot shows the distribution is fairly symmetric with no outliers. There are more than 150 highly trained athletes, which is ten times the sample size of 15. The conditions are met for computing a confidence interval for the mean.

b. $\bar{x} = 6.67$, $s = 1.88$, $t^* = 2.145$ for 14 degrees of freedom. The interval is

$$\bar{x} \pm t^* \cdot \frac{s}{\sqrt{n}} = 6.67 \pm 2.145 \cdot \frac{1.88}{\sqrt{15}} \approx 6.67 \pm 1.04 \text{, or } (5.63, 7.71).$$

You are 95% confident that if you were to ask all highly-trained athletes how long they sleep, that the mean number of hours highly-trained athletes sleep each night is between 5.63 and 7.71 hours.

E47. a. The hypotheses are:
H_0: The mean number of hours highly-trained athletes sleep each night μ is 8.
H_a: $\mu \neq 8$.

b. $t = \dfrac{\overline{x} - \mu_0}{s/\sqrt{n}} = \dfrac{6.67 - 8}{1.88/\sqrt{15}} \approx -2.74$.

c. The *P*-value is approximately 0.0157. If the mean number of hours highly-trained athletes sleep each night is 8, the probability that a random sample of 15 athletes would result in a *t*-statistic at least 2.74 or at most –2.74 is 0.0157.

E49. a. Here, $n = 10$, and using technology we see that
$$\overline{x} \approx 16{,}088, \ s \approx 2852.$$
So, the 95% confidence interval for mean in-state cost for 100 best buy colleges is
$$\overline{x} \pm t^* \dfrac{s}{\sqrt{n}} \approx 16{,}088 \pm 2.262\left(\dfrac{2852}{\sqrt{10}}\right),$$
or about ($14,048, $18,128).

b. Here, $n = 10$, and using technology we see that
$$\overline{x} \approx 17{,}878, \ s \approx 2559.$$
So, the 90% confidence interval for mean debt at graduation for 100 best buy colleges is
$$\overline{x} \pm t^* \dfrac{s}{\sqrt{n}} \approx 17{,}878 \pm 1.833\left(\dfrac{2559}{\sqrt{10}}\right),$$
or about ($16,395, $19,361).

E51. a. The two-sided *P*-value corresponds to the pair of shaded areas.

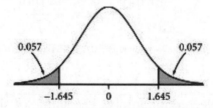

b. The right-hand area ($t > 1.645$) corresponds to evidence that $\mu > \mu_0$. For any given value of the test statistic, *t*, the one-sided *P*-value will be exactly half as large as the two-sided *P*-value.

c. The *P*-value will be larger for a two-sided test.

E53. A

Chapter 11

Practice Problem Solutions

P1. **a.** The population from which the samples were taken is native beggars and Roma beggars; the parameters being estimated are mean amount given to native beggars and mean amount given to Roma beggars.

b. 0.03 euros is the observed difference in the mean amounts given to the samples of beggars, while 0.24 is the margin of error related to a 95% confidence interval.

c. You are 95% confident that the difference in mean amounts given to the populations of native beggars and Roma beggars, with the second subtracted form the first, is between −0.21 and 0.27 euros.

d. Yes, because a difference of 0 is well inside the interval of plausible values.

P2. *Check conditions.* The boxplots of the data show that there is one unusually small heart rate for the males. This may cause problems with the analysis and will be checked out. Otherwise, the distributions look fairly symmetric. These are random samples and presumably independent. Also, both populations are more than ten times the respective sample sizes, so the conditions for inference are met.

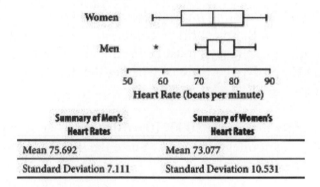

Summary of Men's Heart Rates	Summary of Women's Heart Rates
Mean 75.692	Mean 73.077
Standard Deviation 7.111	Standard Deviation 10.531

Do computations. With 95% confidence, $-4.71 < \mu\,(men) - \mu\,(women) < 9.94$, where $\mu\,(men)$ is the mean heart rate for all men in this population and $\mu\,(women)$ is the mean heart rate for all women in this population.

Write a conclusion in context. Because zero is in the confidence interval, there is insufficient evidence to say that the mean heart rate for men differs from the mean heart rate for women.

A test of significance can be used to answer the question as well.

P3. **a.** *Check conditions:* You are told to assume that this class is a random sample of all students taking the course. It is also probably reasonable to assume that the males who are taking this course is independent of the females taking this course. Both distributions are moderately skewed toward the higher values. With sample sizes of 15 and 46, this moderate skewness would probably not be a big concern, but both distributions also have one outlier on the high end. The analysis should be done both with and without these outliers to see if the conclusion is the same. It is safe to assume that there are more than ten times these sample sizes of each gender in the population of students who take this course.

b. *Compute the interval:* For the formula, consider females to be group 1 and males to be group 2.

Using the summary statistics given, the 90% confidence interval is,

$$(\bar{x}_1 - \bar{x}_2) \pm t^* \sqrt{\frac{s_1^2}{n_1} + \frac{s_2^2}{n_2}} = (10.93 - 8.20) \pm t^* \sqrt{\frac{6.22^2}{46} + \frac{5.94^2}{15}},$$

or $-0.32 < \mu_F - \mu_M < 5.78$.

Using the data with no rounding, you get the interval $(-0.32, 5.79)$.

To see the effect of the outliers, a re-analysis should be conducted with these points removed. Such an analysis gives the 90% confidence interval as

$$1.27 < \mu_F - \mu_M < 5.75$$

where $\bar{x}_F \approx 10.51$, $s_F \approx 5.58$ and $\bar{x}_M = 7$, $s_M \approx 3.84$. This confidence interval no longer overlaps 0.

c. *Give interpretation in context and linked to computations:*

Because the lower bounds of these confidence intervals are so different, one containing 0 and one not containing 0, these results must be used with caution. Ideally, it would be good to get more data to resolve the issue of whether the population means for males and females actually differ.

P4. a. There are certainly more than $1000*10 = 10,000$ 4-year college students and more than $800*10 = 8,000$ 2-year college students. Since the sample sizes are large, the normality condition is met.

b. Here, we have

$$\bar{x}_1 = 330, \ s_1 = 111, \ n_1 = 1000, \ \bar{x}_2 = 310, \ s_2 = 79, \ n_2 = 800.$$

Also, using either technology (which is preferable) or the formula

$$\frac{\left(\frac{s_1^2}{n_1} + \frac{s_2^2}{n_2}\right)^2}{df} = \frac{\left(\frac{s_1^2}{n_1}\right)^2}{n_1 - 1} + \frac{\left(\frac{s_2^2}{n_2}\right)^2}{n_2 - 1},$$

we see that df = 1775. So, $t^* = 1.960$. Thus, the 95% confidence interval is

$$(\bar{x}_1 - \bar{x}_2) \pm t^* \sqrt{\frac{s_1^2}{n_1} + \frac{s_2^2}{n_2}} = 20 \pm 1.960(4.486),$$

or about $(11.2, 28.8)$.

212

c. You are 95% confident that the difference in mean quantitative literacy scores between the population of 4–year college students and the population of 2–year college students is between 11.2 and 28.8 points.

P5. a. State your hypotheses.

H_0: $\mu_F - \mu_M = 0$

H_a: $\mu_F - \mu_M \neq 0$

where μ_F is the mean study time of all female students taking this course and μ_M is the mean study time of all male students taking this course.

b. *Check conditions.* These can be thought of as independent random samples of students taking an introductory statistics course (but not as random samples of all students at the university). Each sample contains one unusually large value, which may have great influence on the results. The data will be analyzed both with and without these values. Also, it is safe to assume that there are more than ten times these sample sizes of each gender in the population of students who take this course.

c. *Compute the test statistic and draw a sketch.*
For the formula, consider females to be group 1 and males to be group 2.

$$t = \frac{\overline{x}_1 - \overline{x}_2}{\sqrt{\dfrac{s_1^2}{n_1} + \dfrac{s_2^2}{n_2}}} = \frac{10.93 - 8.20}{\sqrt{\dfrac{6.22^2}{46} + \dfrac{5.94^2}{15}}} = 1.53$$

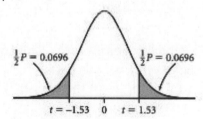

The calculator gives $df \approx 24.8$ and a *P*-value of 0.1387.

d. *Write a conclusion in context.* The *P*-value of 0.1387 means that if it is true that the mean time that all females would report studying is equal to the mean time that all males would report studying, then the probability is 0.1387 of getting a difference in means from samples of these sizes as large (in absolute value) or even larger than 10.93 – 8.20, as obtained in this study. There is insufficient evidence to conclude that there is a difference in mean reported weekly study hours between males and females.

e. However, to see the effect of these outliers, you will re-analyze with these points removed. Such an analysis gives the following results:

H_0: $\mu_F - \mu_M = 0$

H_a: $\mu_F - \mu_M \neq 0$

The test statistic ≈ 2.657 with 31.6544 *df*.

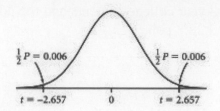

The *P*-value of 0.012 now supports rejection of the hypothesis of equal population means. All things considered, it seems that the analysis without the outliers is the better one for these data. The two outliers inflate the standard deviations to the extent that the test statistic is unduly small. It would be wise to get more data.

P6. a. This is a one-sided test. The null and alternative hypotheses are
$H_o: \mu_1 = \mu_2$, $H_a: \mu_1 > \mu_2$,
where μ_1 denotes mean relative price index for all prices in the poorest neighborhoods and μ_2 denotes the mean relative price index for all prices in the richest neighborhoods

b. No, because both sample sizes are well over 40.

c. The test statistic is $t = \dfrac{\overline{x}_1 - \overline{x}_2}{\sqrt{\dfrac{s_1^2}{n_1} + \dfrac{s_2^2}{n_2}}} = \dfrac{1.22 - 1.11}{\sqrt{\dfrac{(0.03)^2}{1912} + \dfrac{(0.07)^2}{1575}}} \approx 58.12$.

d. The p-value is P(T>58.12), which is nearly zero. This means that it is almost impossible to get a mean for the poorest neighborhoods this much larger than the mean for the richest neighborhoods from samples of these sizes if it is true that the two means are equal for all prices.

P7. Note: The experimental units are the panels, not the individual members. So, the size of each experimental group is 8 because there were 8 panels assigned to each texture.

Check conditions: This is an experiment. Panel members were randomly assigned to the treatment groups as they were recruited. Looking at the distribution of scores for each texture, both distributions are fairly symmetric with no outliers.

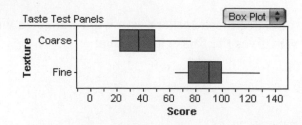

State your hypotheses:

H_0: $\mu_{Coarse} = \mu_{Fine}$, where μ_{Coarse} is the mean palatability score if all the panels had been given food with coarse texture and μ_{Fine} is the mean palatability score if all the panels had been given food with fine texture.

H_a: $\mu_{Coarse} \neq \mu_{Fine}$

Compute the test statistic, find the P*-value, and draw a sketch:*

$$t = \frac{(\overline{x}_{Coarse} - \overline{x}_{Fine}) - (\mu_{Coarse} - \mu_{Fine})}{\sqrt{\dfrac{s^2_{Coarse}}{n_{Coarse}} + \dfrac{s^2_{Fine}}{n_{Fine}}}} = \frac{(38.875 - 90.375) - 0}{\sqrt{\dfrac{20.629^2}{8} + \dfrac{21.159^2}{8}}} \approx -4.929,$$

where the calculator reports 13.991 degrees of freedom.

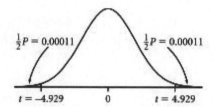

The *P*-value is about 0.00022.

Write your conclusion in context, linked to your computations:

If there is no difference between the mean palatability score that would have been given had all of the panels sampled food with a coarse texture and the mean palatability score had all of the panels sampled food with a fine texture, a *t*-score above 4.929 or below −4.929 would occur only about 0.022% of the time. Because this is such a small probability, you would reject the null hypothesis. You have statistically significant evidence that there is a difference in the mean palatability score between the two texture levels.

P8. Because the previous data indicate that the measurements from the bottom of the river have a greater spread, you would be better off using more of the new measurements at the bottom of the river. This would give your test greater power.

P9. a. If you consider men to be population 1 and women to be population 2, the 95% confidence interval for the difference between the population mean body temperature for men (group 1) and for women (group 2) is:

$$(\overline{x}_1 - \overline{x}_2) \pm t^* \cdot \sqrt{\frac{s^2_1}{n_1} + \frac{s^2_2}{n_2}} = (97.88 - 98.52) \pm 2.101\sqrt{\frac{0.555^2}{10} + \frac{0.527^2}{10}}$$

where the degrees of freedom as given by the calculator is 17.95.

The plausible values for the true difference in mean body temperatures lie in the interval (−1.149, −0.1314).

b. Because this interval is entirely on the negative side of zero, you can conclude that there is a statistically significant difference between the mean body temperature for men and for women.

c. Constructing one confidence interval for the difference resulted in a significant difference, and constructing two separate intervals did not. This indicates that one interval for the difference has more power.

P10. **a.** Because you have data for the first launch and tenth launch for each team, this is a Repeated Measures design. You should use a onc-sample test of the mean of the differences. A one-sided test should be used because you want to test whether the teams improved, which would mean the distance on the tenth launch would be longer.

b. *Check conditions:* This could be considered a random sample of teams. The boxplot of the differences is basically symmetric. The number of teams who have done the Bears in Space activity is probably more than 6•10 or 60 teams. The conditions have been met for a significance test for paired differences.

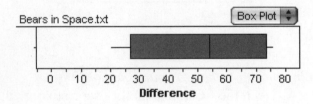

State your hypotheses:
H_0: There is no difference between mean distances on the first launch and mean distances on the tenth launch.
H_a: The mean difference on the tenth launch is greater than the mean distance on the first launch.

Compute the test statistic, find the P*-value, and draw a sketch.*

$$t = \frac{\bar{d} - \mu_d}{s_d / \sqrt{n}} = \frac{50.833 - 0}{23.49 / \sqrt{6}} \approx 5.301$$

with 5 degrees of freedom.

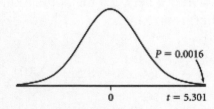

Write your conclusion in context, linked to your computations:
Because the *P*-value of 0.0016 is less than 0.01 you would reject the null hypothesis. If these groups can be considered a random sample of all possible teams of students, you

216

have statistically significant evidence that the mean distance on the tenth launch is greater than the mean distance on the first launch.

c. Yes, the result seems sensible. It was suspected at first that teams would improve with practice. A Type I error could have been made, because if practice actually has no effect you have rejected a true null hypothesis.

d. The conditions are not satisfied for a two-sample test because the sets of distances are not independent samples.

P11. a. You do not have two independent samples because you have data for the fifth launch and tenth launch for each team, this is a Repeated Measures design. You should use a one-sample test of the mean of the differences. A one-sided test should be used because you want to test whether the teams improved, which would mean the distance on the tenth launch would be longer.

b. *Check conditions:* This could be considered a random sample of teams. The boxplot of the differences is basically symmetric. The number of teams who have done the Bears in Space activity is probably more than 6•10 or 60 teams. The conditions have been met for a significance test for paired differences.

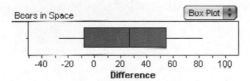

State your hypotheses:
H_0: There is no difference between mean distances on the fifth launch and mean distances on the tenth launch.
H_a: The mean difference on the tenth launch is greater than the mean distance on the fifth launch.

Compute the test statistic, find the P-value, and draw a sketch.

$$t = \frac{\bar{d} - \mu_d}{s_d / \sqrt{n}} = \frac{25.83 - 0}{42.84 / \sqrt{6}} \approx 1.477$$

with 5 degrees of freedom.

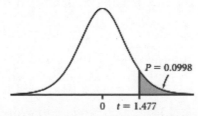

Write your conclusion in context, linked to your computations:
Because the *P*-value of 0.0998 (with no rounding) is greater than 0.01 you would not

reject the null hypothesis. If these groups can be considered a random sample of all possible teams of students, you do not have statistically significant evidence that the mean distance on the tenth launch is greater than the mean distance on the fifth launch.

c. A Type II error could have been made, because if practice still has an effect after the fifth launch you have failed to reject a false null hypothesis.

d. The result may not match your intuition. You would still expect some improvement with practice, although you would expect less improvement from the fifth launch to the tenth than from the first launch to the tenth. At some point you might expect that more practice is not going to cause any more noticeable improvement but you might not expect that to occur already after the fifth launch.

P12. a. You should test the mean of the differences or do one-sample procedures because you have the result of both treatments for each team. This is a repeated measures design.

b. Randomization in this situation means that the order of the treatments was randomly assigned, which is the case here. The distribution of distances has no outliers and no strong skewness.

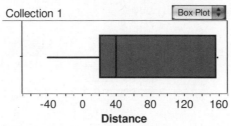

The conditions for inference are met.

Construct the interval: For 5 degrees of freedom, $t^* = 2.571$.

$$\bar{d} \pm t^* \cdot \frac{s_d}{\sqrt{n}} = 61.83 \pm 2.571 \cdot \frac{80.91}{\sqrt{6}} \approx 61.83 \pm 84.92,$$

or $(-23.09, 146.75)$.

Write your interpretation in context, linked to your computations:
You are 95% confident that the mean difference between the distances of bears launched using four books and the distances of bears launched using one book is between -23.09 and 146.75 inches. Because 0 is in the interval, there is not statistically sufficient evidence to support the claim that launch angle makes a difference in how far gummy bears soar.

P13. First, creating the data set comprised of the differences of the form

d = out-state cost − in-state cost

yields

{13894, 6260, 10167, 19142, 9582, 14712, 14236, 11074, 6260, 20608}

Now, observe that using technology yields

218

$$\bar{d} = 12593.8, \ s_d = 4870.48, \ df = 10 - 1 = 9, \ \alpha = 0.10, \ t* = 1.833.$$

Hence, the 90% confidence interval is

$$\bar{d} \pm t* \cdot \frac{s_d}{\sqrt{n}} = 12593.8 \pm 1.833 \cdot \frac{4870.48}{\sqrt{10}},$$

or about (9770.65, 15416.95).

This means that you are 90% confident that the mean difference between typical out-of-state and in-state costs for the 100 best-buy public colleges is between $9,770.65 and $15,416.95.

P14. There is no reason to pair a particular bird with a particular fish, so these are independent samples.

P15. a. The two most striking features of this data set are clear from this scatterplot.

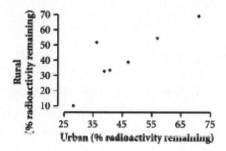

First, points for six of the seven twin pairs fall close to a line, a pattern that strongly suggests that some combination of heredity and early childhood environment is a major determining factor in the rate of tracheobronchial clearance. Second, one twin pair is a clear outlier in the scatterplot. This outlier is the same outlier at about −18 in the *Difference* boxplot below. The other outlier in the boxplot is the point in the lower left corner of the scatterplot.

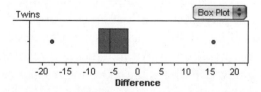

Are the formal tests justified here? There is certainly no randomization. Conditions were not assigned, and so could not be randomized; nor were the twin pairs randomly selected from some population. Indeed, it is hard to specify the population that would be a reasonable target for the inferences: the set of all twin pairs with one twin living in each environment and both twins willing to inhale radioactive Teflon? (Surely not a large or representative component of society!)

In addition, the distribution of the differences, while fairly symmetric, has an outlier on each end. This is more than you would expect from an approximately normal population and a sample size of 7.

b. A test of hypotheses is performed as follows:

State your hypotheses.
H_0: $\mu_d = 0$, and H_a: $\mu_d \neq 0$ where μ_d is the mean difference in the percentage of original radioactivity still remaining 1 hour after inhaling if each treatment could have been given to both twins.

Do computations. The summary statistics with and without the outlying pair are:

	Outliers Included			Outliers Excluded		
	Rural	Urban	Diff.	Rural	Urban	Diff.
Mean	41.5	45.5	4.03	45.74	50.9	5.16
SD	19.0	14.4	10.15	15.69	13.29	2.82
t			1.05			4.10
P-value			0.33			0.015

Without the outliers, you would reject the null hypothesis that the difference in lung clearance between the twins can not reasonably be attributed to chance alone. However, with the outliers, you would fail to reject the null hypothesis and conclude that the difference in lung clearance between the twins can reasonably be attributed to chance alone. Here it would be a wise idea to do the experiment again with more data and randomization.

Observe that for the rural readings in isolation, or the urban readings in isolation, removing the outliers has only a modest effect on the summaries. The place where excluding the outlier has the biggest effect is the *SD* for the differences, which goes from over 10 to under 3.

Overall, there are many reasons not to take the test results at face value. Nevertheless, the result can not be explained by chance variation. One of the alternative possible explanations is that where you live has an effect on how efficiently your lungs clear, with rural environments somewhat healthier, on average. But another possible explanation is that the healthier twin tends to choose to live in the country for some reason. None of this is exactly headline-making news! Perhaps the more interesting results come not from comparing means but from the pattern in the scatterplot.

c. A paired study is probably the better design here, because the analysis of differences reduces the person-to-person variation. In general, the pairs would have to be matched on some other criterion related to breathing capacity.

d. Independent samples would require the independent selection of random samples from clearly defined rural and urban populations of interest. Then, each subject would agree to breathe Teflon particles. Clearance rates could then be measured and the means compared by a two-sample procedure.

The main advantage of using two independent samples is that you would not have to find a criterion on which to pair the subjects. In addition, the two-sample design could be better if there is little person-to-person variability in clearance rates. Twins might be difficult to find for most studies.

P16. In a literal sense, comparing means is meaningful, but in a practical sense it's a waste of time. You don't need statistics to know that hens' eggs are longer than they are wide. (The reason Dempster collected the data was to test the fit of an ellipsoidal model for the volume. In the logarithmic scale, the volume is a weighted sum of the length and width.)

P17. a. No. Asking for volunteers is a useful way of obtaining subjects for an experiment; that is essentially what is done in medical experiments, for example. The main difficulty here is that there is no random assignment into treatment groups. This is an observational study, so you can not attribute any difference in scores to the effect of the music.

b. This is better than the design in part a because it partially accounts for the abilities of the students and can reduce student-to-student variability in the results by looking at differences between pairs. But there still is no random assignment of treatments to subjects.

c. A good design would require a group of students (volunteer or otherwise selected) to be randomly divided between the two treatment groups ("music on" and "music off") and then followed over a period of time to see that they were staying on the treatment. Scores on an exam taken at the end of the time period could be used for assessing the treatment effect. The students could be paired on the basis of past performance in the course.

Exercise Solutions

E1. a. The subjects were randomly assigned to treatments. Both distributions are moderately skewed, but neither has any outliers. Since the distribution of the difference reduces skewness, and the t-procedure is robust against non-normality, the conditions for inference are adequately met to proceed with caution.

b. The interval is

$$(\bar{x}_{\text{Short}} - \bar{x}_{\text{Long}}) \pm t^* \cdot \sqrt{\frac{s_{\text{Short}}^2}{n_{\text{Short}}} + \frac{s_{\text{Long}}^2}{n_{\text{Long}}}} = (13.906 - 8.925) \pm t^* \sqrt{\frac{2.985^2}{4} + \frac{1.669^2}{4}},$$

or (0.50, 9.46), where the degrees of freedom given by the calculator is 4.709. (t^* can be found by **invT(0.975, 4.709)**.)

c. The difference between the mean enzyme concentration of all eight hamsters had they

221

all been raised in short days and the mean concentration had they all been raised in long days.

d. Because 0 is not in the confidence interval, and the treatments were randomly assigned, Kelly has statistically significant evidence that the difference in enzyme concentrations between the two groups of hamsters is due to the difference in daylight.

E3. a. *Check conditions:* You have no information about the shapes of the distributions. With 24 and 27 subjects in the two treatment groups, only a strong skewness or the presence of strong outliers would cause concern, especially when looking at the difference between the means. The subjects were randomly assigned to treatments.

Compute the interval:

$$(\overline{x}_{\text{Leech}} - \overline{x}_{\text{Gel}}) \pm t^* \sqrt{\frac{s^2_{\text{Leech}}}{n_{\text{Leech}}} + \frac{s^2_{\text{Gel}}}{n_{\text{Gel}}}} = (53.0 - 51.5) \pm 2.010 \sqrt{\frac{13.7^2}{24} + \frac{16.8^2}{27}} = 1.5 \pm 8.592,$$

or (–7.092, 10.092), where the degrees of freedom reported by the calculator is 48.66. t^* can be found using **invT(0.975, 48.66)**.

Give interpretation in context: A 95% confidence interval for the difference in initial mean pain levels is (–7.092, 10.092), which includes 0. It is plausible that there is no difference in the means. Therefore the answer is no, there is not statistically significant evidence that the randomization failed to yield groups with comparable means.

b. *Check conditions:* Conditions were checked in part a.

Compute the interval:

$$(\overline{x}_{\text{Leech}} - \overline{x}_{\text{Gel}}) \pm t^* \sqrt{\frac{s^2_{\text{Leech}}}{n_{\text{Leech}}} + \frac{s^2_{\text{Gel}}}{n_{\text{Gel}}}} = (27.6 - 27.1) \pm 2.010 \sqrt{\frac{3.7^2}{24} + \frac{3.7^2}{27}} = 0.5 \pm 2.0864,$$

or (–1.587, 2.5867) with no rounding, where the degrees of freedom reported by the calculator is 48.30. t^* can be found using **invT(0.975, 48.30)**.

Give interpretation in context: You are 95% confident that the difference between the mean body mass indices of the two groups is in the interval (–1.587, 2.5867). The interval includes 0. There is no statistically significant evidence to suggest that randomization failed to yield groups with comparable means.

E5. a. Both populations are fairly symmetric with no outliers, so it is reasonable to assume that both samples are taken from populations that are approximately normally distributed. Also, both populations are more than ten times the respective sample sizes, so the conditions for inference are met.

b. Even though you use the calculator to calculate the interval, you should still write the formula and show the formula with numbers substituted.

For the formula, consider left-handed volunteers to be group 1 and right-handed volunteers to be group 2.

$$(\bar{x}_1 - \bar{x}_2) \pm t^* \sqrt{\frac{s_1^2}{n_1} + \frac{s_2^2}{n_2}} = (59.57 - 58) \pm t^* \sqrt{\frac{14.77^2}{7} + \frac{15.71^2}{23}}$$

or about $(-12.760, 15.899)$ with $df \approx 10.50$.

c. The difference between the mean distance all left-handed volunteers could walk before crossing a sideline and the mean distance all right-handed volunteers could walk before crossing a sideline.

d. Because 0 is near the center of the interval, you have almost no statistically significant evidence that left-and right-handed volunteers differ in the mean number of yards they can walk before crossing a sideline.

E7. *Check conditions.* These data sets are slightly skewed toward the larger values, but the sample sizes are so large that the sampling distributions of the sample means would not reflect this skewness; the sampling distributions would appear nearly normal. The source of the data indicates that these are independent random samples. Also, both populations are more than ten times the respective sample sizes, so the conditions for inference are met.

a. *Do computations.* With 95% confidence, $-0.79 < \mu$ (*length of stay for Insurer A*) $- \mu$ (*length of stay for Insurer B*) < -0.39.

Write a conclusion in context. Because zero isn't a plausible difference, there is evidence that the true mean lengths of stay differ for the two insurers.

b. *Do computations.* With 95% confidence, $-1.79 < \mu$ (*staff per bed for Insurer A*) $- \mu$ (*staff per bed for Insurer B*) < -1.21.

Write a conclusion in context. Because zero isn't a plausible difference, there is evidence that the mean number of full-time staff per bed differs for the two insurers. (A significance test gives $t \approx -10.14$ with p-value nearly 0.) Insurer A serves hospitals that appear to have smaller staff-per-bed ratios. It might be, then, that these hospitals have patients who require less care because they are not as seriously ill as those in hospitals served by Insurer B. That could explain why the length of stay also appears to be smaller for hospitals served by Insurer A.

c. *Check conditions.* You are now back in the realm of proportions, so this question must be answered by methodology from Chapter 8. Let p_1 be the proportion of private hospitals used by all patients of Insurer A and p_2 be the proportion for Insurer B. The source of the data indicates that these are independent random samples. All of $n_1\hat{p}_1 = 367.848$, $n_1(1-\hat{p}_1) = 25.152$, $n_2\hat{p}_2 = 289.08$ and $n_2(1-\hat{p}_2) = 106.92$ are at least 5. Also, both populations are more than ten times the respective sample sizes, so the conditions for inference are met.

Do computations. The interval

$$(\hat{p}_1 - \hat{p}_2) \pm z^* \cdot \sqrt{\frac{\hat{p}_1(1-\hat{p}_1)}{n_1} + \frac{\hat{p}_2(1-\hat{p}_2)}{n_2}}$$

yields the 95% confidence interval (0.16, 0.26).

Write a conclusion in context. The evidence shows that the proportions of all patients who are in private hospitals differ for the two companies. (A significance test gives $z \approx$ 7.77 with p-value nearly 0.) Public hospitals are often general hospitals that have a wide range of facilities and accept almost all kinds of cases, whereas private hospitals are often more specialized and accept a narrower range of illnesses. Thus, more of the serious illnesses, often among patients with a history of illness, may show up at public hospitals. Again, this points to the possibility that Company A may be insuring a preponderance of less ill patients.

E9. a. From the display, the confidence interval is (–0.00646, 0.02254) when estimating the difference between the first half mean and the second half mean. This does not confirm that the heat transfer is decreasing because 0 is still a plausible value for the difference in mean heat transfer.

b. From the display, the confidence interval is about (0.00413, 0.02805) when estimating the difference between the first half mean and the second half mean. This does help confirm that the heat transfer is decreasing because 0 is not a plausible value for the difference in mean heat transfer.

E11. a. The absolute values of the data must be used. The new table would appear as follows:

Absolute Value of Average Error (cm) in Locating Midpoint of Segment	
Vertical	Horizontal
1.6	0.9
1.2	0.4
1.1	0.4
1.1	0.2
1.0	0.2
0.8	0.0
0.4	0.0
0.4	0.0
0.4	0.0
0.4	0.2
0.2	0.4
0.2	0.4
0.0	0.4
0.0	0.8
0.4	

b. *Check conditions:* You are told you could assume that the assignment of treatments (direction of the lines) to subjects was random. The distributions of both treatment groups are skewed right.

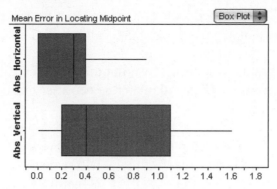

A sample size of 15 is not large enough to compensate for this large amount of skewness if we were examining these data sets by themselves. However, because the difference of means is a bit more lenient, this amount of skewness should be okay for inference.

State your hypotheses:

H_0: The mean absolute value of the error in drawing midpoints on vertical lines, μ_{vert}, is equal to the mean absolute value of the error in drawing midpoints on horizontal lines, μ_{horiz}.

H_a: $\mu_{vert} > \mu_{horiz}$

Compute the test statistic, find the P-value, and draw a sketch:

$$t = \frac{(\bar{x}_{vert} - \bar{x}_{horiz}) - (\mu_{vert} - \mu_{horiz})}{\sqrt{\dfrac{s_1^2}{n_1} + \dfrac{s_2^2}{n_2}}} = \frac{(0.613 - 0.307) - 0}{\sqrt{\dfrac{0.485^2}{15} + \dfrac{0.284^2}{14}}} \approx 2.09,$$

where the degrees of freedom reported by the calculator is 22.85.

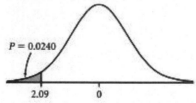

Write your conclusion in context, linked to your calculation: If there were no difference in the mean absolute value of errors between the vertical line group and the horizontal line group, a *t*-statistic at least as large as 2.09 would occur only about 2.4% of the time. With a *P*-value this low, you would reject the null hypothesis. There is sufficient evidence to support the claim that students marking the midpoint on vertical segments have a larger mean absolute value of the error than the students marking on horizontal lines.

E13. a. Yes, when looking at gain in heart rate these can be considered single measurements on each of two randomly assigned treatments. The sample sizes are both

15 (and so don't sum to 40), so the normality condition cannot be checked directly.

b. We test:

$H_o: \mu_1 = \mu_2$, $H_a: \mu_1 > \mu_2$, where μ_1 denotes the mean heart rate gain if all subjects had used the high step and μ_2 denotes the mean gain if all subjects had used the low step.

To do so, we first form two new columns by computing *Exercise – Resting* for the Low Step and the High Step groups. Then, calculate the necessary sample statistics using technology. Indeed, we have:

Low Step Difference (Group 1)	High Step Difference (Group 2)
15	66
21	39
24	45
9	33
42	60
24	27
27	15
6	51
48	24
15	39
15	15
9	15
0	39
18	6
18	57

The summary statistics are:

$$\bar{x}_1 = 19.4, \ s_1 = 12.721, \ n_1 = 15, \ \bar{x}_2 = 35.4, \ s_2 = 18.326, \ n_2 = 15.$$

Also, using either technology (which is preferable) or the formula

$$\frac{\left(\frac{s_1^2}{n_1} + \frac{s_2^2}{n_2}\right)^2}{df} = \frac{\left(\frac{s_1^2}{n_1}\right)^2}{n_1 - 1} + \frac{\left(\frac{s_2^2}{n_2}\right)^2}{n_2 - 1},$$

we see that df = 24.9. The test statistic is $t = 2.778$ and hence the p-value is P(T>2.778) = 0.005. As such, there is strong evidence to conclude that the mean heart rate gain is greater for the high steppers.

E15. a. No randomization is mentioned in assigning treatments to experimental units. (The experimental units here are the 16 dishes.) The distributions are not symmetric, and the dish with 10 mg of resin is highly skewed. Experimental group sizes of 8 will not be enough to compensate for this degree of skewness. The conditions are not met for inference.

226

b. *State Hypotheses:*

H_0: The mean number of termites remaining if all dishes had received a dose of 5 mg of resin, μ_5, is equal to the mean number of termites remaining had all dishes received a dose of 10 mg of resin, μ_{10}.

H_a: $\mu_5 \neq \mu_{10}$

Compute the test statistic, find the P-*value, and draw a sketch:*

$$t = \frac{(\overline{x}_5 - \overline{x}_{10}) - (\mu_5 - \mu_{10})}{\sqrt{\dfrac{s_1^2}{n_1} + \dfrac{s_2^2}{n_2}}} = \frac{(9.50 - 4.13) - 0}{\sqrt{\dfrac{2.67^2}{8} + \dfrac{6.53^2}{8}}} \approx 2.153,$$

where the table yields 9 degrees of freedom.

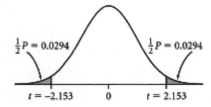

This corresponds to a (two-sided) *P*-value of 0.0597.

Write your conclusion in context, linked to your calculations:

If the mean number of termites surviving if all dishes had received a dose of 5 mg of resin is equal to the mean number of termites surviving had all dishes received a dose of 10 mg of resin (both treatments are equally effective and the difference due only to chance), a test statistic at least as extreme as ± 2.153 would occur in 5.97% of the possible randomizations. This is moderate evidence against the null hypothesis. At the 10% significance level you would reject the null hypothesis, however, you would not do so at the 5% level. You do not have statistically significant evidence at the 5% level that the size of the dose makes a difference in the mean number of termites surviving.

c. The main concerns are the lack of randomization and the skewness of the distribution of termites surviving in these rather small experimental groups—especially in the 10 mg group. The experimenter should repeat the experiment, randomly assigning treatments to the dishes, and using larger number of experimental units to compensate for the skewness and reduce the standard error of the sampling distribution of the mean.

E17. ***Check conditions.*** Conditions for the *t*-procedures were checked for the Special and Control groups in the example in the text, where it was observed that each sample contains a value that is relatively large. A dot plot of all four data sets is shown below. The Weekly and Final data have more symmetric distributions; nevertheless, results here should be viewed with some caution.

The treatments were randomly assigned to the babies, so the condition of random assignment is met.

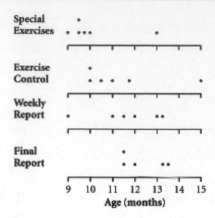

a. State your hypotheses.

H_0: μ *(special)* − μ *(weekly)* = 0

H_a: μ *(special)* − μ *(weekly)* < 0

where μ *(special)* is the mean age of walking if all of the babies could have been given the special exercises treatment and μ *(weekly)* is the mean age of walking if all of the babies could have been given the weekly report treatment.

Compute the test statistic and draw a sketch.

$$t = \frac{\overline{x}_1 - \overline{x}_2}{\sqrt{\dfrac{s_1^2}{n_1} + \dfrac{s_2^2}{n_2}}} = \frac{10.125 - 11.625}{\sqrt{\dfrac{1.447^2}{6} + \dfrac{1.547^2}{6}}} = -1.735$$

($t \approx -1.734$ with no rounding and using the data)

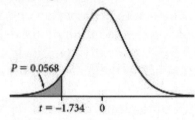

With 9.96 degrees of freedom, the *P*-value is about 0.0568.

Write a conclusion in context.
The *P*-value is just a bit larger than 0.05, so this is not quite sufficient evidence to reject the hypothesis of equal mean walking time, at the 0.05 level of significance, for the special exercises and weekly report groups. An observed difference in means of this size could simply be the result of the randomization of the treatments to the babies.

b. State your hypotheses.

H_0: μ *(exercise)* − μ *(weekly)* = 0

H_a: μ *(exercise)* − μ *(weekly)* < 0

Compute the test statistic and draw a sketch.
For the calculation, consider the exercise control group to be group 1 and the weekly report group to be group 2.

$$t = \frac{\bar{x}_1 - \bar{x}_2}{\sqrt{\dfrac{s_1^2}{n_1} + \dfrac{s_2^2}{n_2}}} = \frac{11.375 - 11.625}{\sqrt{\dfrac{1.896^2}{6} + \dfrac{1.547^2}{6}}} = -0.250$$

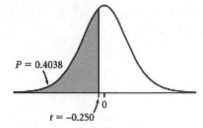

$P = 0.4038$

$t = -0.250$

With 9.61 degrees of freedom, the *P*-value is about 0.4039.

Write a conclusion in context. The *P*-value is large, which makes it clear that there is not sufficient evidence to reject the hypothesis of equal means for the exercise control and weekly report groups. An observed difference in means of this size could quite reasonably be due to the randomization of the treatments to the babies.

c. From the example, a significance test of special exercises versus exercise control yields a *P*-value of 0.1150. The special exercises group is very close to having a significantly lower mean than the weekly report group, with a *P* value that is just slightly larger than 0.05. The next strongest comparison is between the special exercises and exercise control groups, with a *P*-value of 0.1150. The weakest evidence of a difference in means comes from the comparison of exercise control group versus the weekly report group. Good advice to the experimenter might be to suggest that larger sample sizes be obtained for the first two comparisons.

E19. a. Kelly rejected the null hypothesis. If that null hypothesis is actually true, she would have made a Type I error.

b. She could decrease the level of significance say to 0.01%. In fact, the T interval for 99% confidence is (–2.138, 12.101). Because 0 is in the confidence interval, this time, she'd fail to reject H_0.

c. The standard deviation of the experimental groups was small compared to the difference in the means of the two groups.

E21. a. We test the difference between two means. Also, using either technology (which is preferable) or the formula

$$\frac{\left(\dfrac{s_1^2}{n_1} + \dfrac{s_2^2}{n_2}\right)^2}{df} = \frac{\left(\dfrac{s_1^2}{n_1}\right)^2}{n_1 - 1} + \frac{\left(\dfrac{s_2^2}{n_2}\right)^2}{n_2 - 1},$$

we see that df = 107. The test statistic is

$$t = \frac{\overline{x}_1 - \overline{x}_2}{\sqrt{\dfrac{s_1^2}{n_1} + \dfrac{s_2^2}{n_2}}} = 1.954$$

and hence the p-value is 2P(T>1.954) is only slightly higher than 0.05. As such, there is sufficient evidence to conclude that males differ from females with regard to mean NPI scores at the 10% level but not at the 5% level.

b. Now, we test a single mean compared to a standard. This time, the test statistic is given by

$$t = \frac{\overline{x} - \mu}{s/\sqrt{n}} = \frac{17.27 - 15.3}{6.78/\sqrt{144}} = 3.487$$

with df = 144-1=143. So, one–sided P-value is P(T>3.487) is about 0.0003. As such, there is sufficient evidence to conclude that male celebrities have a higher mean NPI score than that of the general public.

c. For each of the seven components, use the test statistic

$$t = \frac{\overline{x}_M - \overline{x}_F}{\sqrt{\dfrac{s_M^2}{144} + \dfrac{s_F^2}{56}}},$$

where $\overline{x}_M$ and $\overline{x}_F$ are the sample means for males and females, respectively, and s_M and s_F are the respective standard deviations. Also, use the standard formula for df, as listed above (or preferably technology) and a p-value that equals 2P(T>t).

Authority: Here, df = 115.46 and $t \approx 1.247$, P-value about 0.2149; there is insufficient evidence to conclude that males differ from females with regard to authority scores.

Exhibitionism: Here, df = 115 and $t \approx 2.436$, P-value about 0.0164; there is insufficient evidence (but close) to conclude that males differ from females with regard to mean exhibitionism scores.

Superiority: Here, df = 107 and $t \approx 1.444$, P-value about 0.1517; there is insufficient evidence to conclude that males differ from females with regard to mean superiority scores.

Entitlement: Here, df = 109 and $t \approx 0.427$, P-value about 0.6703; there is insufficient evidence to conclude that males differ from females with regard to mean entitlement scores.

Exploitativeness: Here, df = 94 and $t \approx 0.134$, *P*-value about 0.8937; there is insufficient evidence to conclude that males differ from females with regard to mean exploitativeness scores.

Self–sufficiency: Here, df = 95 and $t \approx 0.610$, *P*-value about 0.5433; there is insufficient evidence to conclude that males differ from females with regard to mean self–sufficiency scores.

Vanity: Here, df = 97 and $t \approx 3.778$, *P*-value about 0.0003; there is strong evidence to conclude that males differ from females with regard to mean vanity scores.

d. Type I error because, for example, the chance of at least one Type I error in two tests at the 0.05 level is $1 - (0.95)(0.95) = 0.0975$, which is greater than 0.05. The level of each test should be reduced so as to control the overall chance of at least one Type I error.

E23. a. For 9 degrees of freedom, $t^* = 2.262$. The 95% Confidence interval for the difference *Bottom – Mid-depth* is

$$\bar{d} \pm t^* \cdot \frac{s_d}{\sqrt{n}} = 0.99 \pm 2.262 \cdot \frac{0.593}{\sqrt{10}} \approx 0.99 \pm 0.424,$$

or (0.566, 1.414). Because this interval does not contain 0, you have statistically significant evidence that the mean difference is not 0 if the conditions for inference are met.
Note: The paired differences here are a bit skewed with one outlier.

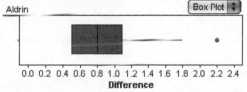

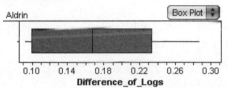

Taking a log transformation of each of the variables and then finding the difference gives a distribution that is much more symmetric with no outlier. The confidence interval is (0.122, 0.221) where the mean difference of the logs is 0.172 and the standard deviation is 0.069. Again, because this interval does not contain 0, you have statistically significant evidence that the mean difference is not 0 if the conditions for inference are met.

Note: Ideally, the differences should be such that they could reasonably have come from a normal distribution, or something close to that, and this still has to be justified. But, two skewed populations can produce a nearly symmetric distribution of differences, so that we need not check the individual populations (or samples) so carefully. The 15/40 rule can still be applied to the single sample of differences. In that regard, the sample under consideration would be highly suspect as a candidate for a t-test.

This problem shows the opposite predicament to the one outlined above; the individual samples behave well but the differences do not.

Why do we not insist on a transformation here? Probably because we did not transform these data in the past, as the individual samples do not look terribly skewed. It is more difficult to compare the result to past results if we transform here and not there.

A side note: The main problem with the 15/40 rule is that for sample sizes under 15 you have no good statistical way of checking to see if they came from a normal population, as almost anything can happen in a sample that small. The best way to check this out in the real world is to talk to the experts about the measurements being made and see if they have any reason to believe the population (whatever that might be) is highly skewed or subject to outliers. So, we do not push the transformation idea too strongly for small samples; the transformation may do exactly the wrong thing.

b. No. In that example the conclusion was that there was not sufficient evidence to conclude the difference was different from 0.

c. The one-sample test of the mean difference has more power. You just saw that because, for the same data, the one-sample test found a statistically significant difference and the two-sample test did not.

E25. a. The repeated measures design shows the strongest relationship as it should because these are measurements made on the same person. The scatterplot of the data for the completely randomized design shows the weakest relationship or very little association between the measurements.

b. *Completely Randomized Design. Check conditions.* You know the randomization condition is met. Boxplots show no reason to be concerned about non-normality here.

State your hypotheses. The null hypothesis is that the true difference between mean pulse rates for standing and sitting (for this group of subjects) is 0; the alternate is that standing increases the mean pulse rate.

H_0: $\mu\,(stand) - \mu\,(sit) = 0$

H_a: $\mu\,(stand) - \mu\,(sit) > 0$

Compute the test statistic and draw a sketch. The two-sample t-statistic is

$$t = \frac{(x_1 - x_2) - 0}{\sqrt{\dfrac{s_1^2}{n_1} + \dfrac{s_2^2}{n_2}}} = \frac{(75.71 - 64.57) - 0}{\sqrt{\dfrac{11.68^2}{14} + \dfrac{9.33^2}{14}}} = 2.788$$

With 24.79 *df,* this produces a *P*-value of 0.005.

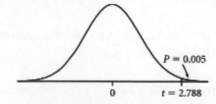

232

Write a conclusion in context. Reject H$_0$. There is sufficient evidence to conclude that standing does increase the mean pulse rate by more than would be expected by chance for this group of subjects.

c. *Matched Pairs Design.* The scatterplot of pairs from the matched pairs design shows a fairly strong positive association, so a one-sample test on the differences should be powerful.

Check Conditions. You know the randomization condition is met. The boxplot of the differences shows a little skewness, but not enough to be a serious problem for this large a sample.

State your hypotheses.
H$_0$: $\mu_d = 0$, and H$_a$: $\mu_d > 0$, where μ_d is the mean difference in pulse rates if each treatment could have been assigned to each subject within each pair.
Compute the test statistic and draw a sketch. The one-sample *t*-statistic calculated from the 14 differences is

$$t = \frac{\bar{d} - \mu_d}{s_d / \sqrt{n}} = \frac{6.0 - 0}{6.563 / \sqrt{14}} = 3.421$$

With 13 *df,* this leads to a *P*-value of 0.0023.

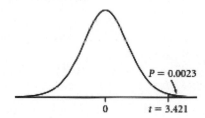

P = 0.0023

0   t = 3.421

Write a conclusion in context. Reject H$_0$. Again, there is evidence to conclude that standing does increase the mean pulse rate by more than would be expected by chance, and the evidence is a little stronger (smaller *P*-value) than for the completely randomized design.

d. *Repeated Measures Design.* The scatterplot of pairs taken from the repeated measures design again shows a very strong positive trend.

Check conditions. You know the randomization condition is met. The boxplot of the differences shows very little skewness.

State your hypotheses.
H$_0$: $\mu_d = 0$, and H$_a$: $\mu_d > 0$ where μ_d is the mean difference in pulse rates if each treatment could have been given in both orders to each subject.

Do computations. The one-sample *t*-statistic calculated from the 28 differences is given by

$$t = \frac{\bar{d} - \mu_0}{s_d / \sqrt{n}} = \frac{4.643 - 0}{3.613 / \sqrt{28}} = 6.800$$

With 27 *df*, the *P*-value is less than 0.0001. The test statistic is quite large and would be difficult to see in a sketch.

Write a conclusion in context, linked to your computations. Reject H_0. Once again, here is evidence to conclude that standing does increase the mean pulse rate by more than would be expected by chance. The evidence is even stronger than in the matched pairs design.

e. In this study, all three designs produced statistically significant results, with decreasing *P*-values of 0.005, 0.0023, and a *P*-value less than 0.0001. So, the evidence in favor of the alternate increased as you reduced variation between pairs by first matching on base pulse rate and then using repeated measures. This is fairly typical of the performance of these three designs and is consistent with the results from the display.

E27. a. This is a matched pairs design when comparing experimental group to control group, and a repeated measures design when comparing scores between Form L and Form M within groups.

b. Analyze the differences between experimental and control scores on Form M. The null and alternative hypotheses are:
H_0: $\mu_d = 0$, H_a: $\mu_d \neq 0$, where μ_d is the mean difference in the scores.

First, compute the differences "Experiment Form M -- Control Form M" to get the new sample data:

$$\{9, 1, 21, 11, 23, 21, 21, -1, 0, 10, 10\}$$

Using technology, we find that $\bar{d} = 11.455$, $s_d = 9.015$. So, the test statistic is given by

$$t = \frac{\bar{d} - 0}{s_d / \sqrt{11}} = 4.214$$

with df = 11-1=10. So, the p-value is 2P(T>04.214) is 0.0018. So, there is strong evidence to declare that the mean difference between the experimental and control group scores on Form M is greater than zero.

E29. a. We use the test performed in E28a here. Since the test is a measure of aptitude and this would likely not change drastically in such a short time, this is what you would expect.

b. Analyze the differences between experimental scores on Forms M and L. The null and alternative hypotheses are:

H_o: $\mu_d = 0$, Ha: $\mu_d \neq 0$, where μ_d is the mean difference in the scores.

First, compute the differences "Experimental Form L -- Experimental Form M" to get the new sample data:

$$\{-10, -12, -19, -13, -19, -12, -20, -11, -7, -10, 0\}$$

Using technology, we find that $\bar{d} = -12.091$, $s_d = 5.839$. So, the test statistic is given by

$$t = \frac{\bar{d} - 0}{s_d / \sqrt{11}} = -6.868$$

with df = 11-1=10. So, the p-value is 2P(T<-6.868) is nearly zero. So, there is strong evidence to declare that the mean difference between the experimental group scores on Forms L and M is different from zero. If you suspect that the M&M treatment is effective, this is exactly what you would expect.

E31. a. *Check conditions:* These are paired data, so the analyses will be done using one-sample techniques. The problem states that the drugs were assigned in random order. The first step of a good analysis is to take a careful look at the data. The boxplots of the differences show one outlier in the values of A – C. Other than that, there is no reason to suspect that these differences could not have come from normally distributed populations.

b. The 98% confidence for the mean of A – B is

$$\bar{d} \pm t^* \cdot \frac{s_d}{\sqrt{n}} = -0.61 \pm 2.896 \frac{1.21}{\sqrt{9}},$$

or (-1.778, 0.558). Because this interval overlaps zero, there is not sufficient evidence to say that the mean flicker frequency for Treatment A differs from that for Treatment B.

c. The 98% confidence for the mean of B – C is

$$\bar{d} \pm t^* \cdot \frac{s_d}{\sqrt{n}} \text{ or } 1.613 \pm 2.896 \cdot \frac{1.274}{\sqrt{9}}$$

or (0.38, 2.84). Because this interval does not overlap zero, there is evidence to say that the mean flicker frequency for Treatment B differs from that for Treatment C, in the direction indicating that C causes more drowsiness.

d. The 98% confidence for the mean of A – C is

$$\bar{d} \pm t^* \cdot \frac{s_d}{\sqrt{n}} \text{ or } 1.003 \pm 2.896 \cdot \frac{1.500}{\sqrt{9}}$$

or (–0.445, 2.451). This indicates no significant difference between the mean flicker frequencies for Treatments A and C, because the interval overlaps zero. These are the differences with the outlier, so it is appropriate to check its influence by doing the analysis without the outlier. The resulting 98% confidence interval is

$$-0.353 < \mu (A - C) < 1.578$$

which is slightly narrower than the first interval but still shows no significant difference between the means for A and C.

e. Taking the above analyses together, you can say that C appears to cause more drowsiness than the placebo B, but none of the other differences shows up as significant. Here you see an example of the commonly occurring bind you get into when comparing more than two means. It looks like B beats C and A is not different from B, but A does not beat C. In general, it is impossible to place more than two means in a complete ordering by a series of tests of significance. There are special procedures to compare three or more means, but they are beyond the scope of this book.

E33. a. It is reasonable to regard the nine occasions as a representative and possibly quasi-random sample of some vaguely defined larger population of occasions, and the careful 4 x 4 arrangement of cotton balls of the two types offers a near guarantee that the bees are not choosing one kind over the other on the basis of location.

Let $d_i = Stung - Fresh$ for occasion i, and let μ_d be the underlying mean difference. The hypothesis that bees have no preference can be written H$_0$: $\mu_d = 0$. The one-sided alternative, that bees are more likely to sting the balls that had been previously stung, is H$_a$: $\mu_d > 0$.

b. It is unclear whether any randomization took place. If Free wanted to keep the cotton balls of the same type together, he could have randomized which side each treatment was on, but there is no mention of that. The distribution of the differences is fairly symmetric with the exception of one rather extreme outlier.

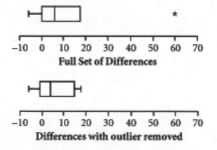

You should do the analysis both with and without the outlier to see if you come to the same conclusion.

c. With the full data set, $\bar{d} = 12$, $s \approx 19.9$.

$$t = \frac{\bar{d} - \mu_d}{s_d / \sqrt{n}} = \frac{12 - 0}{19.92 / \sqrt{9}} = 1.807$$

with 8 degrees of freedom. The *P*-value is about 0.0542.

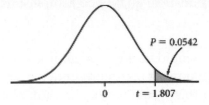

Without the outlier, $\bar{d} = 6$, $s \approx 9.133$.

$$t = \frac{\bar{d} - \mu_d}{s_d / \sqrt{n}} = \frac{6 - 0}{9.133 / \sqrt{8}} = 1.858$$

with 7 degrees of freedom. The P-value is about 0.0528.

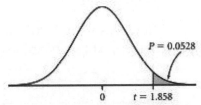

Write your conclusion in context, linked to your computations: With a P-value of 0.0542 or 0.0528, this result is not statistically significant at the 0.05 level. You would not reject the null hypothesis whether you include the outlier or not. There is not statistically significant evidence at the 5% level that there would be a difference between the number of new stingers in the previously stung cotton balls and the fresh cotton balls had Free been able to give all cotton balls both treatments.

Note that removing the outlier cuts the mean difference in half, from 12 to 6 stings, but cuts the standard deviation in half also, so the t-statistic is virtually unchanged. The evidence can hardly be taken as conclusive because the P-values are borderline at best (for one-tailed tests). The data do tend to confirm the theory that smoke masks some odor that a bee leaves behind, along with its stinger, in order to tell other bees to "sting here," but the evidence is not overwhelming.

E35. a. This is a completely randomized design; the data are not paired. The two-sample t-procedure is an appropriate way to compare means. As before, random assignment of treatments to subjects allows the same analyses as random sampling from two different populations.

b. You might pair rooms, one with carpet and one without, that are near each other and have similar uses. This may be difficult, however, unless the experiment allows you to install carpet long before the data are collected. Alternatively, you could use a repeated measures design, giving each room both treatments. You need to randomize which eight rooms get the carpet first.

E37. For this situation, it is not the center but the variability that is of interest. Comparing means might be relevant to make sure that the new settings gave the same mean as the old settings, but that's not what this question is about. (Reducing the variability is very often the main challenge in statistical quality control, just as it often is in design of experiments.)

E39. a. Comparing means would be appropriate if it is reasonable to assume that the students have a somewhat random or else representative set of eruptions. There appears to be no obvious reason why this would not be the case. Note that the data for analysis are the times between eruptions. It turns out that there is a substantial difference: After eruptions that last less than three minutes, the average waiting time is 54 minutes, with a standard deviation of 6.3 minutes; after eruptions that last longer than 3 minutes, the average waiting time is 78 minutes, with a standard deviation of 6.9 minutes.

b. The interruption times are unpaired measurements.

E41. a & b. The distributions of differences with the 0's and without (the latter called the adjusted differences) are shown in the next two boxplots.

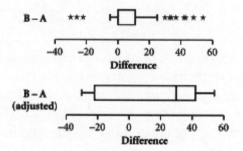

The original data show many outliers because the variability is so small (due to the many 0's) that otherwise ordinary values show up as being far away from the mean. The adjusted data look much better for inference based on t-procedures. For comparison purposes, the summary statistics and both 95% confidence intervals are shown here.

	B – A	(B – A) Adjusted
Count	40	15
Mean	6.38	17
Median	0	30
SD	20.14	30.60
Std Error	3.18	7.90

$$-0.06 < \mu(B-A) < 12.82 \text{ or } (-0.07, 12.82) \text{ with no rounding}$$
$$0.05 < (\mu(B-A) \text{ adj}) < 33.95 \text{ or } (0.06, 33.94) \text{ with no rounding}$$

c. Note that the first interval estimates the mean difference for all accounts and would lead to a conclusion that the two population means do not differ. The second interval estimates the mean difference for all accounts with nonzero differences and would lead to a conclusion that they do differ. The second interval is much longer because of the increased variation and smaller sample size. However, the method that produced the second interval will have nearly the correct capture rate and is the more reliable estimate because the conditions are better met. Whereas because of the outliers, the method of the first interval will have a capture rate that is too low.

E43. a. The probability of a positive difference is 0.5; the probability of 7 positive differences, using the binomial distribution, is $\binom{7}{0}(0.5)^7 = 0.0078$; the probability of 6 out of 7 positive differences is $\binom{7}{6}(0.5)^7 = 0.0547$; the probability of 6 or more positive differences is $0.0547 + 0.0078 = 0.0625$.

b. The value of the test statistic is 6; a large value supports the alternative that urban life results in a higher percentage of radioactivity remaining

c. $P(T \geq 6) = P$-value is 0.0625 (from part a).

d. We do not reject the null hypothesis of no difference in percentage of remaining radioactivity, at the 5% level.

E45. a. The null hypothesis is that the mean time until the next eruption is the same for the two groups, short (following an eruption of 3 minutes or less) and long (following an eruption of more than 3 minutes). The alternate hypothesis is that the means are unequal. You could also argue that a one-sided alternate is appropriate here, because theory predicts a shorter waiting time following a shorter eruption, and vice-versa. If μ_S and μ_L are the mean waiting times following short and long eruptions, the hypotheses are H_0: $\mu_S = \mu_L$ versus either H_a: $\mu_S < \mu_L$ or H_a: $\mu_S \neq \mu_L$.

b. There is no explicit randomization here. Samples were not chosen using a chance device, and the conditions of interest (long or short eruptions) could not be assigned, and so could not be randomly assigned. Moreover, the measurements come from a time series, and it is reasonable to expect the conditions affecting the duration of an eruption or time to the next eruption to be closely related to the conditions affecting the duration of the following eruption and time until the eruption after that. In short, it is almost certain that the data do not satisfy the independence assumption.

c. The standard deviations for the two distributions are quite close, and the distributions are roughly symmetric. (Because of the large sample sizes, the slight skewness in the short group is not an issue.) There are a few outliers, but here, too, the large sample sizes provide reassurance. Note also that removing the four outliers shown in the plot will strengthen the already strong evidence of a difference.

d. After eruptions that last less than 3 minutes ($n = 67$), the average waiting time is 54.46 minutes, with a standard deviation of 6.3; after eruptions that last longer than 3 minutes ($n = 155$), the average waiting time is 78.16 minutes, with a standard deviation of 6.89 minutes. A two-sample t-statistic computed from the summary statistics is huge ($t = 25.0$ with 136.3 degrees of freedom and P-value essentially 0), and H_0 is rejected.

e. The P-value is so extremely small that issues of one-sided versus two-sided alternates and some adjustment for observations that are correlated in time are of no concern. You don't need a statistical test to conclude that there is a "real" difference in waiting times for the two groups of eruptions.

Note that this is an example where it would be hard to satisfy the independent random sampling that the *t*-test needs in order to be valid, but the evidence is so clear cut that you don't need fine-tuned statistics to decide that the difference is real.

E47. a. As the description makes clear, there is one set (sample) of subjects, with two measurements on each subject—control and alcohol.

b. A scatterplot shows one clear outlier, Subject 10, with readings of 900 and 900: an endurance champion as well as an outlier. The remaining nine points lie scattered about a line with slope near zero, with no evidence of any association, positive or negative. Side-by-side boxplots confirm the presence of the outlier; what makes Subject 10 unusual is the long period of useful consciousness after drinking whiskey.

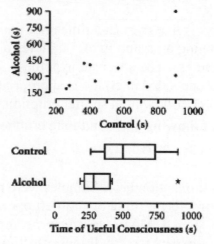

The two distributions are roughly symmetric, with only a slight skewness toward high values. However, the spreads are unequal, with the larger spread belonging to the group with the larger values, as is so often the case for waiting-time data.

A stemplot of the differences shows a distribution skewed toward the high values:

```
−0   05   55
 0   00   65   76
 1   75   90
 2
 3   90
 4
 5   30   90
```

Notice that the outlying Subject 10 contributes a difference of 0, which falls in the middle part of the data: Removing the outlier won't have much effect on the *P*-value.
Here the null hypothesis is that alcohol has no effect on the mean duration of useful consciousness. On the basis of physiology, you can rule out the possibility that alcohol increases the mean duration, which leaves as the alternate hypothesis H$_a$: $\mu_A < \mu_C$,

240

where μ_A and μ_C are the mean durations after alcohol and under the control conditions, respectively. This is the same as saying the mean of the differences, μ_{C-A}, is greater than zero.

c. The summary statistics for these data are shown next.

	Control	Alcohol	C − A
n	10	10	10
$\bar{x}$	546.6	351.0	195.6
s	238.81	210.88	230.53

The 95% confidence interval for μ_{C-A} is

$$\bar{d} \pm t^* \cdot \frac{s_d}{\sqrt{n}} \text{ or } 195.6 \pm 2.262 \cdot \frac{230.53}{\sqrt{10}} \approx 195.6 \pm 164.90 \approx (30.70, 360.50)$$

or (30.69, 360.51) without rounding.

As usual μ_{C-A} refers to a "true" mean of differences, here the difference in duration of useful consciousness, but because there is no explicit randomization, you can't pin down to what "true difference" refers. Because the subjects are not a random sample from a population, there is no straightforward way to figure out to whom, other than the subjects themselves, the results apply. Moreover, all the subjects received the control conditions first, treatment second, so the observed difference could, in principle, be due to something other than alcohol that was linked to the order, although this seems doubtful. A reasonable conclusion is that alcohol appears to reduce the duration of useful consciousness at high altitude, but that it remains to be established how true this is in general.

E49. a. Each infant was tested on both sets of stimuli, so it is a repeated measures, one-sample test.

b. $SE = s / \sqrt{n}$ so $s = SE \cdot \sqrt{n} = SE \cdot \sqrt{24}$, which gives 2.01 and 2.20 for the familiar items and novel items, respectively.

c. There are 23 degrees of freedom in the one–sample test with 24 observations.

d. The standard error is $\dfrac{s_d}{\sqrt{n}}$. The value of t is given and we know that

$$t = \frac{\bar{d} - 0}{s_d / \sqrt{n}} \text{ so } \frac{s_d}{\sqrt{n}} = \frac{\bar{d}}{t} = \frac{(8.85 - 7.97)}{2.3} \approx 0.38$$

e. The result is statistically significant at the 5% level. There is evidence to conclude that the mean listening time is larger for the non-words than for the words for these infants, or for the population of infants that these might represent.

E51. **a.** This is a repeated measures, one–sample test since each patient received both treatments.

b. It could be made double-blind by having neither the patient nor the examining physician know in which order the treatments were applied.

c. $SE = s_d / \sqrt{n} = s_d / \sqrt{10}$

d. $t = \dfrac{\bar{d}}{SE(differences)} = \dfrac{(84 - 72)}{4.7} \approx 2.55$, correct except for rounding; the (two-sided) P-value is correct for $df = 8$.

e. The measures for which the result is statistically significant are "number of correct responses" and "percent conceptual level." The treatment was effective in improving performance on these two cognitive tasks.

Chapter 12

Practice Problem Solutions

P1. a. For a tetrahedral die, we'd expect each face to come up one-fourth of the time, so we'd expect in each category to have a count of $\frac{50}{4}$, or 12.5.

b. The value of χ^2 would be

$$\chi^2 = \frac{\sum(O-E)^2}{E} = \frac{(14-12.5)^2}{12.5} + \frac{(17-12.5)^2}{12.5} + \frac{(9-12.5)^2}{12.5} + \frac{(10-12.5)^2}{12.5} = 3.28$$

P2. 0. The numerator will be 0 in each term of the chi-square calculation.

P3. Using the histogram for the 12-sided die, the heights of the bars of the histogram to the right of 21.3 look to be about $60 + 32 + 22 + 22 + 10$ or about 146. So, the *P*-value is about 146/5000, or 0.0292 with $df = 11$. From the table, the *P*-value is about 0.025 and from the calculator it is about 0.030. Reject the null hypothesis at the 0.05 level.

P4. Using the histogram for an eight-sided die, the heights of the bars to the right of 21.3 look to be very small and the *P*-value would be close to zero. In fact, there are only 14 cases out of 5000 where the chi-square value was at least 21.3 so the simulated *P*-value is 14/5000 or 0.0028. From the table with $df = 7$, the *P*-value is between 0.0025 and 0.005 and the calculator gives a *P*-value of 0.0034. Reject the null hypothesis at the 0.05 level.

P5. a. From Table C, the critical value of χ^2 with 3 degrees of freedom is 7.81. Since 8.24 is greater than 7.81, the result is statistically significant at the 0.05 level.

b. The critical value of χ^2 with 19 degrees of freedom is 30.14. Because 8.24 is less than 30.14, the result is not statistically significant at the 0.05 level.

P6. a. 0.1006   **b.**   0.0304

P7. a. Expect each number to occur the same number of times, namely $\frac{20,000}{6} = 3,333\frac{1}{3}$

b. To use a chi-square goodness-of-fit test, set up the table as shown:

	Outcome					
	1	2	3	4	5	6
Observed Frequency	3407	3631	3176	2916	3448	3422
Expected Frequency	3333.3	3333.3	3333.3	3333.3	3333.3	3333.3

Check conditions: This situation meets the criteria for a chi-square goodness-of-fit test. There were 20,000 independent rolls. Each roll is a 1, 2, 3, 4, 5, or 6. You have a model that gives the hypothesized proportion of outcomes in the population that fall into each category from which the expected frequencies can be derived. The expected frequency in each category is 5 or greater as shown in the table above.

State your hypotheses:
H_0: The die is fair; that is, the probability the die lands on each face is $\frac{1}{6}$.

H_a: The die is not fair.

Compute the test statistic and draw a sketch: The test statistic is

$$\chi^2 = \sum \frac{(O-E)^2}{E} \approx 94.189$$

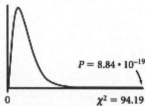

Comparing the test statistic to the χ^2 distribution with $6-1=5$ degrees of freedom, you find that the value of χ^2 from the sample, 94.19, is very far out in the tail. The *P*-value from the calculator is $8.84 \cdot 10^{-19}$.

Write a conclusion in context, linked to your computations: You should reject the null hypothesis. You can not attribute the differences to random variation. A value of χ^2 this large is extremely unlikely to occur if the die is fair. This die (or the method of tossing or recording the results) appears to avoid 4's and favor 2's.

P8. a. Expect each to occur the same number of times, namely $\frac{700}{12} = 58\frac{1}{3}$

b. Because these data are tied to months of the year, you should first look at a plot over time of the number of births.

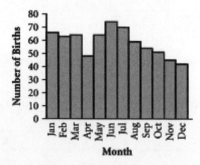

The number of births generally decreases throughout the year, with the most births in the late spring/early summer.

To use a chi-square goodness-of-fit test, we set up this table.

244

Month	Observed Number of Births	Expected Number of Births
January	66	$\left(\frac{31}{365}\right)700 \approx 59.452$
February	63	$\left(\frac{28}{365}\right)700 \approx 53.699$
March	64	$\left(\frac{31}{365}\right)700 \approx 59.452$
April	48	$\left(\frac{30}{365}\right)700 \approx 57.534$
May	64	$\left(\frac{31}{365}\right)700 \approx 59.452$
June	74	$\left(\frac{30}{365}\right)700 \approx 57.534$
July	70	$\left(\frac{31}{365}\right)700 \approx 59.452$
August	59	$\left(\frac{31}{365}\right)700 \approx 59.452$
September	54	$\left(\frac{30}{365}\right)700 \approx 57.534$
October	51	$\left(\frac{31}{365}\right)700 \approx 59.452$
November	45	$\left(\frac{30}{36}5\right)700 \approx 57.534$
December	42	$\left(\frac{31}{365}\right)700 \approx 59.452$
Total	700	700

Check conditions: There were 700 births, but they all were in one hospital during one particular year. So, this is not a random sample but more of an observational study. Each birth is classified into exactly 1 of 12 categories. You have a model that gives the hypothesized proportion of outcomes in the population that fall into each category. The expected frequency in each category is 5 or greater as shown in the table above.

State your hypotheses:
H_0: The number of births in a month is proportional to the number of days in the month.
H_a: The number of births in a month is not proportional to the number of days in the month for at least one month.

Compute the test statistic and draw a sketch: The test statistic is

$$\chi^2 = \sum \frac{(O-E)^2}{E} = \frac{(66-59.452)^2}{59.452} + \frac{(63-53.699)^2}{53.699} + \frac{(64-59.452)^2}{59.452} + \cdots + \frac{(42-59.452)^2}{59.452}$$
$$\approx 20.468$$

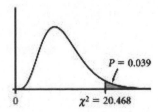

Comparing the test statistic to the χ^2 distribution with $12 - 1 = 11$ degrees of freedom, you find that $\chi^2 = 20.468$ gives a *P*-value of 0.0393.

Write a conclusion in context: You can reject the null hypothesis at the 5% level, but not the 1% level. It would be a rare event (with $\alpha = 0.05$) to get a value of χ^2 this large or larger if the number of births in a month is proportional to the number of days in the month. The differences between the observed frequencies and expected frequencies can not easily be attributed to chance variation in the timing of births.

Apparently, there are more births in the late spring/early summer and in the first months of the year. However, you have to be suspicious of the conclusion that the number of births varies by season because you don't have a random sample of births—you have all the births in this particular year for this particular hospital. Births appear generally to decrease over the year, but perhaps women were leaving this part of Switzerland or perhaps a new hospital opened and women started going to it; or there may be some other explanation why the decrease occurred in this hospital at this time.

P9. We will perform a chi-square goodness-of-fit test on the data shown in the table below. The expected frequencies are in the right column.

Age	Countrywide Percentage	Observed Number of Grand Jurors	Expected Number of Grand Jurors
21–40	42	5	0.42(66), or 27.72
41–50	23	9	0.23(66), or 15.18
51–60	16	19	0.16(66), or 10.56
61 or older	19	33	0.19(66), or 12.54
Total	100	66	66

Check conditions: This situation meets the criteria for a chi-square goodness-of-fit test. There were 66 jurors selected, but as usual you can't be sure they were independently or randomly selected. This isn't a problem because this is essentially what you want to test. Each juror falls into exactly one of the age groups. You have a model that gives the hypothesized proportion of outcomes in the population that fall into each category. The expected frequency in each category is 5 or greater as shown in the table above.

State your hypotheses:
H_0: The distribution of the ages of the grand jurors looks like a typical random sample from the ages of all adults in this county.
H_a: The distribution of ages of the grand jurors doesn't look like a typical random sample from the ages of all adults in this county.

Compute the test statistic, find the P-*value, and draw a sketch:*
    The test statistic is

$$\chi^2 = \sum \frac{(O-E)^2}{E} = \frac{(5-27.72)^2}{27.72} + \frac{(9-15.18)^2}{15.18} + \frac{(19-10.56)^2}{10.56} + \frac{(33-12.54)^2}{12.54} \approx 61.27$$

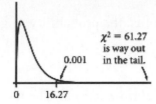

$\chi^2 = 61.27$
is way out
in the tail.

0.001

0   16.27

Comparing the test statistic to the χ^2 distribution with $4 - 1 = 3$ degrees of freedom, you find that the value of χ^2 from the sample, 61.27, is very far out in the tail. The *P*-value is 3.15×10^{-13}, or 0.000000000000315.

Write a conclusion in context: You should reject the null hypothesis. You can not attribute the differences to random sampling. A value of χ^2 this large or larger is extremely unlikely to occur if grand jurors are selected without regard to age. Either older people are more likely to be available for grand juries or they are deliberately selected.

P10. Choices A and B only. Some notes as to why are as follows:

A – Switching the order of the categories does not matter since they are summed and summing is a commutative property.

B – The observed frequencies are always whole numbers since the frequencies represent unique counts of observed events.

C – The expected frequencies are NOT always whole numbers since they are calculated by dividing the total number of observations by the number of possible distinct outcomes. If one rolls a 2-sided coin 51 times, the expected frequencies for a fair coin = 25.5.

D – The number of degrees of freedom is NOT necessarily 1 less than the sample size. The df is calculated as the number of categories – 1.

E – A high value of χ^2 indicates a LOW level of agreement between the observed frequencies and the expected frequencies. In fact a sufficiently large χ^2 indicates that the proportions of the different outcomes are not equal to the expected/hypothesized proportions.

P11. a. The technique in Chapter 8 would be a test of significance for a proportion.

Check conditions: This is a random sample from a potentially infinite population. Both $np_0 = 50$ and $n(1 - p_0) = 50$ are at least 10, where n is the sample size and p_0 is the probability of getting tails if spinning a coin is fair. The conditions are met for a significance test for a proportion.

State your hypotheses:

H_0: Spinning a coin is fair; that is, the probability of getting a tail when spinning this coin is 0.5.

H_a: Spinning a coin is not fair.

Compute the test statistic and draw a sketch: The test statistic is

$$z = \frac{\hat{p} - p_0}{\sqrt{\dfrac{p_0 \cdot (1 - p_0)}{n}}} = \frac{0.64 - 0.5}{\sqrt{\dfrac{0.5 \cdot 0.5}{100}}} = 2.8$$

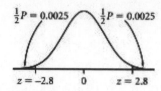

The *P*-value for a two-tailed test is about 0.005.

Write a conclusion in context, linked to your computations: You should reject the null hypothesis that spinning a coin is fair. A fair coin spun 100 times would produce a proportion of tails of at least 0.64 or at most 0.36 only 0.5% of the time. You have evidence that this coin favors tails.

b. To use a chi-square goodness-of-fit test, we set up this table.

Outcome	Observed	Expected
Tails	64	50
Heads	36	50
Total	100	100

Check conditions: This situation meets the criteria for a chi-square goodness-of-fit test. There were 100 independent coin spins. Each coin spin is either a head or a tail. You have a model that gives the hypothesized proportion of outcomes in the population that fall into each category. The expected frequency in each category is 5 or greater as shown in the table above.

State your hypotheses:

H_0: Spinning a coin is fair; that is, the probability the coin lands tails when spinning is 0.5, and the probability it lands heads when spun is 0.5.

H_a: Spinning a coin is not fair.

Compute the test statistic and draw a sketch: The test statistic is

$$\chi^2 = \sum \frac{(O - E)^2}{E} = \frac{(64 - 50)^2}{50} + \frac{(36 - 50)^2}{50} = 7.84$$

Comparing the test statistic to the χ^2 distribution with $2 - 1 = 1$ degree of freedom, you find that the value of χ^2 from the sample, 7.84, is far out in the tail. The *P*-value from the table is about 0.005.

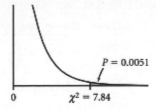

$P = 0.0051$

$0 \qquad \chi^2 = 7.84$

Write a conclusion in context, linked to your computations: You should reject the null hypothesis. You can not attribute the difference from the expected proportions to random variation. A value of χ^2 this large is extremely unlikely to occur if spinning a coin is fair. This coin appears to favor tails when spun.

c. The *P*-values for the two different techniques are the same. Also, note that the test statistic $z = 2.8$ and $\chi^2 = 7.84$ are related in the following way: $z^2 = \chi^2$. This is always the case and is not hard to prove algebraically.

P12. a. The technique in Chapter 8 would be a test of significance of a proportion.

Check conditions: This is a random sample from a potentially infinite population. Both $np_0 = n(1-p_0) = 50$ are at least 10. Where p_0 is the probability of getting heads when spinning a checker, if spinning a checker is fair.

State your hypotheses:
H_0: Spinning a checker with putty on the crown side is fair; that is, the probability of getting a "crown" by spinning this checker is 0.5.
H_a: Spinning this checker is not fair. The probability of getting a "crown" is not 0.5.

Compute the test statistic and draw a sketch:
The test statistic is

$$z = \frac{\hat{p} - p_0}{\sqrt{\dfrac{p_0 \cdot (1 - p_0)}{n}}} = \frac{0.23 - 0.5}{\sqrt{\dfrac{0.5 \cdot 0.5}{100}}} = -5.4$$

$\frac{1}{2}P = 3.35 \cdot 10^{-8}$ $\qquad$ $\frac{1}{2}P = 3.35 \cdot 10^{-8}$

$z = -5.4 \qquad 0 \qquad z = 5.4$

The *P*-value for a two-tailed test is $6.7 \cdot 10^{-8}$.

Write your conclusion in context, linked to your computations: A *P*-value this small is statistically sufficient evidence to reject the null hypothesis. Spinning a checker and having it land heads only 23 times or less or 77 times or more out of 100 spins would be extremely unlikely if spinning the checker was fair.

249

b. To use a chi-square goodness-of-fit test, we set up this table.

Outcome	Observed	Expected
Heads	23	50
Tails	77	50
Total	100	100

Check conditions: This situation meets the criteria for a chi-square goodness-of-fit test. There were 100 independent checker spins. Each checker spin is either a head or a tail. You have a model that gives the hypothesized proportion of outcomes in the population that fall into each category. The expected frequency in each category is 5 or greater as shown in the table above.

Compute the test statistic, find the P-value, and draw a sketch: The test statistic is:

$$\chi^2 = \sum \frac{(O-E)^2}{E} = \frac{(23-50)^2}{50} + \frac{(77-50)^2}{50} \approx 29.16$$

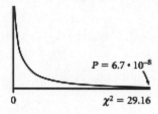

$P = 6.7 \cdot 10^{-8}$

$\chi^2 = 29.16$

Comparing the test statistic to the χ^2 distribution with $2 - 1 = 1$ degrees of freedom, you find that $\chi^2 = 29.16$ gives a *P*-value of $6.7 \cdot 10^{-8}$.

Write your conclusion in context, linked to your computations: A *P*-value this small is statistically sufficient evidence to reject the null hypothesis. Spinning a checker and having it land heads only 23 times or less or 77 times or more out of 100 spins would be extremely unlikely if spinning the checker was fair and ended up "heads" 50% of the time.

c. The *P*-values for the two different techniques are the same. Also, note that the test statistic $z = -5.4$ and $\chi^2 = 29.16$ are related in the following way: $z^2 = \chi^2$. This is always the case and is not difficult to prove algebraically.

P13. a.

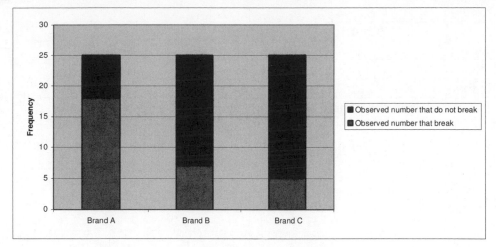

b. From this plot alone, it is clear that those in Brand A are more likely to break than are the other two brands.

P14. a.

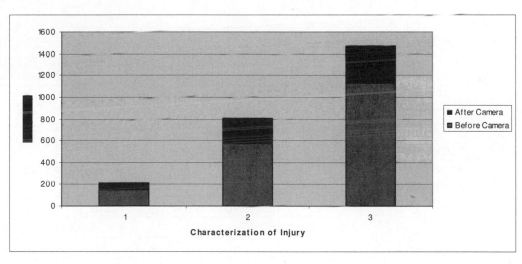

Here:

1 = Death or Disabling Injury

2 = Evident Injury

3 = Possible Injury

b. The number of right-angle crashes decreased by about two-thirds, and the distribution of the severity of injuries appears somewhat different. The proportions of more serious injuries are a bit larger in the after camera crashes.

P15.

	Brand A	Brand B	Brand C	Total
Yes	10	10	10	30
No	15	15	15	45
Total	25	25	25	75

P16. First, we compute row and column totals for the observed frequencies:

	Before Camera	After Camera	Total
Death or Disabling Injury	152	59	211
Evident Injury	571	237	808
Possible Injury	1131	338	1469
Total	1854	634	2488

As such, the table of expected frequencies is given by

	Before Camera	After Camera	Total
Death or Disabling Injury	$\frac{(1854)(211)}{2488} = 157.23$	53.77	211
Evident Injury	602.1	205.9	808
Possible Injury	1094.7	374.34	1469
Total	1854	634	2488

P17. a. Here are the computations used to compute the test statistic:

Outcome	Observed Frequency, O	Expected Frequency, E	O-E	$(O-E)^2 / E$
A breaks	18	10	8	6.4
A does not break	7	15	-8	4.27
B breaks	7	10	-3	0.9
B does not break	18	15	3	0.6
C breaks	5	10	-5	2.5
C does not break	20	15	5	1.67
Total	75	75	0	16.34

Summing the values in the right column yields

$$\chi^2 = \sum \frac{(O-E)^2}{E} = 16.34 \,.$$

b. df = (2-1)(3-1) = 2

c. Using the calculator to determine the p-value for the test with the test statistic and df given above, we get approximately 0.00028. Alternatively, using Table C, we find that this value of χ^2 is significant at the 0.001 level.

d.

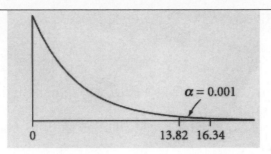

P18. **a.** The test statistic is given by

$$\chi^2 = \sum \frac{(O-E)^2}{E} = \frac{(152-157.23)^2}{157.23} + \ldots + \frac{(338-374.34)^2}{374.34} = 11.72.$$

b. df = (3-1)(2-1) = 2

c. About 0.0028 (or, 0.0025 < P-value < 0.005 from Table C); yes, this is significant at the 5% level.

d. Reject Ho. If it is true that the distribution of severity of accidents is the same before and after the cameras, there is less than a 0.005 chance of getting proportions of observed frequencies that are as different or more different in at least one category of severity than those in Display 12.38. In other words, the differences in the two distributions cannot reasonably be attributed to chance alone.

P19. The new table appears below with expected frequencies included.

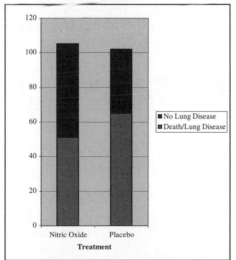

	Nitric Oxide	Placebo	Totals
Death/Survival with chronic lung disease	Obs = 51 Exp = 58.84	Obs = 65 Exp = 57.16	116
Survival without chronic lung disease	Obs = 54 Exp = 46.16	Obs = 37 Exp = 44.84	91
Totals	105	102	207

Check conditions: Treatments were randomly assigned to subjects and all expected frequencies are at least five as shown in the table above. Each subject falls into exactly one category.

State your hypotheses:
H_0: If both treatments could be assigned to all subjects, the resulting distributions of outcomes would be the same.
H_a: If both treatments could have been assigned to all subjects, the resulting distributions of outcomes would differ.

Compute the test statistic, find the P-value, and draw a sketch:

$$\chi^2 = \sum \frac{(O-E)^2}{E} = \frac{(51-58.84)^2}{58.84} + \frac{(65-57.16)^2}{57.16} + \frac{(54-46.16)^2}{46.16} + \frac{(37-44.84)^2}{44.84} \approx 4.823$$

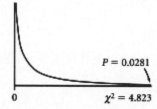

Comparing the test statistic to a χ^2 distribution with one degree of freedom, you see that a χ^2 value of 4.823 is in the tail and the *P*-value is 0.0281. Using the table, the *P*-value would be between 0.025 and 0.05.

Write your conclusion in context, linked to your computations: The *P*-value of 0.0281 is less than 0.05 so you would reject the null hypothesis. There is now a statistically significant difference between the proportions that fall into each category in the two distributions.

P20. The table with expected frequencies included is shown:

Degree of Hemorrhage	Nitric Oxide	Placebo	Totals
Severe	Obs = 13 Exp = 18.77	Obs = 24 Exp = 18.23	37
Not Severe	Obs = 27 Exp = 18.77	Obs = 10 Exp = 18.23	37
None	Obs = 65 Exp = 67.46	Obs = 68 Exp = 65.54	133
Total	105	102	207

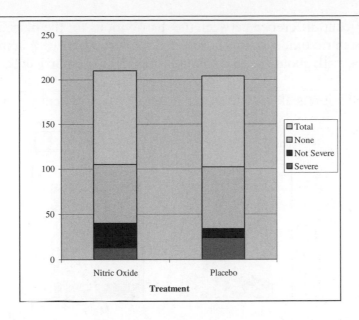

Check conditions: Treatments were randomly assigned to subjects and all expected frequencies are at least five as shown in the table above. Each subject falls into exactly one category.

State your hypotheses:
H_0: If both treatments could be assigned to all subjects, the resulting distributions of outcomes would be the same.
H_a: If both treatments could have been assigned to all subjects, the resulting distributions of outcomes would differ.

Compute the test statistic, find the P-value, and draw a sketch:

$$\chi^2 = \sum \frac{(O-E)^2}{E} = \frac{(13-18.77)^2}{18.77} + \frac{(24-18.23)^2}{18.23} + \frac{(27-18.77)^2}{18.77} + \frac{(10-18.23)^2}{18.23} + \frac{(65-67.46)^2}{67.46}$$

$$+ \frac{(68-65.54)^2}{65.54} \approx 11.108$$

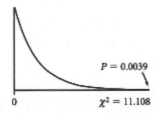

Comparing the test statistic to a χ^2 distribution with two degree of freedom, you see that a χ^2 value of 11.108 is in the tail and the *P*-value is 0.0039. The *P*-value from the table is between 0.0025 and 0.005.

Write your conclusion in context, linked to your computations: The *P*-value of 0.0039 is less than 0.05. You would reject the null hypothesis that if both treatments could be assigned to all subjects, the distributions of outcomes would be the same. There is a statistically significant difference between the distributions for the placebo and nitric oxide groups. The nitric oxide treatment resulted in more *Not severe* hemorrhages and fewer *Severe* ones, with about the same number having no hemorrhages.

P21. a. The population is all female students and all male students. Observe that

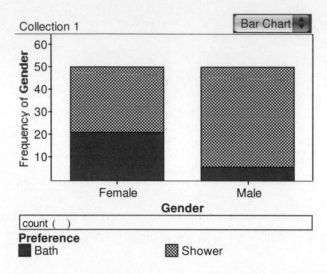

from the bar chart, the proportion of females who prefer a bath to a shower is much greater than the proportion of males who prefer baths to showers. It appears that females are more likely than males to prefer a bath.

b. We test:

H_0: $p_1 = p_2$ against H_a: $p_1 \neq p_2$, where p_1, p_2 are the proportions of males and females who prefer baths over showers, respectively.

Here, we have

$$\hat{p}_1 = 0.12, \quad \hat{p}_2 = 0.42, \quad \hat{p} = \frac{6+21}{50+50} = 0.27.$$

Hence, the test statistic is given by

$$z = \frac{(\hat{p}_1 - \hat{p}_2) - 0}{\sqrt{\hat{p}(1-\hat{p})\left(\frac{1}{n_1} + \frac{1}{n_2}\right)}} = 3.38.$$

The p-value is $2P(Z > 3.38) = 0.0008$.

c. It is helpful to put the observed and expected frequencies into one table.

	Male	Female	Total
Bath	Obs = 6 Exp = 13.5	Obs = 21 Exp = 13.5	27
Shower	Obs = 44 Exp = 36.5	Obs = 29 Exp = 36.5	73
Total	50	50	100

256

$$\chi^2 = \sum \frac{(O-E)^2}{E} = \frac{(6-13.5)^2}{13.5} + \frac{(21-13.5)^2}{13.5} + \frac{(44-36.5)^2}{36.5} + \frac{(29-36.5)^2}{36.5} \approx 11.42$$

This table has 1 degree of freedom. The *P*-value estimated from Table C is less than 0.001. (Table C only goes up to $\chi^2 = 10.83$ for I *df*. If male and female students have the same preferences, the probability of getting a test statistic of 11.42 or larger is less than 0.001 (or about 0.00073).

d. The two tests give the same *P*-value and so the same conclusion: If the same proportion of male as female students prefer a bath, the probability is only 0.0007 of getting a difference in proportions as large or larger than that from these two samples.

P22. a. This is the marginal relative frequency of all those sampled that prefer a bath, *P(Bath)*.

b. This is the conditional relative frequency of those who are female given that they prefer a bath, *P(Female | Bath)*.

c. This could represent either the marginal relative frequency of those sampled who are female or those who are male, *P(Female)* or *P(Male)*.

d. This is neither a conditional nor a marginal relative frequency. It is the frequency of those sampled that are male and prefer a bath, *P(Male and Bath)*.

P23. a. Expected frequencies for right-handed are 955.86 and 1048.14.

	Men	Women	Total
Right-handed	955.86	1048.14	**2004**
Left-handed	97.78	107.22	**205**
Ambidextrous	13.36	14.64	**28**
Total	**1067**	**1170**	**2237**

b. Opinions may vary as to independence. P25 will shed some light on this question.

P24. a. $\chi^2 = \sum \frac{(O-E)^2}{E} = 1.997$

b. df = 2 -1 = 1. So, the p-value is 0.1576. So, even at the 10% level, there is insufficient evidence to conclude that the variables are associated.

c. There isn't statistically significant evidence that the proportion of all San Francisco bars with glamorized images of smoking that are compliant with the smoking law is different from the proportion of all San Francisco bars without such images that are compliant.

P25. Because the data came from a single survey which recorded (among other things) gender and handedness as variables, a chi-square test of independence is the appropriate test. The expected frequencies were calculated in P23. The bar chart and ribbon chart below show the observed frequencies.

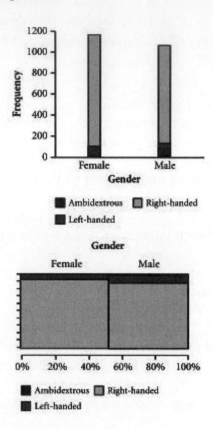

Check conditions: The sample was a random sample. Each member of the population does fall into exactly one of the cells determined by the two variables, and each expected frequency is at least 5 as shown earlier.

State your hypotheses:
H_0: Gender and handedness are independent.
H_a: Gender and handedness are not independent.

Compute the test statistic, find the P-value, and draw a sketch:

$$\chi^2 = \sum \frac{(O-E)^2}{E} = \frac{(934-955.86)^2}{955.86} + \frac{(1070-1048.14)^2}{1048.14} + \frac{(113-97.78)^2}{97.78} + \cdots + \frac{(8-14.64)^2}{14.64} \approx 11.806$$

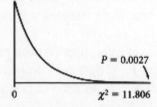

$P = 0.0027$

$0 \qquad\qquad\qquad \chi^2 = 11.806$

258

Comparing the test statistic to a chi-square distribution with 2 degrees of freedom, you find that this value of χ^2, 11.81, is out in the tail of the distribution and the P-value is 0.0027. Using the table, you get a P-value between 0.0025 and 0.005.

Write your conclusion in context, linked to your computation: You should reject the null hypothesis. If gender and handedness were independent, getting a value of χ^2 as large as or larger than 11.806 from a sample of 2237 is unlikely. You have evidence that gender and handedness are associated. In particular, it appears that women are more likely to be right-handed and left-handed or ambidextrous persons are more likely to be men.

P26. a. The expected frequencies are given by

	Appl. Abandoned	Invention Patented	Appl. Pending	Total
No IDS filed	$\frac{(750)(856)}{3000} = 214$	300.5	235.5	750
IDS filed with application	226.56	312.52	244.92	784
IDS filed after appl. filing date	419.44	588.98	461.58	1470
Total	860	1202	942	3004

b. The following conditions are indeed met:

- You should have independent random samples of fixed (but not necessarily equal) sizes taken from two or more large populations (or two or more treatments are randomly assigned to subjects).

- Each outcome falls into exactly one of several categories, with the categories being the same in all populations.

- The expected frequency in each cell is 5 or greater.

The test statistic is given by

$$\chi^2 = \sum \frac{(O-E)^2}{E} = 80.38.$$

Since df = (3-1)(3-1)=4, we see that the p-value is nearly zero. So, we reject the null hypothesis.

c. There is statistically significant evidence that if you examine the population of all patent applications, the proportion of applications that fall into at least one of the three categories is not equal across all IDS categories.

P27. a. independence; From this middle school, get a random sample of children who ate breakfast this morning and an independent random sample of children who didn't (this would be harder to do).

b. The table below shows the observed and expected frequencies, and the segmented bar chart and ribbon chart show the observed frequencies.

		Paying Attention		
		Yes	**No**	**Totals**
Breakfast This Morning	**Yes**	Obs: 22 Exp: 18.15	Obs: 19 Exp: 22.85	**41**
	No	Obs: 5 Exp: 8.85	Obs: 15 Exp: 11.15	**20**
	Totals	**27**	**34**	**61**

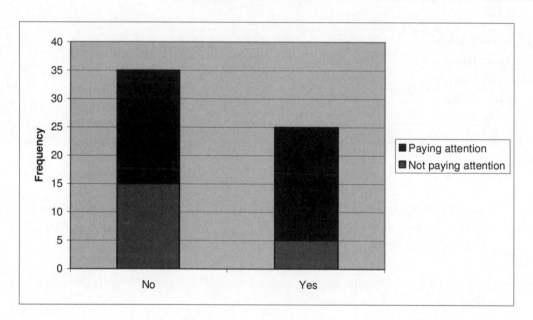

Check conditions: You do not know whether this was a random sample. Each person surveyed does fall into exactly one cell based on the two variables, and all the expected frequencies are at least 5 as shown above. If the sampling was done randomly the conditions are met for a chi-square test of independence.

State your hypotheses:

H_0: Eating breakfast this morning and paying attention are independent.
H_a: These variables are associated (not independent).

Compute the test statistic, find the P-value, and draw a sketch:

$$\chi^2 = \sum \frac{(O-E)^2}{E} = \frac{(22-18.15)^2}{18.15} + \frac{(19-22.85)^2}{22.85} + \frac{(5-8.85)^2}{8.85} + \frac{(15-11.15)^2}{11.15} \approx 4.475$$

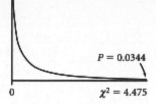

$P = 0.0344$

0 $\chi^2 = 4.475$

Comparing the test statistic to a chi-square distribution with 1 degree of freedom, you can see that the value of χ^2, 4.475, is in the tail. Using the table, the *P*-value is between 0.025 and 0.05. Using the calculator, the *P*-value is 0.0344.

Write your conclusion in context, linked to your computations: You should reject the null hypothesis. If, among all students at this middle school, having breakfast in the morning and paying attention are independent, the probability of getting a chi-square of 4.475 or higher from a random sample of this size is 0.0344. (Recall that you don't know whether this was a random sample.) You have evidence that these variables are associated if the sample was taken randomly.

P28. a. Note that $\chi^2 = 1067.404$ with df = (4-1)(2-1)=3. So, the p-value is nearly zero. Hence, it is statistically significant. The proportions in each region with complete kitchen facilities aren't all that different (all about 99%), so this is strong evidence of a weak association.

b. The expected frequencies are as follows:

	Northeast	Midwest	South	West
Complete Kitchen	$\frac{(1948)(397)}{1969} = 392.77$	389.80	669.78	495.66
Incomplete Kitchen	4.23	4.20	7.22	5.34

So, the test statistic is given by

$$\chi^2 = \sum \frac{(O-E)^2}{E} = 0.75 \ .$$

Since df = (2-1)(4-1)=3, we see that the p-value is > 0.25, and so the result is not significant.

c. No.

Exercise Solutions

E1. a.

Blood Type	Expected Frequency
O	0.49(254)=124.46
A	0.27(254)=68.58
B	0.20(254)=50.8
AB	0.04(254)=10.16

b. No, because the test statistic is

$$\chi^2 = \sum \frac{(O-E)^2}{E} = \frac{(134-124.46)^2}{124.46} + \cdots + \frac{(5-10.16)^2}{10.16} = 4.04$$

with df = 3. So, the p-value is 0.2572.

c. H_0: The distribution of blood types in the population is equal to the distribution of blood types in the model.
H_a: The distribution of blood types is not consistent with the model. At least one ratio in the distribution is not equal to the ratio in the model.

Check conditions: While the expected frequency in each category is at least 5, it's not clear whether this is a random sample of black homicide victims.

Write your conclusion in context, linked to your computation: Because the *P*-value is higher than any reasonable significance level, you would not reject the null hypothesis. As such, we conclude that the difference between the proportions in the sample of homicide victims with the various blood types and proportions in the donor population isn't statistically significant.

E3. *Check Conditions:* This is not a simple random sample of patient's seizures, instead it is an observational study. The seizures are all from patients at Tampa General Hospital. However, each seizure did occur during exactly one of the four periods, you do have a model of hypothesized proportions that you can use to compare to the sample frequencies, and all expected frequencies are at least 5 and are shown below. Only the random sample condition is not met.

State your hypotheses:
H_0: Seizures are equally likely during all four given phases of the moon.
H_a: Seizures are more likely during some lunar phases than others.

Compute the test statistic, find the P-*value, and draw a sketch:*
The expected count in each category is 470 / 4 = 117.5.
The test statistic is

$$\chi^2 = \sum \frac{(O-E)^2}{E} = \frac{(103-117.5)^2}{117.5} + \frac{(121-117.5)^2}{117.5} + \frac{(94-117.5)^2}{117.5} + \frac{(152-117.5)^2}{117.5} \approx 16.723$$

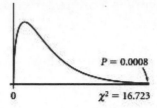

$$P = 0.0008$$
$$0 \qquad \chi^2 = 16.723$$

Comparing the test statistic to the χ^2 distribution with $4 - 1 = 3$ degrees of freedom, you find that $\chi^2 = 16.723$ gives a P-value of 0.0008.

Write your conclusion in context, linked to your computations: Because a P-value of 0.0008 is so small you would reject the null hypothesis. If this was a random sample, then these data do not fit the model of equal likelihood of seizures during the four phases of the moon. If this was a random sample and the model was correct that it is equally likely to have an epileptic seizure during the different phases of the moon, then, there would be only an 0.08% chance of getting a result as extreme as or more extreme than the one in the sample. You'd have statistically sufficient evidence that seizures are not equally likely over the four phases of the moon.

The largest deviations from the model can be looked at in a couple ways. The deviations are −14.5, 3.5, −23.5, 34.5. You can also look at the contribution of each element to the chi-square calculation. The TI-84+ calculator automatically calculates these and stores them in a list called CNTRB. The contributions are 1.789, 0.104, 4.700, and 10.130. The largest deviation occurs during the last quarter of the moon. There were many more seizures during the last quarter, and there were actually *fewer* during the full moon than the model predicts.

E5. a. To use a chi-square goodness-of-fit test, we set up this table. The expected frequencies are in the right column.

Type of Pea	Observed	Expected
Smooth Yellow	315	$\left(\frac{9}{16}\right)556 = 312.75$
Wrinkled Yellow	101	$\left(\frac{3}{16}\right)556 = 104.25$
Smooth Green	108	$\left(\frac{3}{16}\right)556 = 104.25$
Wrinkled Green	32	$\left(\frac{1}{16}\right)556 = 34.75$
Total	556	556

b. In such case, $O - E$ is approximately zero. So, the test statistic is close to zero.

c. *Check conditions:* You have a sample of 556 peas. It is not a random sample nor do you know whether there was a random assignment of treatments but you are trying to determine whether the peas fit the probability model proposed by Mendel's theory. Each pea is classified into exactly one of four categories. You have a model that gives the hypothesized proportion of outcomes in the population that fall into each category. The expected frequency in each category is 5 or greater as shown in the table above.

State your hypotheses:
H₀: Peas are produced so that $\frac{9}{16}$ are smooth yellow, $\frac{3}{16}$ are wrinkled yellow, $\frac{3}{16}$ are smooth green, and $\frac{1}{16}$ are wrinkled green.
Hₐ: Peas are produced in some other ratio.

Compute the test statistic and draw a sketch: The test statistic is

$$\chi^2 = \sum \frac{(O-E)^2}{E} = \frac{(315-312.75)^2}{312.75} + \frac{(101-104.25)^2}{104.25} + \frac{(108-104.25)^2}{104.25} + \frac{(32-34.75)^2}{34.75} \approx 0.47$$

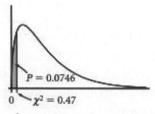

$P = 0.0746$

$0 \qquad \chi^2 = 0.47$

Comparing the test statistic to the χ^2 distribution with $4 - 1 = 3$ degrees of freedom, you find that the value of χ^2 from the sample, 0.47, is close to 0. Because you are trying to determine whether Mendel's results are *closer* to the theory than you would expect, you want the probability of a chi-square statistic *less than* or equal to 0.47. The probability of getting a chi-square statistic this close or even closer to 0 is only 0.0746.

Write a conclusion in context: This P-value in and of itself is a bit suspicious, but not conclusive. However, Mendel performed several such experiments and in each case, the P-value was suspiciously low, leading statisticians to believe that the data may have been "fudged" to fit his theory. In other words, the results are too close to Mendel's theory to look like a random sample or a random assignment of treatments.

E7. To do a chi-square test you must work with frequencies, not percentages. The table must be converted using 3000 as the total number in each sample.

	Crossley	Gallup	Roper	Actual Result
Truman	1350	1320	1140	1500
Dewey	1500	1500	1590	1350
Thurmond	60	60	150	90
Wallace	90	120	120	60

Check conditions: You do not know that these polling organizations used random sampling techniques. Statistical methods were not as well developed in 1948 as they are today so it is entirely possible that random sampling was not used. Each voter belongs to one of the four categories (candidates), you have a model of hypothesized proportions (the actual results), and each expected count is at least 5 as shown in the table above.

State your hypotheses:

H_0: The proportions of voters in the population who favored each candidate in a poll by a particular organization are equal to the proportions reflected in the actual election results.
H_a: The proportions of voters in the population who favored each candidate in a poll by a particular organization is different from those reflected in the election results.

Compute the test statistic, find the P-*value, and draw a sketch:* Each poll will have its own test statistic:

Crossley:

$$\chi^2 = \sum \frac{(O-E)^2}{E} = \frac{(1600-1350)^2}{1350} + \cdots + \frac{(60-90)^2}{90} \approx 56.67$$

P-value: 3×10^{-12}

Gallup:

$$\chi^2 = \sum \frac{(O-E)^2}{E} = \frac{(1300-1320)^2}{1320} + \cdots + \frac{(60-120)^2}{120} \approx 84.55$$

P-value: 3×10^{-23}

Roper:

$$\chi^2 = \sum \frac{(O-E)^2}{E} = \frac{(1500-1140)^2}{1140} + \cdots + \frac{(60-120)^2}{120} \approx 203.91$$

P-value: 2×10^{-49}

Comparing each test statistic to the χ^2 distribution with $4 - 1 = 3$ degrees of freedom, you find that the value of χ^2 from each sample gives an extremely small P-value.

Write your conclusion in context, linked to your computations: None of the polls are even close to a reasonably good fit to the actual election results. The Crossley poll is the best fit of the three, but even its P-value of $3 \cdot 10^{-12}$ is strong evidence to reject the null hypothesis (or the condition of random sampling – those conditions are important!).

E9. a. There were a total of 1598 fatalities in the study. The model says twice as many fatalities should occur on the two control Sundays as on Superbowl Sunday. Put these numbers in a table:

Sunday	Observed Number of Fatalities	Expected Number of Fatalities
Super Bowl Sunday	662	$\frac{1}{3} \cdot 1598 = 532.667$
Control Sunday	936	$\frac{2}{3} \cdot 1598 = 1065.333$
Total	1598	1598

Check Conditions: Here you do not have a random sample, you have the population of all fatalities within the first four hours after the Superbowl telecast, as well as the fatalities

265

during the same hours on the previous and the following Sundays. There is no greater population about which to make an inference. What you want in this case is to see whether the difference between the observed numbers and the model could be reasonably attributed to chance. In other words, if the true probability of a fatality within four hours after the Superbowl is the same as during those hours on any other Sunday, are the observed numbers reasonably likely? Each fatality falls into one of the two categories (Superbowl Sunday or control Sunday), you do have a hypothesized proportion for each category, and the expected frequencies are each at least 5 as shown in the table above. The conditions are met for a chi-square goodness of fit test.

State your hypotheses:
H_0: The difference between the observed number of traffic fatalities during the first four hours after the Superbowl and the traffic fatalities on the previous and the following Sunday can be attributed to chance variation.

H_a: The difference between the observed number of traffic fatalities during the first four hours after the Superbowl and the traffic fatalities on the previous and the following Sunday can not be attributed to chance variation.

Calculate the test statistic, find the P-*value, and draw a sketch:*
This test could be done as a z-test for a proportion, but we will do a chi-square goodness of fit test.

$$\chi^2 = \sum \frac{(O-E)^2}{E} = \frac{(662-532.667)^2}{532.667} + \frac{(936-1065.333)^2}{1065.333} \approx 47.10$$

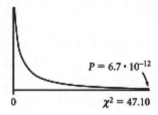

Comparing the test statistic to the χ^2 distribution with $2 - 1 = 1$ degree of freedom, you find that a χ^2 value of 47.10 gives a P-value of $6.7 \cdot 10^{-12}$.

Write your conclusion in context, linked to your computations: With a P-value so close to zero, you would reject the null hypothesis that the difference in the number of traffic fatalities during the first four hours after a Superbowl and (half) the number of fatalities during the same four hours on the previous and following Sundays can reasonably be attributed to chance variation. If the probabilities were equal, numbers as far from the predicted frequencies as observed in the data would be extremely unlikely. You have evidence that the difference can not be reasonably attributed to chance alone.

b. There were a total of 406 accidents in the study. The model says twice as many fatalities should occur on the two control Sundays as on Superbowl Sunday. Put these

266

numbers in a table:

Sunday	Observed Number of Accidents	Expected Number of Accidents
Super Bowl Sunday	141	$\frac{1}{3} \cdot 406 = 135.33$
Control Sunday	265	$\frac{2}{3} \cdot 406 = 270.67$
Total	406	406

Check Conditions: Here you do not have a random sample, you have the population of all accidents within the home state of the winning Super Bowl team within the first four hours after the Superbowl telecast, as well as the accidents during the same hours on the previous and the following Sundays. There is no greater population about which to make an inference. What you want in this case is to see whether the difference between the observed numbers and the model could be reasonably attributed to chance. In other words, if the true probability of an accident within four hours after the Superbowl is the same as during those hours on any other Sunday, are the observed numbers reasonably likely? Each accident falls into one of the two categories (Superbowl Sunday or control Sunday), you do have a hypothesized proportion for each category, and the expected frequencies are each at least 5 as shown in the table above. The conditions are met for a chi-square goodness of fit test.

State your hypotheses:
H_0: The difference between the observed number of traffic accidents during the first four hours after the Superbowl and the traffic accidents on the previous and the following Sunday can be attributed to chance variation.
H_a: The difference between the observed number of traffic accidents during the first four hours after the Superbowl and the traffic accidents on the previous and the following Sunday can not be attributed to chance variation.

Calculate the test statistic, find the P-*value, and draw a sketch:*
This test could be done as a z-test for a proportion, but we will do a chi-square goodness of fit test.

$$\chi^2 = \sum \frac{(O-E)^2}{E} = \frac{(141-135.333)^2}{135.333} + \frac{(265-270.667)^2}{270.667} \approx 0.356$$

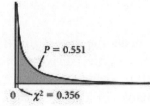

Comparing the test statistic to the χ^2 distribution with $2 - 1 = 1$ degree of freedom, you find that a χ^2 value of 0.356 gives a *P*-value of 0.551.

Write your conclusion in context, linked to your computations: With a *P*-value so large, you would not reject the null hypothesis that the difference in the number of traffic accidents during the first four hours after a Super Bowl and (half) the number of accidents during the same four hours on the previous and following Sundays can reasonably be attributed to chance variation. If the probability of an accident fits the 1:2 model, numbers as close to the predicted frequencies as observed in the data would be very likely. The observed difference in the number of accidents can be reasonably attributed to chance alone.

E11. a. Assuming that the 18 college-educated participants were randomly selected from the population of college-educated persons, then this condition is met. However, this is not explicitly stated. Each outcome falls into exactly one of a fixed number of categories. The null hypothesis suggests that the participants randomly selected which item was the dog food. As such the expected frequencies for selecting dog food should be equal across all five foods. However, the fourth assumption of having an expected frequency of at least 5 in each category is not met as shown below.

Food Type	Observed Frequency, O	Expected Frequency, E	O-E	$(O-E)^2 / E$
Duck liver mousse	1	3.6	-2.6	1.877778
Spam	4	3.6	0.4	0.044444
Dog food	3	3.6	-0.6	0.1
Park liver pâté	2	3.6	-1.6	0.711111
Liverwurst	8	3.6	4.4	5.377778
Total	18	18	0	8.111111

b. Summing the values in the right column in the above table, we see that $\chi^2 = 8.11$ with df $= 5 - 1 = 4$. As such, the p-value > 0.05, which is not significant.

c. The researchers treated the total observed frequencies and total expected frequencies as 1 and then calculated the percentage of 1 that each group represented. Instead, they should have treated the total observed frequencies as 18 – the total number of observations.

E13. a. Each digit should occur $\frac{1}{10}$ of the time.

b. You will perform a chi-square goodness-of-fit test on the data. Each expected frequency is $0.1(104) = 10.4$.

Check conditions: This situation meets the criteria for a chi-square goodness-of-fit test. There are 104 measurements of the lengths of rivers, presumably independent measurements. However, note that the rivers were not randomly selected—they are 104 of the most important rivers. Perhaps you can view their final digits, however, as a random sample. Each measurement ends in exactly one of the digits. You have a model that gives the hypothesized proportion of outcomes in the population that fall into each category. The expected frequency in each category is 10.4 which is at least 5.

State your hypotheses:

H_0: The probability that the length of a river ends in any given digit is $\frac{1}{10}$.

H_a: The probability is not $\frac{1}{10}$ for all of the digits.

Compute the test statistic, find the P-*value, and draw a sketch:* The test statistic is

$$\chi^2 = \sum \frac{(O-E)^2}{E} = \frac{(44-10.4)^2}{10.4} + \frac{(6-10.4)^2}{10.4} + \frac{(9-10.4)^2}{10.4} + \cdots + \frac{(9-10.4)^2}{10.4} \approx 131.77$$

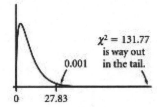

Comparing the test statistic to the χ^2 distribution with $10-1=9$ degrees of freedom, you find that the value of χ^2 from the sample, 131.77, is very far out in the tail. The P-value is 5.13×10^{-24}.

Write a conclusion in context: You should reject the null hypothesis. You can not attribute the differences to chance variation. A value of χ^2 this large or larger is extremely unlikely to occur if rivers are accurately measured to the nearest mile and each digit is equally likely to be the final digit. It appears that the measurements were approximated in some cases and not measured to the nearest mile, as there are too many that end in 0 and in 5.

E15. a. These are the four populations of all ratings given to the nominees of the four presidents. The samples of presidential nominations to judgeships are independent, however not random samples. Each outcome/rating falls into exactly one of the rating categories. The expected frequency in each cell is at least 5 as shown below.

b.

Expected Frequencies

President	Not Qualified or Split Between Qualified/Not Qualified	Qualified or Qualified/Well-Qualified	Well-Qualified or Well-Qualified/Qualified	Total
Reagan	6.769716088	14.27129338	36.95899054	58
G.H.W. Bush	6.186119874	13.04100946	33.77287066	53
Clinton	10.50473186	22.14511041	57.35015773	90
G.W. Bush	13.53943218	28.54258675	73.91798107	116
Total	37	78	202	317

The calculations needed to compute the test statistic are:

Chi-square Calculations -- (O-E)2 / E			
President	Not Qualified or Split Between Qualified/Not Qualified	Qualified or Qualified/Well-Qualified	Well-Qualified or Well-Qualified/Qualified	Total
Reagan	5.73383538	0.037208495	1.310304978	7.081349
G.H.W. Bush	0.772555366	1.885711931	0.227662368	2.88593
Clinton	5.361488618	0.033002148	0.771059819	6.165551
G.W. Bush	0.884492754	1.499704346	0.128505142	2.512702
Total	12.75237212	3.45562692	2.437532307	18.64553

Summing the values in the right column of the table below, we see that

$$\chi^2 = \sum \frac{(O-E)^2}{E} = 18.65$$

with df = (4-1)* (3–1) = 6. As such, the p-value < 0.005.

Based on the chi-square results, the differences in the proportions of nominees who received ratings for the different presidents cannot reasonably attributed to chance alone. However, the conditions for using the chi-square test of homogeneity were not met as discussed above.

E17. a. In the table below, the new category appears entirely in CAPS.

Observed Frequencies:

	Domestic Veg.	Imported Veg.	Total
No Residue	0.738(672)=496	0.604(2447)=1478	1974
Residue in Violation	0.024(672)=16	0.054(2447)=132	148
SOME RESIDUE, BUT NOT IN VIOLATION	0.238(672)=160	0.342(2447)=837	997
Total	672	2447	3119

b. The expected frequencies are given by

	Domestic Veg.	Imported Veg.	Total
No Residue	$\frac{(1974)(672)}{3119} = 425.31$	1548.69	1974
Residue in Violation	31.89	116.11	148
SOME RESIDUE, BUT NOT IN VIOLATION	214.81	782.19	997
Total	672	2447	3119

c. The population is all domestic vegetables sent to market and all imported vegetables sent to market. Also, the following conditions are indeed met:

- You should have independent random samples of fixed (but not necessarily equal) sizes taken from two or more large populations (or two or more treatments are randomly assigned to subjects).

• Each outcome falls into exactly one of several categories, with the categories being the same in all populations.

• The expected frequency in each cell is 5 or greater.

d. The value of the test statistic is given by

$$\chi^2 = \sum \frac{(O-E)^2}{E} = 42.89.$$

Since df = (3-1)(2-1)=2, the p-value is nearly zero.

e. The proportion of domestic vegetables and the proportion of imported vegetables are different for at least one category of pesticide residue. The imported vegetables show higher amounts of residue.

E19. a. The populations are all children aged 2 through 6 whose parents when to high school or had job training; all children where one parent went to college; all children where both parents went to college.

b. The observed frequencies are:

	Purchased	Did Not Purchase	Total
Parents have HS or job training	9	7	16
One parent went to college	10	16	26
Both parents went to college	15	63	78
Total	34	86	120

c.

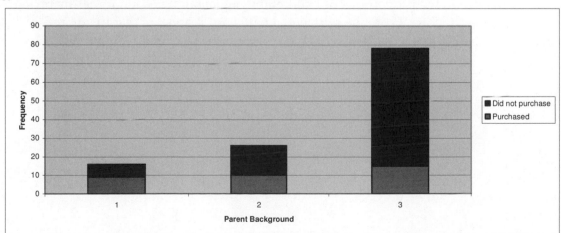

Here: 1 = HS or job training, 2 = One attended college, 3 = Both attended college

d. The expected frequencies are given by

	Purchased	Did Not Purchase	Total
Parents have HS or job training	$\frac{(16)(34)}{120}=4.53$	11.47	16
One parent went to college	7.37	18.63	26
Both parents went to college	22.1	55.9	78
Total	34	86	120

We test:

Ho: The proportion of all children in each category of parent education who would buy cigarettes is the same.

Ha: The proportion who would buy cigarettes in at least one category of parent education is different from the proportion in one of the other categories.

Check Conditions: This isn't a random sample from all children in these three categories, but are the children available for the study. One expected count is 4.5, but this isn't a serious problem.

The test statistic is given by

$$\chi^2 = \sum \frac{(O-E)^2}{E} = 10.64.$$

Since df = (3-1)(2-1)=2, we see that the p-value is between 0.0025 and 0.005.

e. These aren't random samples from the three populations, so you can conclude only that the difference in the proportions of children buying cigarettes cannot reasonably be attributed to chance alone. From the segmented bar chart, it appears that the more education the parents have, the less likely the children are to purchase cigarettes.

E21. a. The table of observed frequencies is given here.

	Carbolic Acid Used	Carbolic Acid Not Used	Total
Patient Lived	34	19	53
Patient Died	6	16	22
Total	40	35	75

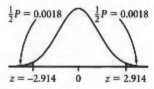

From this plot, it seems clear that there should be a smaller proportion of deaths in operations where carbolic acid is used. This difference looks unlikely to have occurred by chance. However, there is an important caveat here. You do not know how Lister selected the patients who would go into each treatment.

b. The test statistic z is

$$z = \frac{(\hat{p}_1 - \hat{p}_2) - (p_1 - p_2)}{\sqrt{\hat{p}(1-\hat{p})\left(\frac{1}{n_1} + \frac{1}{n_2}\right)}} = \frac{(0.85 - 0.543) - 0}{\sqrt{0.707(1-0.707)\left(\frac{1}{40} + \frac{1}{35}\right)}} \approx 2.914$$

(or $z = 2.915$ with no rounding)

$$\tfrac{1}{2}P = 0.0018 \qquad \tfrac{1}{2}P = 0.0018$$

$$z = -2.914 \qquad 0 \qquad z = 2.914$$

With a test statistic of 2.915, the *P*-value for a two-tailed test is 2(0.0018) = 0.0036. If all subjects could have been assigned each treatment and which treatment the subject received made no difference in survival, the probability of getting a difference in sample proportions as large or larger in absolute value as from the actual experiment is only 0.0036 (provided some randomization was used).

c. The table of expected frequencies is given here.

	Carbolic Acid Used	Carbolic Acid Not Used	Total
Patient Lived	28.27	24.73	53
Patient Died	11.73	10.27	22
Total	40	35	75

The test statistic is given by

$$\chi^2 = \sum \frac{(O-E)^2}{E} = \frac{(34-28.27)^2}{28.27} + \frac{(19-24.73)^2}{24.73} + \frac{(6-11.73)^2}{11.73} + \frac{(16-10.27)^2}{10.27} \approx 8.495$$

Comparing the test statistic to a chi-square distribution with $df = 1$, you see that this χ^2 value is in the tail. The P-value is 0.0036. If all subjects could have been assigned each treatment and which treatment the subject received made no difference in survival, the probability of getting a χ^2 statistic as large as or larger than from the actual experiment is only 0.0036.

d. The two P-values are the same. The values of z^2 and χ^2 are the same as well if nothing is rounded.

E23. The observed and expected frequencies are given in the next two tables.

Observed Frequencies

	No Bandage	Skin-Colored Bandage	Brightly Colored Bandage	Total
Pain Gone	0	9	15	24
Almost Gone	5	9	2	16
Still There	15	2	3	20
Total	20	20	20	60

Expected Frequencies

	No Bandage	Skin-Colored Bandage	Brightly Colored Bandage	Total
Pain Gone	8	8	8	24
Almost Gone	5.33	5.33	5.33	16
Still There	6.67	6.67	6.67	20
Total	20	20	20	60

Here is the segmented bar graph of observed frequencies.

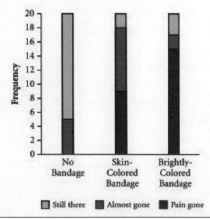

274

Check conditions: This situation fits the criteria for a chi-square test of homogeneity. The treatments were randomly assigned to the available subjects. Each answer falls into exactly one of three categories. The expected number of answers in each cell is 5 or more as seen in the table.

State your hypotheses:

H_0: Suppose all 60 children could have been given the no-bandage treatment, all could have been given the flesh-colored bandage treatment, and all could have been given the brightly colored bandage treatment. Then the proportion of answers that would have fallen into each of the three categories would have been the same for all three types of bandage treatments.

H_a: For at least one answer, the proportion of children who give that answer is not the same for all three types of bandage.

Compute the test statistic and draw a sketch: The test statistic is

$$\chi^2 = \sum \frac{(O-E)^2}{E} = \frac{(0-8)^2}{8} + \frac{(9-8)^2}{8} + \frac{(15-8)^2}{8} + \cdots + \frac{(3-6.67)^2}{6.67} \approx 34.575$$

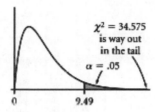

Comparing the test statistic to the χ^2 distribution with 4 degrees of freedom, you find that the value of χ^2 from the sample, 34.575, is extremely far out in the tail and certainly greater than 9.49 ($\alpha = 0.05$). The *P*-value is close to 0. The calculator gives a *P*-value of $5.7 \cdot 10^{-7}$.

Write a conclusion in context: You reject the null hypothesis. You can not reasonably attribute the differences in the answers to the random assignment of treatments to children. A value of χ^2 this large or larger is extremely unlikely to occur if the distribution of answers for all 60 children would have been the same for each type of bandage. You conclude that these children given different bandages tend to give different answers to the questions.

E25. The data are not in the right format for a chi-square test of homogeneity. The numbers in the cells are the percentages by year, not frequencies. The estimated observed frequencies and the corresponding expected frequencies are shown in the table below. (NOTE: You need only consider the first and last rows – the other data is included just as a fyi.)

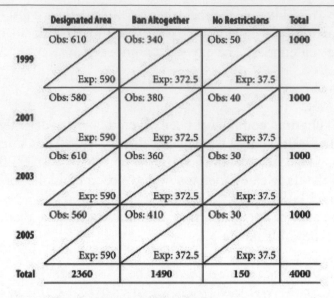

	Designated Area	Ban Altogether	No Restrictions	Total
1999	Obs: 610 / Exp: 590	Obs: 340 / Exp: 372.5	Obs: 50 / Exp: 37.5	1000
2001	Obs: 580 / Exp: 590	Obs: 380 / Exp: 372.5	Obs: 40 / Exp: 37.5	1000
2003	Obs: 610 / Exp: 590	Obs: 360 / Exp: 372.5	Obs: 30 / Exp: 37.5	1000
2005	Obs: 560 / Exp: 590	Obs: 410 / Exp: 372.5	Obs: 30 / Exp: 37.5	1000
Total	2360	1490	150	4000

Here is a segmented bar chart.

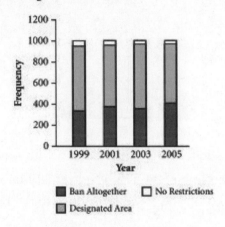

Our visual impression from this chart is that there is little difference in the proportions who give each answer over the various years this question was asked. However, again the sample sizes are quite large, so small differences in percentages may be statistically significant.

Check conditions: The conditions essentially are met in this situation for a chi-square test of homogeneity. You can think of each year's survey as basically four large populations and that a random sample of size 1000 was taken independently from each population. Each response falls into exactly one of four categories. All of the expected frequencies are at least 5 as shown above.

State your hypotheses:
H_0: If in 1999 and 2005, you had asked all adults in the United States this question, the distribution of responses would be the same for each year.
H_a: The distribution of responses would not be the same in each year. That is, in at least one year, the proportion of all adults who would give one of the responses is different from the proportion in other year.

276

Compute the test statistic: The test statistic is

$$\chi^2 = \sum \frac{(O-E)^2}{E} = \frac{(610-590)^2}{590} + \cdots + \frac{(30-37.5)^2}{37.5} \approx 13.67$$

Comparing the test statistic to the χ^2 distribution with 2 degrees of freedom, you find that the value of χ^2 from the sample, 13.67, is reasonably far out in the tail. From the table, the *P*-value is about 0.001.

Write a conclusion in context: You should reject the null hypothesis. You cannot attribute the differences in the answers to the fact that you have only a sample of adults for each year and not the entire population. A value of χ^2 this large or larger is extremely unlikely to occur in samples of this size if the distribution of responses is the same for each year. You conclude that the proportion of responses in at least one category changed in from 1999 to 2005.

E27. a. A one-sided *z*-test for a difference in proportions is most appropriate in this case. Let p_1 refer to the proportion of men in the U.S. who report drinking at least several alcoholic drinks per week and p_2 the same proportion for women.

$$z = \frac{(\hat{p}_1 - \hat{p}_2) - (p_1 - p_2)}{\sqrt{\hat{p}(1-\hat{p})\left(\frac{1}{n_1} + \frac{1}{n_2}\right)}} = \frac{(0.26 - 0.13) - 0}{\sqrt{0.195 \cdot 0.805\left(\frac{1}{500} + \frac{1}{500}\right)}} \approx 5.188$$

You could also do a chi-square test of homogeneity if you're careful. A chi-square test does not technically apply to a one-sided hypothesis. However, since you are only comparing two proportions you could use a chi-square test if you divide the *P*-value by 2 and your alternative hypothesis is in the same direction the data indicate. In this case, the alternative hypothesis is that men drink more, and the data in the sample show men drinking more. You must first convert the percents to frequencies.

	Report Drinking at Least Several Drinks per Week	Do Not Report Drinking at Least Several Drinks per Week	Total
Men	130	370	500
Women	65	435	500
Total	195	805	1000

The test statistic is given by

$$\chi^2 = \sum \frac{(O-E)^2}{E} = \frac{(130-97.5)^2}{97.5} + \frac{(370-402.5)^2}{402.5} + \frac{(65-97.5)^2}{97.5} + \frac{(435-402.5)^2}{402.5} \approx 26.92$$

Note that $z^2 = \chi^2$. The *P*-value for the one-sided *z*-test is half the *P*-value for the χ^2 test.

b. Here you must use a chi-square test of homogeneity. The percents must be converted to frequencies.

	Great Britain		Canada		United States		Total
Men Who Drink at Least Several Alcoholic Beverages per Week	Obs: 240	Exp: 185	Obs: 185	Exp: 185	Obs: 130	Exp: 185	555
Men Who Do Not Drink at Least Several Alcoholic Beverages per Week	Obs: 260	Exp: 315	Obs: 315	Exp: 315	Obs: 370	Exp: 315	945
Total	500		500		500		1500

$$\chi^2 = \sum \frac{(O-E)^2}{E} = \frac{(240-185)^2}{185} + \frac{(185-185)^2}{185} + \frac{(130-185)^2}{185} + \frac{(260-315)^2}{315} + \frac{(315-315)^2}{315}$$
$$+ \frac{(370-315)^2}{315} \approx 51.91$$

c. Here, as in part a, you could use either a chi-square test of homogeneity or a z-test for the difference of two proportions. In this case the test is two-sided so no adjustment to the P-value is needed.

Chi-square:

	Canada		United States		Total
Women Who Drink Regularly	Obs: 85	Exp: 75	Obs: 65	Exp: 75	150
Women Who Do Not Drink Regularly	Obs: 415	Exp: 425	Obs: 435	Exp: 425	850
Total	500		500		1000

The test statistic is given by

$$\chi^2 = \sum \frac{(O-E)^2}{E} = \frac{(85-75)^2}{75} + \frac{(65-75)^2}{75} + \frac{(415-425)^2}{425} + \frac{(435-425)^2}{425} \approx 3.14$$

z-test:

$$z = \frac{(\hat{p}_1 - \hat{p}_2) - (p_1 - p_2)}{\sqrt{\hat{p}(1-\hat{p})\left(\frac{1}{n_1} + \frac{1}{n_2}\right)}} = \frac{(0.17 - 0.13) - 0}{\sqrt{0.15 \cdot 0.85 \left(\frac{1}{500} + \frac{1}{500}\right)}} \approx 1.77$$

E29. a. No. Not every member of each population falls into exactly one of the categories. Some Internet users will fall into more than one category and some users will fall into none of them.

b. Yes. You need to also add a row for those that have not used the internet for sending and reading e-mail if you plan to use a chi-square test, or you could use a z-test for the difference of two proportions. Here is the chi-square test:

	2003	2005	Total
Used the Internet for Sending and Reading Email	Obs: 630 / Exp: 641.5	Obs: 653 / Exp: 641.5	1283
Did Not Use the Internet for Sending and Reading Email	Obs: 120 / Exp: 108.5	Obs: 97 / Exp: 108.5	217
Total	750	750	1500

The test statistic is given by

$$\chi^2 = \sum \frac{(O-E)^2}{E} = \frac{(630-641.5)^2}{641.5} + \frac{(503-641.5)^2}{641.5} + \frac{(120-108.5)^2}{108.5} + \frac{(247-108.5)^2}{108.5}$$
$$\approx 2.85$$

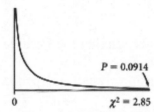

$P = 0.0914$

$\chi^2 = 2.85$

Comparing the test statistic from these samples, 2.85, with a chi-square distribution with 1 degree of freedom, this value of χ^2 lies in the tail, but not far into the tail. The P-value is 0.0914. This is not statistically significant at the 0.05 significance level. You do not have strong evidence that the proportion of the population that uses the internet for sending and reading e-mail has changed. (Note that, had you done a one-sided z-test, with the alternative hypothesis that the proportion has increased, you would have statistically significant evidence against the null hypothesis.)

E31. a. The observed frequencies are

	In a Romantic Relationship	Not in a Romantic Relationship	Total
Muscular	0.37(121)=45	0.63(121)=76	121
Non-muscular	0.16(38)=6	0.84(38)=32	38
Total	51	108	159

The proportion of all males in a romantic relationship is $\frac{51}{159} = 0.32$, or about 32%.

b. H_0: If all male characters in these types of films were analyzed, the proportion of muscular characters who were in a romantic relationship would be equal to the proportion of non-muscular characters who were in a romantic relationship (and the proportions not in a romantic relationship also would be equal). Alternatively, muscularity and romantic status are independent.

H_a: The proportions are different. (Because the chi-square test is two-sided.)

c. The expected frequencies are

	In a Romantic Relationship	Not in a Romantic Relationship	Total
Muscular	$\frac{(121)(51)}{159} = 38.81$	82.19	121
Non-muscular	12.19	25.81	38
Total	51	108	159

The test statistic is given by

$$\chi^2 = \sum \frac{(O-E)^2}{E} = 6.08.$$

d. Since $df = (2-1)(2-1) = 1$, we see that the P-value is 0.0137 from the calculator or between 0.01 and 0.02 from Table C. So, the result is significant and we conclude that muscularity and romantic relationship are not independent.

E33. a. The proportions in each category under the assumption of independence are given by

	Business	Humanities	Social and Behavioral Sciences	Other	Total
Men	0.0946	0.0774	0.0731	0.1849	0.43
Women	0.1254	0.1026	0.0969	0.2451	0.57
Total	0.22	0.18	0.17	0.43	1.00

b. The observed frequencies are given by

	Business	Humanities	Social and Behavioral Sciences	Other	Total
Men	144,179	177,965	111,411	281,804	615,360
Women	191,121	156,372	147,684	373,555	868,732
Total	335,300	334,337	159,095	655,360	1,484,092

c. The variables are associated, not independent. For example, men are less likely to major in business than are women.

E35. a. First, to get the observed frequencies, multiply the decimal version of each cell by 556:

	Men	Women	Total
Advantage Being a Man	178	267	445
Advantage Being a Woman	50	22	72
No Advantage	322	256	578
Unsure	6	11	17
Total	556	556	1112

Now, the expected frequencies are as follows:

	Men	Women	Total
Advantage Being a Man	$\frac{(445)(556)}{1112} = 222.5$	222.5	445
Advantage Being a Woman	36	36	72
No Advantage	289	289	578
Unsure	8.5	8.5	17
Total	556	556	1112

We test:

H_0: Gender and opinion on this issue are independent.

H_a: Gender and opinion on this issue are associated.

The following conditions are indeed met:

- You should have independent random samples of fixed (but not necessarily equal) sizes taken from two or more large populations (or two or more treatments are randomly assigned to subjects).

- Each outcome falls into exactly one of several categories, with the categories being the same in all populations.

- The expected frequency in each cell is 5 or greater.

The test statistic is given by

$$\chi^2 = \sum \frac{(O-E)^2}{E} = 37.70.$$

Since df = (4-1)(2-1)=3, we see that the p-value is close to zero. Thus, we reject the null hypothesis and conclude that gender and opinion on this issue are associated.

b. Yes, because the H_o was rejected.

c. Take a random sample of 556 men and then an independent random sample of 556 women.

E37. a.

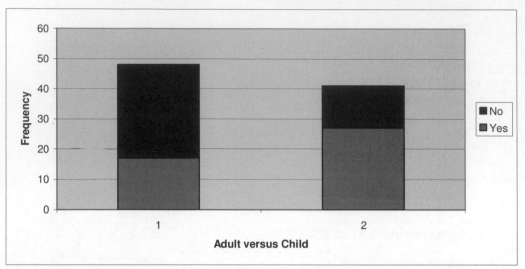

Here: 1 = Adult, 2 = Child

Observe that a far lower proportion of adults survived than children.

b. The expected frequencies are given by

	Adult	**Child**	**Total**
Yes	$\frac{(48)(44)}{89} = 23.73$	24.27	48
No	20.27	20.73	41
Total	44	45	89

Note that this is not a random sample, but the expected number in each cell is 5 or more. So, we will run the test anyway.
The test statistic is given by

$$\chi^2 = \sum \frac{(O-E)^2}{E} = 8.20 .$$

Since df = (2-1)(2-1)=1, we see that the p-value is 0.0042. So, we reject Ho that the difference in the proportions of adults and children who survived can reasonably be attributed to chance alone.

E39. a. Observational study.

b. Homogeneity. There are two populations, Super Bowl Sundays and Control Sundays. The idea is to see if the proportions are equal between these two populations. If it were a test of independence, they would have taken a single sample of Sundays and sorted them into non-Super Bowl and Super Bowl. Because the data is not for a sample of Sundays, but for all Super Bowl Sundays and Control Sundays, you must carefully state your hypotheses and conclusions. You are not inferring about a larger population, but trying to determine if the difference in the proportion of fatalities in the two populations can

282

reasonably be attributed to chance.

c. The table below includes the observed and expected frequencies for a chi-square test of homogeneity.

	Super Bowl Sundays	Control Sundays	Total
Killed	Obs: 284 Exp: 208.14	Obs: 163 Exp: 238.86	447
Survived	Obs: 48 Exp: 123.86	Obs: 218 Exp: 142.14	266
Total	332	381	713

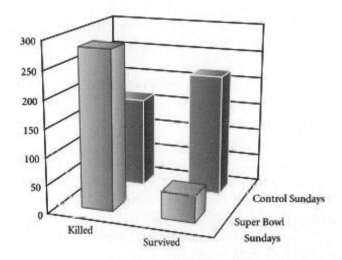

Examining the column chart, you find that the rows do not appear to have the same shape. Super Bowl Sundays have far more that didn't survive proportionally. It is not clear whether the differences will turn out to be statistically significant from looking at this chart.

Check conditions: You do not have a random sample of Sundays, or even of Super Bowl Sundays. You have all Super Bowl Sundays and all Sundays before and after Super Bowl Sundays. The question is whether the observed difference in frequencies can reasonably be attributed to chance under the hypothesis that fatalities in alcohol related injuries are equally likely on the two types of Sundays. It is the case that all people involved in alcohol related accidents on one of the two types of Sundays fall into one of the two categories, and all expected frequencies are at least 5 as shown above.

State your hypotheses:
H_0: The difference in the proportions of fatal alcohol related accidents for a Super Bowl Sunday and the Sunday before or the Sunday after the Super Bowl can reasonably be attributed to chance.
H_a: The difference in the proportions of fatal alcohol related accidents for a Super Bowl

Sunday and the Sunday before or the Sunday after the Super Bowl can not reasonably be attributed to chance and you should look for another explanation.

Compute the test statistic, find the P-*value, and draw a sketch:*

$$\chi^2 = \sum \frac{(O-E)^2}{E} = \frac{(284-208.14)^2}{208.14} + \frac{(163-238.86)^2}{238.86} + \frac{(48-123.86)^2}{123.86}$$

$$+ \frac{(218-142.14)^2}{142.14} \approx 138.69$$

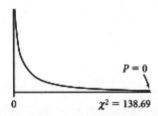

Comparing the test statistic to a chi-square distribution with 1 degree of freedom, you see that the value of χ^2 from this sample, 138.69, is extremely far out in the tail. The *P*-value is extremely close to zero.

Write your conclusion in context, linked to your computations: If the probability that a person in an alcohol related accident is killed is the same the first four hours after a Super Bowl as it is during the same four hours on a Sunday before or after a Super Bowl, a chi-square statistic as high as or higher than 138.69 is extremely unlikely. You should reject the null hypothesis. Something other than random variation must account for the high incidence of fatalities following a Super Bowl.

E41. **a.** The test should be a one-sided two-sample z-test of the difference of two proportions, because the conclusion is based on a one-sided test,

b. We test:
H$_0$: $\hat{p}_1 = \hat{p}_2$ against H$_a$: $\hat{p}_1 < \hat{p}_2$, where $\hat{p}_1, \hat{p}_2$ are the proportion who survived when defibrillation began after 2 minutes or no later than 2 minutes.

Here,
$$\hat{p}_1 = \tfrac{455}{2045} = 0.222, \quad \hat{p}_2 = \tfrac{1863}{4744} = 0.393, \quad \hat{p} = \tfrac{1863+455}{4744+2045} = 0.341, \quad n_1 = 4744, \quad n_2 = 2045 .$$

The test statistic is given by

$$z = \frac{(\hat{p}_1 - \hat{p}_2) - 0}{\sqrt{\hat{p}(1-\hat{p})\left(\frac{1}{n_1} + \frac{1}{n_2}\right)}} = -13.57 .$$

So, P(Z<-13.57) is nearly zero. Hence, we reject the H$_0$ that the proportion of all such patients who would survive to discharge is the same whether defibrillation begins in 2 minutes or less or not. This assumes that the patients can be considered a random sample of all such patients. If not, conclude only that the difference in the two proportions who survive cannot reasonably be attributed to chance alone.

E43. a.

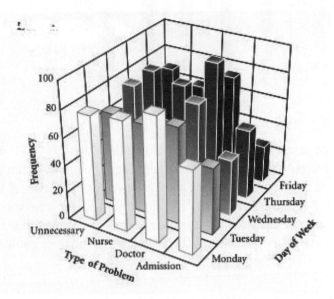

b. The infirmary was used most heavily on Monday and Thursday. There were relatively few admissions on Friday. Perhaps students saved their problems until Thursday after classes were over and perhaps the infirmary wasn't open on the weekends, so students who got sick on the weekend had to save their problems until Monday.

c. Yes, you could use a test of independence because there was one large population that was classified according to day and severity of problem. However, some students may consider it a test of homogeneity even though that test requires that independent random samples be drawn from two or more populations. They could argue that the five populations of students were all of the students who came on the five days and they then were classified according to severity of problem. The design of the study involves no random sampling so does not fit the conditions for a chi-square test.

d. A test of independence is the more appropriate test.

First, examining the column chart above or the segmented bar chart here, you see that there are indeed more visits on Monday and Thursday. However, the pattern of admissions appears roughly to be the same from day to day, which is what a test of independence will test.

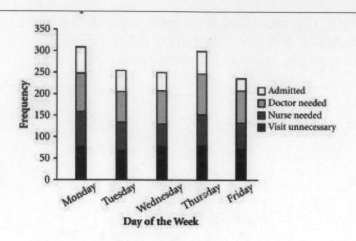

Check conditions: Because the table includes all students who visited the infirmary during a school year, these data can not reasonably be considered a simple random sample taken from one large population. Thus, you will test whether it is reasonable to attribute the difference in the proportions to chance or whether you should look for some other explanation. Each person who visited the infirmary was classified according to day of the week and severity of the problem. The expected number in each cell is 5 or more. (See the printout below for expected frequencies.)

State your hypotheses:
H_0: The differences in the proportions of the various problems from day to day can reasonably be attributed to chance.
H_a: The differences in the proportions can not reasonably be attributed to chance, and you should look for some other explanation.

Compute the test statistic and draw a sketch: This Minitab printout shows expected frequencies under the assumption of independence. It also gives the computation of the test statistic χ^2 and the *P*-value.

Chi-Square Test: Expected frequencies are printed below observed frequencies.

```
          Mon     Tues    Wed     Thurs   Fri     Total
1         77      67      77      78      70      369
          84.69   69.57   68.19   81.94   64.62
2         80      66      53      73      62      334
          76.66   62.97   61.72   74.17   58.49
3         90      71      76      95      75      407
          93.41   76.73   75.21   90.38   71.27
4         61      49      42      52      28      232
          53.25   43.74   42.87   51.52   40.63
Total     308     253     248     298     235     1342

ChiSq = 0.698 + 0.095 + 1.138 + 0.189 + 0.449 +
0.146 + 0.146 + 1.233 + 0.018 + 0.211 + 0.124 +
0.428 + 0.008 + 0.236 + 0.195 + 1.129 + 0.633 +
0.018 + 0.005 + 3.924 = 11.023

df = 12, P = 0.527
```

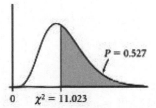

Comparing the test statistic to the χ^2 distribution with 12 degrees of freedom, you can see from the printout that the value of χ^2 from the sample, 11.023, is not out in the tail. The *P*-value is 0.527.

Write a conclusion in context: You should not reject the null hypothesis. These are typical of the results you would expect to see if there was no association between severity of problem and day of the week. No further explanation for these results is needed.

E45. To use a chi-square test, set up a table. For the remaining four outcomes (2, 3, 4, or 5), there are a total of 78 rolls.

Outcome	Observed Count	Expected Count
2	18	78 / 4 = 19.5
3	28	78 / 4 = 19.5
4	10	78 / 4 = 19.5
5	22	78 / 4 = 19.5
Total	78	78

Check conditions: You are considering only the die rolls that are not a 1 or a 6. They will each fall into one of the other four categories. You have a model that gives the hypothesized proportion of outcomes that fall in each category. Rolling the die is a random event, so you have an independent random sample, and each expected frequency is more than 5 as shown in the table above.

State your hypotheses:
H_0: The shaved die is fair with respect to the outcomes 2, 3, 4, and 5.
H_a: The die is not fair with respect to these four outcomes.

Calculate the test statistic, find the P-*value, and draw a sketch:*
The test statistic is

$$\chi^2 = \sum \frac{(O-E)^2}{E} = \frac{(18-19.5)^2}{19.5} + \frac{(28-19.5)^2}{19.5} + \frac{(10-19.5)^2}{19.5} + \frac{(22-19.5)^2}{19.5} \approx 8.77$$

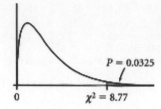

$P = 0.0325$

0 $\chi^2 = 8.77$

Comparing the test statistic to the χ^2 distribution with $4 - 1 = 3$ degrees of freedom, you find that $\chi^2 = 8.77$ gives a *P*-value of 0.0325.

Write your conclusion in context, linked to your computations: The *P*-value is less than the commonly used 0.05 so you would reject the null hypothesis at the 5% significance level. If the die was indeed fair with respect to these values, there would only be a 3.25% chance of getting a chi-square value as extreme as or more extreme than 8.77. So, there is statistically sufficient evidence that the die was unfair with regard to the outcomes 2, 3, 4, and 5 in addition to the 1 and 6.

E47. a.

	Right	**Left**	**Total**
Married	$\frac{(1678)(2150)}{2477} = 1456.5$	221.5	1678
Not Married	693.5	105.5	799
Total	2150	327	2477

b.

	Right	**Left**	**Total**
Married	1462	216	1678
Not Married	688	111	799
Total	2150	327	2477

c. We test:
H_0: *Handedness* and *Marital status* are independent.
H_a: *Handedness* and *Marital status* are associated.

The test statistic is given by

$$\chi^2 = \sum \frac{(O-E)^2}{E} = 0.49.$$

Since df = (2-1)(2-1)=1, we see that the p-value is about 0.4833. So, we do not reject the null hypothesis and conclude that *Handedness* and *Marital status* are independent.

E49. For this goodness-of-fit test, the probabilities are hypothesized to be 0.5, 0.25, and 0.25 for the respective categories. The χ^2 statistic is 0.88, with 2 degrees of freedom, producing a *P*-value of 0.644. There is not statistically sufficient evidence to reject the percentages claimed.

E51. *Check conditions:* This sample can not reasonably be considered a simple random sample taken from one large population. Thus, you should test whether it is reasonable to attribute the difference in the proportions to chance or whether you should look for some other explanation. Each person on the *Titanic* was classified according to class of travel and survival status. The expected number in each cell is 5 or more. (See the next printout for expected frequencies.)

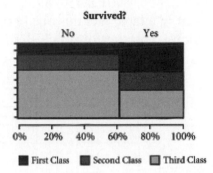

State your hypotheses:

H$_0$: The differences in the proportions of passengers of various classes of travel who were saved can reasonably be attributed to chance.

H$_a$: The differences in the proportions of passengers of various classes of travel who were saved can not reasonably be attributed to chance, and you should look for some other explanation.

Compute the test statistic, find the P*-value, and draw a sketch:* The following printout shows expected frequencies under the assumption of independence. It also gives the computation of the test statistic χ^2 and the *P*-value.

Chi-Square Test: Expected frequencies are printed below observed frequencies.

```
              First     Second    Third     Total
    1         203       118       178       499
              123.23    108.07    267.70
    2         122       167       528       817
              201.77    176.93    438.30

    Total     325       285       706       1316

    ChiSq = 51.632 + 0.913 + 30.057 + 31.535 +
            0.558 + 18.358 = 133.052

    df = 2, P = 0.000
```

Comparing the test statistic to the χ^2 distribution with 2 degrees of freedom, you see from the printout that the value of χ^2 from the sample, 133.052, is extremely far out in the tail. The *P*-value is close to 0. The test statistic is quite large and would be difficult to see in a sketch.

Write a conclusion in context: You should reject the null hypothesis. These are not the results you would expect if people were placed on lifeboats without regard to class of travel. The explanation apparently is that first-class passengers were indeed the first to be allowed into lifeboats and third-class passengers were last.

E53. a. Results of the test will vary according to the sample taken. It is likely that the null hypothesis can not be rejected.

b. 0123456789012345678901234567890123456789 . . . Take a large sample of pairs of random digits. Record each pair in a table like this one (a blackline master of this table appears at the end of this section). Then perform a chi-square test of independence.

First Digit in the Pair

	0	1	2	3	4	5	6	7	8	9
0										
1										
2										
3										
4										
5										
6										
7										
8										
9										

Second Digit in the Pair (rows 0–9)

E55. We test:
H_0: Passing final exam and type of writing instrument are independent.
H_a: Passing final exam and type of writing instrument are associated.

The following are the expected frequencies:

	Pencil	Ballpoint Pen	FeltTip Marker	Total
Yes	12.8	7.2	10	30
No	19.2	10.8	15	45
Total	32	18	25	75

The test statistic is given by

$$\chi^2 = \sum \frac{(O-E)^2}{E} = 7.70.$$

Since df = (3-1)(2-1)=2, we see that the p-value is 0.0213. Thus, we reject the null hypothesis and conclude that passing final exam and type of writing instrument are associated.

Chapter 13

Practice Problem Solutions

P1. a. By definition of slope, the slope of the line is 9 calories per gram of fat.

b. From the graph, we estimate the y-intercept to be 200. It tells you the number of calories in a serving containing no grams of fat.

c. Using technology, we find that the regression line has equation $\hat{y} = 191.6 + 13.4x$, where x is fat grams and y is calories. Note that a salad with one more gram of fat than another will have about 13.4 more calories, on average. Also, the calculated 13.4 is somewhat close to the theoretical 9.

d. Calories in pizza come from carbohydrates and protein, as well as from fat. So, there will naturally be variation in the values not reflected in the line itself. Also, the measurement process is likely not 100% accurate.

e. Here, $x = 30$. So, $\mu_y = 200 + 9(30) = 470$. So, $\varepsilon = 600 - 470 = 130$.

f. Note that $\hat{y}(30) = 593.6$, and since the observed data value is $y(30)=600$, the residual is $600 - 593.6 = 6.4$.

g.

x	y_i	$\hat{y}_i = 191.6 + 13.4x$	$(y_i - \hat{y}_i)^2$
57	960	955.4	4.6^2
59	950	982.2	32.2^2
43	820	767.8	52.2^2
22	460	486.4	26.4^2
30	600	593.6	6.4^2

So,

$$s = \sqrt{\frac{4.6^2 + 32.2^2 + 52.2^2 + 26.4^2 + 6.4^2}{3}} \approx 38.8$$

P2. $s = \sqrt{\dfrac{\sum(y_i - \hat{y}_i)^2}{n-2}} \approx \sqrt{\dfrac{0.349593}{5-2}} \approx 0.341366$

P3.

a. Plot I has σ=3; it has less spread in the y-values at each level of x.

b. Plot II has σ=3; it has less spread in the y-values at each level of x, especially noticeable at x=0.

c. Smaller spread of x

d. Plot III because of the smaller spread on x and greater variability in y

P4. a. 5, because you saw in an earlier activity that a greater variability in the conditional distributions of y results in a larger value in the numerator of s_{b_1}.

b. 3, because you saw in an earlier activity that a smaller spread in the values of x results in a smaller value in the denominator of s_{b_1}.

c. 10, because, all else being equal, a larger (random) sample size tends to result in less variability in the estimates of parameters.

Each of these two histograms show 100 sample slopes from regression lines computed from repeated random samples taken from a population of (x, y) pairs with regression slope 0.8. The top plot is for samples of size 10, and the bottom plot is for samples of size 20. Note that both distributions center at about 0.8, but there is much more variability in the distribution of the sample slopes for the samples of size 10.

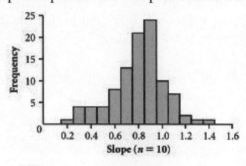

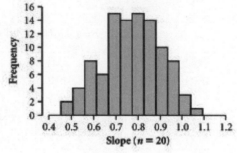

d. The theoretical slope does not matter, everything else being equal.

e. The theoretical intercept does not matter because, all else being equal, all it does is indicate whether one cloud of points is higher or lower than another.

P5. a. Using the value from P2 and $\sqrt{\sum\left(x_i - \overline{x}\right)^2} \approx \sqrt{5.3814} = 2.3197845$;

$$s_{b_1} = \frac{\sqrt{\sum\left(y_i - \hat{y}_i\right)^2 / (n-2)}}{\sqrt{\sum\left(x_i - \overline{x}\right)^2}} \approx \frac{0.341366}{2.3197845} \approx 0.1472$$

b. The standard error for the slope appears in the row "Sulfate" and column "Stdev" (On some printouts, "Sulfate" will be replaced with x. This means the standard error of the coefficient of x — the standard error of the slope.) The estimate s of the variability in y

292

about the line is found in the fourth row of the display as $s = 0.3414$. The equation of the regression line from the printout is

$$\hat{y} = 1.7153 + 0.5249x.$$

c. The first value is the slope of the regression line, b_1. The second value is the estimated standard error of that estimated slope, as calculated in part a. The third value is the first divided by the second,

$$t = \frac{b_1}{s_{b_1}} = \frac{0.5249}{0.1472} \approx 3.57$$

The final value, 0.038, is a two-sided P-value for $t = 3.57$, computed with $df = 3$.

P6. a. The "true" line should pass through the points $(0, 10)$ and $(4, 18)$, so its equation is $y = \beta_0 + \beta_1 x = 10 + 2x$. From the regression analysis in the display, $\hat{y} = b_0 + b_1 x = 9.91 + 2.03x$. The estimated slope is very close to 2, and the estimated intercept is close to 10.

Regression Analysis

```
The regression equation is
y = 9.91 + 2.03 x

Predictor    Coef     Stdev    t-ratio    p
Constant     9.9100   0.9889   10.02      0.000
x            2.0300   0.3496   5.81       0.000
s = 3.127  R-sq = 65.2%  R-sq(adj) = 63.3%

Analysis of Variance

SOURCE       DF    SS       MS       F       p
Regression   1     329.67   329.67   33.71   0.000
Error        18    176.01   9.78
Total        19    505.68

Unusual Observations

                                Stdev.
Obs.  x       y       Fit       Fit      Residual   StResid
8     0.00    2.600   9.910     0.989    -7.310     -2.46R
R denotes an obs. with a large st. resid.
```

b. The standard deviation, s, of the residuals is 3.04, very close to the theoretical value of 3.

c. The standard deviation of b_1 is 0.3496. In theory, the slope should have standard deviation

$$\frac{\sigma}{\sqrt{\sum(x_i - \bar{x})^2}} = \frac{3}{\sqrt{80}} = 0.335$$

d. The mean of the responses at $x = 0$ is 9.91, and the mean at $x = 4$ is 18.03. The regression line with intercept 9.91 and slope 2.03 passes through these means at $x = 0$ and $x = 4$, respectively.

P7. V, II, I, IV, III

P8. The test statistic is

$$t = \frac{b_1 - 0}{s_{b_1}} = \frac{0.133399 - 0}{0.3617} \approx 0.3688 .$$

From the TI-84+, with $df = 4$, the P-value is 0.731. From Table B, you can say only that the P-value is larger than 2(0.25) or 0.50. You cannot reject the null hypothesis that the slope is 0. There is not statistically sufficient evidence here of a non-zero linear relationship between the percentage of sulfate and the redness for soil samples from Mars.

P9. The scatterplot in the display shows very weak correlation, so the slope will be close to zero. The t-value will be small (in absolute value) and the P-value will be large.

P10. The t-statistic is

$$t = \frac{0.6268 - 0}{0.11} \approx 5.6982 .$$

With $11 - 2 = 9$ degrees of freedom, the P-value from the calculator is 0.00029. With a P-value this small, you reject the hypothesis that there is no non-zero linear relationship between percentage of sulfur and redness.

P11. The scatterplot with regression line is shown next. It looks like a straight line will serve as a good model of the relationship between average temperature and number of chirps per second.

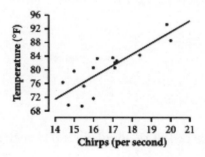

a. The equation is $\hat{y} = 25.23 + 3.291x$ where x is the number of chirps per second and y is the temperature. You can expect the temperature to rise about 3.291 degrees if the number of chirps per second increases by 1.

b. The plots appear below.

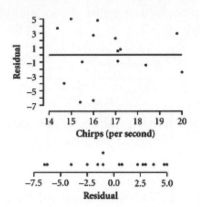

The residual plot shows no obvious pattern, so a linear model fits the data well. There is no evidence that the residuals tend to change in size as *x* increases. The dot plot of the residuals shows no outliers or obvious skewness or any other indications of non-normality. Of course, these plots can't check the condition that you have a random sample of cricket chirping.

c. From the printout, the test statistic is 5.47 and the *P*-value is close to 0. You reject the hypothesis that there is no linear relationship between rate of chirping and temperature. If this was a random sample of cricket chirps, then there is a very small chance (about 0.00011) of getting a test statistic as extreme as or more extreme than the one in the sample if there is really no linear relationship between temperature and the number of cricket chirps.

P12. *Check Conditions:* Because the caption tells you that this is from a study of identical twins, it is clearly not a random sample of people, and is probably not a random sample of identical twins.

The plot of these data does not show a linear relationship, but one that is quite curved (sideways). There is a wider spread of *y*-values for smaller values of *x* and the residual plot shows an upward linear trend if you ignore the two points in the upper left-hand corner. If you go ahead and fit the regression line anyway, the equation is

$$IQ = 0.9969 \cdot head\_\text{circumference} + 45.050$$

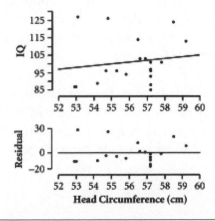

295

The boxplot of the residuals themselves does not appear to be from a normal distribution; rather it is skewed right, a pattern influenced by the sideways curvature of the plot. There is also the outlier that is a problem.

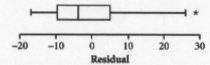

These issues make inference problematic and these data do not meet the conditions of inference. However, if you should proceed with the test:

State your hypotheses:
H_0: $\beta_1 = 0$, where β_1 is the slope of the true linear relationship between *Head Circumference* and *IQ* for all individuals.
H_a: $\beta_1 \neq 0$

Calculate the test statistic and find the P-*value:*

$$t = \frac{b_1 - \beta_{1_0}}{s_{b_1}} = \frac{0.9969 - 0}{1.68838} \approx 0.5904$$

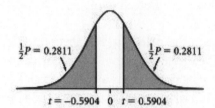

The *P*-value for a test statistic $t \approx 0.5904$ with $20 - 2 = 18$ degrees of freedom is approximately 0.5622.

Write your conclusion in context, linked to your computations:
Do not reject the null hypothesis. There is no statistically significant evidence of a linear relationship between *head circumference* and *IQ*. If there is no linear relationship between head circumference and IQ you are still reasonably likely to get a sample with a slope that differs from zero by 0.9969 or more simply due to chance.

P13. a. The 90% confidence interval is
$$b_1 \pm t^* \cdot s_{b1} = -0.033718 \pm 2.353(0.0039)$$
You are 90% confident that the slope of the true regression line for predicting titanium dioxide content from silicon dioxide content for Mars rocks is in the interval -0.0337 ± 0.0092, or -0.0429 to -0.0245.

b. The value $s = 0.0257$ is an estimate of σ, the standard deviation of the residuals from the true regression line. In other words, it is an estimate of a typical distance of the points from that line. It can also be interpreted as an estimate of the variability in the titanium dioxide percentage at each fixed silicon dioxide percentage.

P14. a. The 90% confidence interval is
$$b_1 \pm t^* \cdot s_{b1} = 0.026856 \pm 2.132(0.0461)$$
You are 90% confident that the slope of the true regression line for predicting titanium dioxide content from silicon dioxide content for Mars soil is in the interval 0.026856 ± 0.09829, or -0.0714 to 0.1251.

b. Because the confidence interval for the soil samples entirely overlaps the one for the rocks, you can't conclude that the slopes are different.

P15. a. The samples of sheets may not be random and there is a cyclical pattern in the residuals, so the results of the analysis must be taken with some caution.

b. First, we calculate s_{b_1} :

y_i	$\hat{y}_i = 0.072x + 2.7$	$(y_i - \hat{y}_i)^2$	$x_i - \bar{x} = x_i - 150$
6	6.3	0.09	-100
11	9.9	1.21	-50
12.5	13.5	1	0
17	17.1	0.01	50
21	20.7	0.09	100

So, $s_{b_1} = \sqrt{\frac{2.4}{3(25,000)}} = 0.00566$. Since $b_1 = 0.072$ and df $= 3$, we see that the confidence interval is about $(0.054, 0.090)$. If one sheet is added to the stack being measured the thickness increases from 0.054 to 0.090 mm, on average. The slope is an estimate of the thickness of a sheet.

P16. The scatterplot with regression line and residual plot beneath it is:

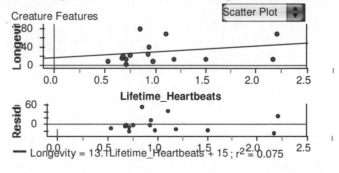

297

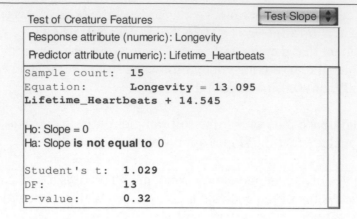

Note that the regression line is $\hat{y} = 13.1x + 15$, where x is heartbeats and y is longevity. As seen from the above plots, the residuals are patternless and the points in the scatterplot conform reasonably well to a linear fit. Further, note from the analysis that P-value is about 0.32. As such, we do not reject the hypothesis of a zero slope when using lifetime heartbeats to predict longevity. The result can be trusted because the data show lots of scatter with little in the way of a linear trend.

b. The scatterplot with regression line and residual plot beneath it is:

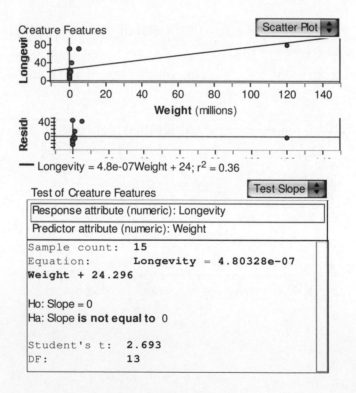

Note that the regression line is $\hat{y} = \left(4.8 \times 10^{-7}\right) x + 24.3$, where x is weight and y is longevity. As seen from the above plots, the residuals are not exactly patternless, being stacked near a single x-value, and the points in the scatterplot are not described well by a linear fit. Running the analysis anyway, we find that the P-value is about 0.018. Hence,

298

we reject the hypothesis of a zero slope when using weight to predict longevity. The result cannot be trusted because the shape of the plot is not linear and dominated by the whale as a very influential point.

c. The scatterplot with regression line and residual plot beneath it is:

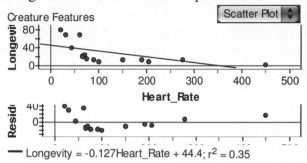

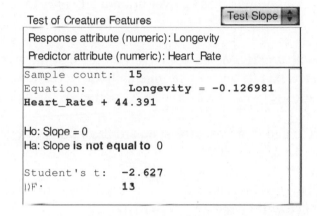

Note that the regression line is $\hat{y} = -127x + 44.4$, where x is heart rate and y is longevity. As seen from the above plots, the residuals are patternless and the points in the scatterplot conform reasonably well to a linear fit. Further, note from the analysis that the P-value is about 0.021. So, we reject the hypothesis of a zero slope when using heart rate to predict longevity. The result cannot be trusted because the data show a definite curvature.

Exercise Solutions

E1. a. Use the form $\hat{y} = b_0 + b_1 x$. The slope and intercept were estimated from the data because the "true" values, β_1 and β_0, are unknown.

b. The response is the opening day; the predictor is the number of inches of *swe* at Flattop Mountain; the equation is

$$opening\ day = 150 + \left(\frac{1}{0.57}\right) \cdot (swe - 30) = 97.37 + 1.75 \cdot swe$$

c. If *swe* = 31.0, *opening day* = 151.62. So, the prediction is day 152 or June 1.

d. The random variation, ε, should be relatively large. Numerous and stronger conditions other than *swe* affect the opening date and vary widely from year to year. These include temperature, park management concerns, maintenance needs, road crew safety, and how many days of precipitation occur after they begin trying to plow the road. The estimate given on the website for a typical random variation, ε, is about 9 days. Note that if the conditional distributions are approximately normal, the middle 95% of the opening dates for a given level of *swe* has a large range of *swe* $\pm$ 1.96(9), or about 36 days.

E3. a. A line is a suitable summary of the relationship between "HGM and GGE" and "ES and GGE." There is a lack of pattern in the data values in the scatterplot for "PCS and GGE," but it is possible that a linear relationship describes it. However, for "CY and GGE" the haps and lack of homogeneity of variation likely prevent a linear description of the data.

b. Using technology with x = HGM and y = GGE, we see that the slope of the regression line is -0.36. This means that for every one-mile an hour increase in HGM, the GGE decreases about 0.36 tons per year, on average. Also, the SE of slope is about 0.025.

c. s = 0.51 tons per year and represents the standard deviation of the residuals around the regression line relating GGE and HGM.

E5. a. π or about 3.1416. This means that for every 1 cm increase in the distance across, the circumference will increase by 3.1416 cm.

b. In this case, there is a linear model that is known to fit the situation perfectly: $C = \pi d$. However, it is difficult to measure C and d with much accuracy.

E7. a. The soil samples should have the larger variability in the slope because the distance of *y* from the regression line tends to be larger compared to the spread in *x*.

b. The standard error for the soil samples is 0.36165. (Again, you may decide to have students do this by reading a computer printout.) From P6, the standard error for the rock samples was 0.1472. As predicted, the standard error for the soil samples is much larger.

E9. A. The plot shows a linear trend with large but homogeneous variation in heights across values of age. This matches Plot III.

B. The plot shows a linear trend with homogeneous variation that is smaller than the variation in A. This matches Plot I.

C. The plot shows a linear trend with variability in height increasing with age. This matches Plot II.

Overall, the conditions for the inference methods of this chapter are met for the plots in A and B but not for C.

E11. From largest to smallest: V, I, II, III, IV

E13. a. First, you need an estimate of both the mean and the variability of the conditional distribution of redness for a sulfur content like Half Dome's of 2.72%. The regression line gives you the estimate of the mean. That line is $\hat{y} = 1.71525 + 0.524901x$, so when $x = 2.72$, $\hat{y} = 1.71525 + 0.524901(2.72) = 3.14298$.

b. The best estimate of σ is s:

y_i	$\hat{y}_i = 0.525x + 1.175$	$(y_i - \hat{y}_i)^2$
2.39	1.595	0.795^2
2.73	2.356	0.374^2
2.75	2.603	0.147^2
3.18	2.656	0.524^2
4.14	3.280	0.86^2

So, $s = \sqrt{\dfrac{\sum (y_i - \hat{y}_i)^2}{n-2}} \approx 0.341$.

c. Choose a value for x.

Predict the mean value of y by finding $\hat{y}$ through the regression line. (This is 3.14298 for Half Dome.)

Generate a value at random from a normal distribution with mean 0 and standard deviation 0.3414.

Add that value to the $\hat{y}$ calculated above. This is your estimate of the redness for a rock with sulfate percentage x. Note that this is the same thing as picking a value at random from $N(\hat{y}, 0.3414)$.

Complete the above four steps for each value of x in the study.

Fit a least squares regression line through the resulting points.

Use the regression line to predict the redness when x is 2.72, the value for Half Dome.

d. This model is reasonable given the limited information we have, but it may not be reasonable given that this is a sample of only a few rocks and we can't be sure that (1) the relationship is linear, (2) the variation in y is constant for all values of x, or (3) the distribution of the errors is normal.

E15. a. From the display, $s_{b_1} = 0.06786$

b. We test: H_o: $\beta_1 = 0$, H_a: $\beta_1 \neq 0$.
From the display, the p-value is 0.165. Hence, there is not significant evidence that the slope differs from 0.

c. For $\alpha = 0.10$ and df = 15-2=13, we have $t^* = 1.771$. So, the 90% confidence interval is
$$b_1 \pm t^* s_{b_1} = 0.0997 \pm 1.771(0.06786),$$
or about -0.204 to 0.2200. This means that if passenger compartment size increases by a square foot, the greenhouse gas emissions are predicted to decrease by up to 0.204 tons per year or increase by up to 0.220 tons per year. In other words, the interval is not informative about the linear relationship between these variables.

E17. We test: H_o: $\beta_1 = 0$, H_a: $\beta_1 \neq 0$. Using technology, we obtain the regression line:

```
The regression equation is
GGE = 2.65 + 1.09 CY


Predictor    Coef   SE Coef      T       P
Constant    2.6500   0.6887    3.85   0.002
CY          1.0860   0.1202    9.03   0.000


S = 0.795213    R-Sq = 86.3%    R-Sq(adj) = 85.2%
```

The scatterplot with regression line is:

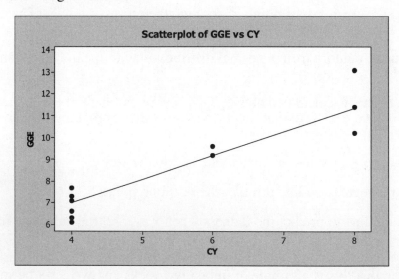

The histogram of the residuals and residual plot are as follows:

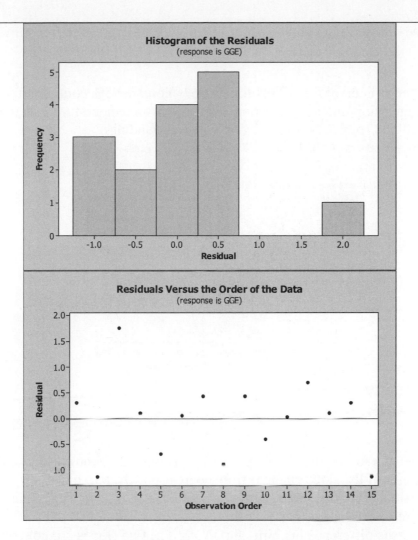

Note that lack of a linear relationship in the scatterplot and the lack of normality in the residuals histogram. This suggests that the conditions necessary to run the regression analysis are questionable. That said, observe from the display that the p-value is nearly zero. Hence, there is strong evidence to conclude that the change in GGE as the number of cylinders increases is greater than 0.

E19. a. i. You checked the conditions in P11. Using the information in the printout in P11 with 13 degrees of freedom,

$$b_1 \pm t^* \cdot s_{b_1} = 3.2911 \pm 2.160(0.6012) = 3.2911 \pm 1.2986 .$$

A 95% confidence interval is 1.9925 to 4.5897. You are 95% confident that the true slope of the regression line for predicting the temperature from the chirp rate is in the interval 1.9925 to 4.5897 (or 1.9924 to 4.5898 with no rounding).

ii. First, note that the assumptions were not checked in P11, given that the residuals will be different when you reverse the roles of *chirp rate* and *temperature*. The residual plot and a dot plot of the residuals appear next with the regression printout. The conditions of linear trend, same spread across *x,* and normal residuals look like they are

met. Using the information in the printout below with 13 degrees of freedom,
$$b_1 \pm t^* \cdot s_{b_1} = 0.21192 \pm 2.160(0.03871) = 0.21192 \pm 0.08361$$

A 95% confidence interval is 0.12831 to 0.29553. You are 95% confident that the true slope of the regression line for predicting chirp rate from temperature is in the interval 0.12831 to 0.29553 (or 0.12829 to 0.29556 with no rounding).
The regression equation is: Chirps = –0.31 + 0.212 Temp

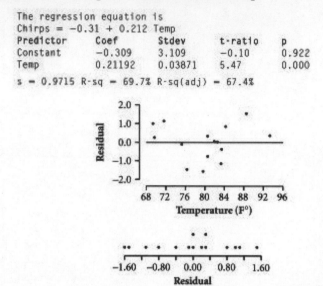

```
The regression equation is
Chirps = -0.31 + 0.212 Temp
Predictor     Coef      Stdev     t-ratio     p
Constant     -0.309     3.109      -0.10     0.922
Temp          0.21192   0.03871     5.47     0.000
s = 0.9715  R-sq = 69.7%  R-sq(adj) = 67.4%
```

b. When you reverse the roles of *chirp rate* and *temperature,* the entire regression line changes. The unit of the slope changes from degrees per chirp to chirps per degree. The sizes of the residuals change too because they are measured from a different line and from a different direction. Further, they are measured in different units (difference in temperature versus difference in chirp rate). *Note:* The two slopes are not reciprocals of one another either.

E21. First, note that it is not appropriate to do a test of significance that the slope of the regression line for predicting height from arm span is 0. You are interested in whether 1 is a plausible value for the slope, not 0. You have two choices. You can do a test that $\beta_1 = 1$ using the *t*-statistic:
$$t = \frac{b_1 - 1}{s_{b_1}} = \frac{0.95166 - 1}{0.03325} \approx -1.45383$$

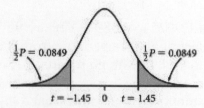

The *P*-value is 2(0.08485) ≈ 0.1697 with no rounding.

With 13 degrees of freedom, you can not reject the hypothesis that the slope is 1. It is plausible that if one person has an arm span 1 cm longer than another, they tend to be 1 cm taller.

Alternatively, a 95% confidence interval for the slope is:
$$b_1 \pm t * \cdot s_{b_1} = 0.95166 \pm 2.160(0.03325) = 0.95166 \pm 0.07182$$

This interval (0.8798, 1.0235) includes 1 as a plausible value for β_1. So these data are consistent with Leonardo's model.

The scatterplot appears in E2, part b, and the residual plot appears in the solution to E2, part d. It shows that a linear trend is reasonable because the residuals show no pattern and stay roughly the same size across all values of x. This plot of the residuals shows that they could reasonably have come from a normal distribution.

E23. a. Consider the following residual plot and univariate plot of the residuals:

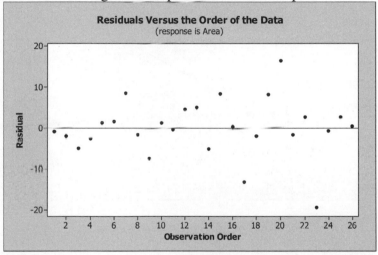

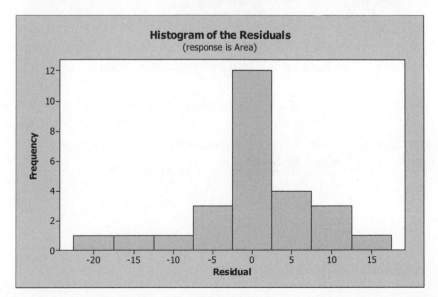

The variation in areas tends to increase with age.

b. The regression analysis is as follows:

```
The regression equation is
Area = - 3.17 + 1.12 Age


Predictor     Coef  SE Coef       T      P
Constant    -3.173    3.063   -1.04  0.311
Age         1.1151   0.1212    9.20  0.000


S = 7.20505   R-Sq = 77.9%   R-Sq(adj) = 77.0%
```

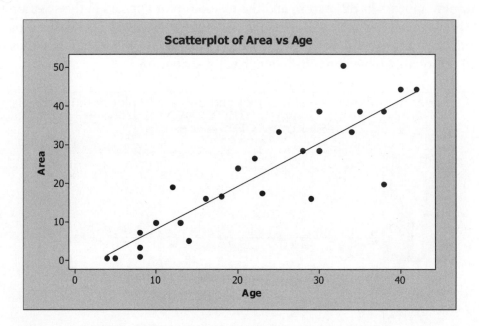

Observe that $b_1 = 1.1151$, $s_{b_1} = 0.1212$, df = 26−2=24, $\alpha = 0.10$, and $t^* = 1.711$. So, the 90% confidence interval is

$$b_1 \pm t^* \cdot s_{b_1} \approx 1.1151 \pm (1.711)(0.1212)$$

or about 0.91 to 1.32. This means that you are 90% confident that the cross-sectional area increases from 0.91 to 1.32 square mm per year, on average.

E25. A regression analysis and needed graphs appear next. The regression equation is *size* = 31.83 − 0.712 *acid*. The slope of −0.712 means that for every increase of 1 μ g/ml in acid concentration, the radius of the fungus colony tends to decrease by 0.712 mm. The test of the significance of the slope follows.

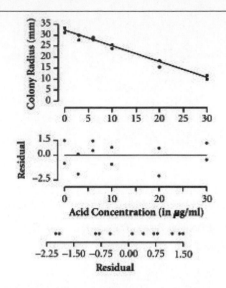

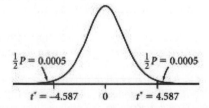

Check conditions: The text does not say whether the petri dishes were randomly selected to receive the different concentrations. But that's a safe bet, given that the experiment was published in *Science*. The residual plot shows that a linear trend is reasonable and the residuals stay roughly constant across all values of *x*. The plot of the residuals shows no reason to suspect that the population of residuals isn't normal.

State your hypotheses:
H_0: $\beta_1 = 0$
H_a: $\beta_1 \neq 0$

where β_1 is the true slope of the linear relationship for predicting colony radius from acid concentration.

Compute the test statistic and draw a sketch:
The printout gives $t = -19.84$, $df = 10$, and a *P*-value of 0. From Table B, the *P*-value is < 0.001. It would be hard to see this in a sketch. The largest critical values in Table B are ±4.587 and the test statistic here is much more extreme.

Write a conclusion in context:
Because the *P*-value is so small, reject the null hypothesis. There is a negative linear relationship between colony radius and acid concentration. The more acid applied, the smaller the radius, at least over the range of acid concentrations applied.

The regression equation is: Size = 31.8 – 0.712 Acid

307

```
The regression equation is
Size = 31.8 - 0.712 Acid

Predictor    Coef       Stdev       t-ratio    p
Constant     31.8298    0.5569      57.15      0.000
Acid         -0.71201   0.03589     -19.84     0.000
s = 1.295 R-sq = 97.5% R-sq(adj) = 97.3%

Analysis of Variance

SOURCE       DF     SS        MS        F         p
Regression   1      660.57    660.57    393.64    0.000
Error        10     16.78     1.68
Total        11     677.35
```

E27. Here are the regression analysis, scatterplot, and residual plot:

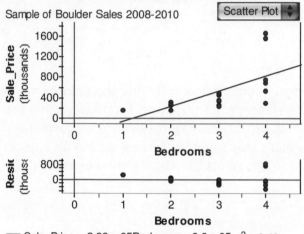

Scatter Plot

Sample of Boulder Sales 2008-2010

— Sale_Price = 2.98e+05Bedrooms - 3.8e+05; r² = 0.43

Test Slope

Test of Sample of Boulder Sales 2008-20

Response attribute (numeric): Sale_Price

Predictor attribute (numeric): Bedrooms

Sample count: **20**

Equation of least-squares regression line:
 Sale_Price = 297684 Bedrooms - 380080
Alternative hypothesis: The slope of the least squares regression line **is not equal to** 0.

The test statistic, Student's t, is **3.649**. There are **18** degrees of freedom (two less than the sample size).

If it were true that the slope of the regression line were equal to 0 (the null hypothesis), and the sampling process were performed repeatedly, the probability of getting a value for Student's t **with an absolute value this great or greater** would be **0.0018**.

Since the p-value is 0.0018, we conclude that there is a significant linear relationship between number of bedrooms and price. In this context, this means that the sales price of a house that has one more bedroom than another house should be about $297.684 greater,

308

on average.

The conditions are not met as the data plot shows curvature and the variation in prices increases with the number of bedrooms. Moreover, there is only 1 one-bedroom house in the analysis, but the others are equally represented. To get a fairer picture of the true relationship, you should sample more one-bedroom houses.

E29. a. Plot I looks excellent. Plot III is also a possibility at least for all points but those at the ends. Plot II is curved, IV shows no linear trend, and V has a very influential data point on the upper right.

b. Plot IV should produce a regression line with slope close to zero.

c. Plot V has correlation closest to 1 with Plot III a very close second. Note that the different scales can make correlation difficult to judge.

E31. The narrowest confidence interval should be for all depths together (the last plot) because the points cluster relatively close to the regression line compared to the other plots *and* the sample size is larger.

The widest confidence interval should be for the mid-depth measurements (second plot) because the outlier will result in a large estimated standard error for the response.

Note that the outlier in the plot for all depths together is neither an outlier among the values of x nor an outlier among the values of y. However, it stands out on the residual plot.

The margins of error for a 95% confidence interval for the slope are

Surface 0.377827
Mid-depth 0.7069
Bottom 0.353189
All 0.195099

E33. a. The scatterplot below shows that the relationship between Pinkerton's estimate and the actual number has a positive linear trend. However, the variation in the residuals is not constant and there is one outlier in both variables that may have strong influence on the analysis. The two outliers in the plot of the residuals is also a reason for concern and perhaps a transformation is in order. The equation of the regression line is *actual number of regiments* = –2.7935 + 1.08021 · *Pinkerton's estimate.*

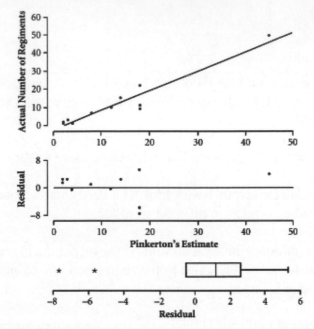

Note: A log-log transformation takes care of both problems in the residuals—although there is still some skewness. Here are the plots:

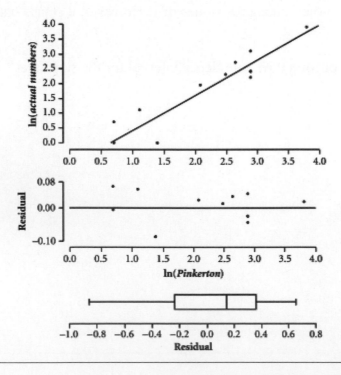

310

The equation of the regression line is:
$$\ln(Actual\_Number) = -0.77956 + 1.18145 \cdot \ln(Pinkerton).$$

b. Conditions were checked in part a. This is not a random sample so you are testing to see if a slope of this size could have happened merely by chance.
The test statistic is for the untransformed data is:
$$t = \frac{b_1 - \beta_1}{s_{b_1}} = \frac{1.0802 - 0}{0.10316} \approx 10.471$$

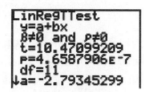

For a value of $t = 10.471$ with $13 - 2 = 11$ degrees of freedom, the P-value is very close to zero.

If Pinkerton could not predict the actual number at all, which would amount to random guessing on his part, the probability of a t-statistic 10.471 or more (in absolute value) with 13 estimates is almost 0. There is strong evidence that Pinkerton's predicting ability was better than chance. However, this result is suspect because of the non-constant variation and outliers in the residuals.

The test statistic is for the transformed data is: $t = \frac{b_1 - \beta_1}{s_{b_1}} = \frac{1.18145 - 0}{0.152456} \approx 7.749$

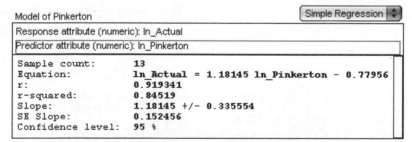

For a value of $t = 7.749$ with $13 - 2 = 11$ degrees of freedom, the P-value is very close to zero.

Write your conclusion in context, linked to your computations:
You should reject the null hypothesis that the slope of the true linear relationship between the natural logarithm of the actual number of regiments and the natural logarithm of Pinkerton's estimate is 0. With a P-value close to zero, if this null hypothesis were true, there would be a very little chance that a random sample of 13 estimates would result in a t-statistic this large or larger in absolute value. Even though this is not a random sample,

there is evidence of a linear relationship between the logarithm of the variables that cannot be explained by chance alone.

c. For every one regiment increase in Pinkerton's estimate, you'd expect the actual number of regiments to increase by about 1.0802.

The 95% confidence interval for the slope of the untransformed data is
$$b_1 \pm t^* \cdot s_{b_1} \approx 1.0802 \pm 2.201 \cdot 0.10316 \approx 1.0802 \pm 0.227055$$
or about (0.85315, 1.307255).

The plausible values for this slope, if these could be considered a random sample of a large population of estimates, are between 0.85315 and 1.3073. If Pinkerton was right on target, the slope of the line would be 1, which does lie in this interval.

Note: For the transformed data, for every one unit increase in the logarithm of Pinkerton's estimate, you'd expect the log of the actual number of regiments to increase by about 1.18145.

The 95% confidence interval for the slope of the transformed data is
$$b_1 \pm t^* \cdot s_{b_1} \approx 1.18145 \pm 2.201 \cdot 0.152456 \approx 1.18145 \pm 0.33556$$
or about (0.84589, 1.51701).

The plausible values for this slope, if these could be considered a random sample of a large population of estimates, are between 0.846 and 1.517. If Pinkerton was right on target, the slope of this line would also be 1, which does lie in this interval.

d. Eliminating Virginia does not change the conclusion of either part. It causes a slight decrease in the slope, (from 1.0802 to 0.8040) and increases the standard error of the slope, thereby widening the confidence interval and increasing the P-value. But the P-value is still small (0.0005) and the confidence interval still contains 1. Eliminating Virginia also does not take care of the uneven variation across values of x either. The plot of the residuals no longer has outliers but is a little skewed to the left.

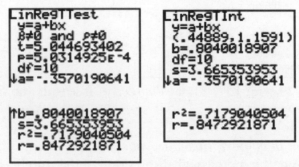

Note: The conclusion does not change for the transformed data either. But in the transformed data, Virginia was not as influential as it is in the original data.

E35. a. The scatterplot with regression line and statistical analysis is provided below.

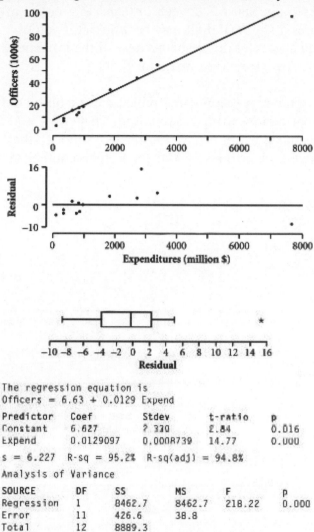

The regression equation is
Officers = 6.63 + 0.0129 Expend

Predictor	Coef	Stdev	t-ratio	p
Constant	6.627	2.330	2.84	0.016
Expend	0.0129097	0.0008739	14.77	0.000

s = 6.227   R-sq = 95.2%   R-sq(adj) = 94.8%

Analysis of Variance

SOURCE	DF	SS	MS	F	p
Regression	1	8462.7	8462.7	218.22	0.000
Error	11	426.6	38.8		
Total	12	8889.3			

The slope of the regression line is 0.0129 thousand officers per million dollars of expenditure. So an increase on one million dollars means 0.0129 · 1000 = 12.9 new officers. You may want to ask your students to construct a confidence interval around this estimate.

All conditions are not met for inference in this problem. Although this is a random sample of states and the relationship appears linear, there is a pattern in the residual plot. With the exception of the outlier, the residual plot has an upward linear trend. The residual do not look to be reasonably constant across values of x. The boxplot looks fairly symmetric, but has an outlier.

Proceeding anyway and using the values for the estimated slope and its standard error on the printout, a 95% confidence interval for the expected increase in the number of police (in thousands) for a \$1 million increase in expenditure is
$$0.01291 \pm (2.201)(0.000874) \approx 0.01291 \pm 0.00192$$

or (0.01099, 0.01483). Because the increase in police is measured in thousands, this is equivalent to a predicted increase of between 11.0 and 14.8 police officers per $1 million increase in expenditures. (The final digit may be different if you use a calculator.) However, you should be leery of this result because of the influential point, California and because the data did not meet the conditions.

b. The right end of the regression line is pulled down a bit by the influential point (California) and will spring upward if it is removed. That effect is seen in the next analysis, which shows much better behaved residuals. Although there is increasing variation in the residuals as *x* increases and there is still an outlier in the residual plot.

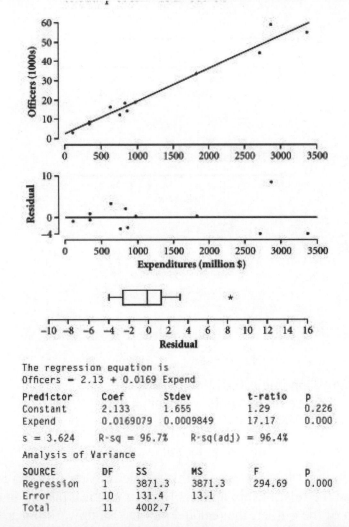

```
The regression equation is
Officers = 2.13 + 0.0169 Expend

Predictor    Coef        Stdev        t-ratio    p
Constant     2.133       1.655        1.29       0.226
Expend       0.0169079   0.0009849    17.17      0.000

s = 3.624    R-sq = 96.7%    R-sq(adj) = 96.4%

Analysis of Variance

SOURCE       DF    SS        MS        F         p
Regression   1     3871.3    3871.3    294.69    0.000
Error        10    131.4     13.1
Total        11    4002.7
```

The expected increase in the number of police officers is now 0.016908 thousand, or 16.908 new officers for every one million dollar increase in expenditures.

The 95% confidence interval for this slope is
$$0.016908 \pm (2.228)(0.0009849) \approx 0.016908 \pm 0.002194,$$
or (0.01471, 0.01910), which translates into an increase of about 14.7 to 19.1 police officers per $1 million increase in expenditures. The governments will get fewer police for their dollars if they attempt to keep up with California!

314

c. The analysis of the relationship between number of police and the population of the states yields:

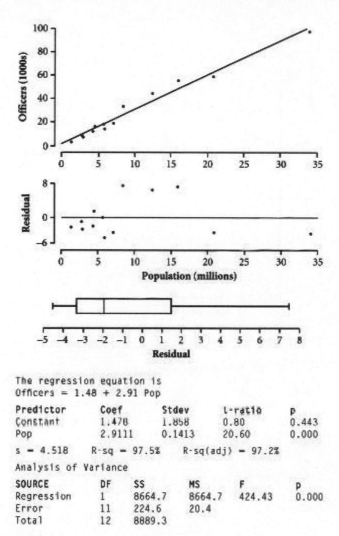

```
The regression equation is
Officers = 1.48 + 2.91 Pop

Predictor       Coef        Stdev       t-ratio     p
Constant        1.478       1.858       0.80        0.443
Pop             2.9111      0.1413      20.60       0.000

s = 4.518       R-sq = 97.5%       R-sq(adj) = 97.2%

Analysis of Variance

SOURCE          DF      SS        MS        F         p
Regression      1       8664.7    8664.7    424.43    0.000
Error           11      224.6     20.4
Total           12      8889.3
```

You expect the first state to have 2.91115 thousand, or 2911.15, more police officers. Again, you may want to ask your students to construct a confidence interval for this estimate: A 95% confidence interval on the expected increase in the number of police officers per million increase in the population is

$$2.91115 \pm (2.201)(0.1413),$$

or (2.600, 3.222). You are 95% confident that a state with a million more people than another will have between about 2600 and 3222 more police officers.

d. With California removed, the results are:

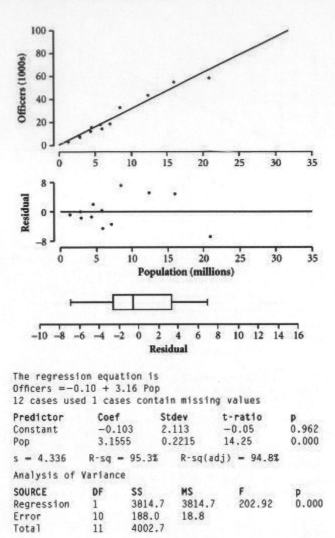

```
The regression equation is
Officers =-0.10 + 3.16 Pop
12 cases used 1 cases contain missing values

Predictor     Coef        Stdev        t-ratio      p
Constant      -0.103      2.113        -0.05        0.962
Pop           3.1555      0.2215       14.25        0.000

s = 4.336     R-sq = 95.3%     R-sq(adj) = 94.8%

Analysis of Variance

SOURCE        DF          SS           MS           F         p
Regression    1           3814.7       3814.7       202.92    0.000
Error         10          188.0        18.8
Total         11          4002.7
```

The predicted increase in the number of police officers is now 3.15548 thousand, or 3155.48 instead of 2911.15 for every increase of one million people in the population.

The resulting 95% confidence interval on the slope is
$$3.15548 \pm (2.228)(0.2215),$$
or (2.662, 3.649). The increase in the number of police officers per million increase in population is expected to be between 2562 and 3774, which is not much different from the interval found in part c. Note that the influence of California on the estimated slope is greater in the case of expenditures, where California is farther away from the linear pattern generated by the rest of the data, than in the case of population, where California lies closer to the linear pattern.

E37. If "standard deviation of the responses" is interpreted as meaning "for each fixed x," then this is a description of σ and so both values are equal. To see this, note that each error can be written as $y - \mu_y$. For a fixed x, μ_y is a constant, so the standard deviation of the errors, $y - \mu_y$, is equal to the standard deviation of the values of y, or σ. Because this is true for each value of x, it is true for the entire population of errors.

If "standard deviation of the responses" is interpreted as meaning "for all values of y, regardless of x," then this value will be larger than the standard deviation of the errors, σ. In the unusual case where the regression line has 0 slope, the two values will be equal.

Chapter 14

Practice Problem Solutions

P1. a. $\bar{x} = \dfrac{n_1\bar{x}_1 + n_2\bar{x}_2}{n_1 + n_2} = \dfrac{6(10.125) + 6(11.367)}{6+6} = 10.75$

b. $s_p^2 = \dfrac{(n_1-1)s_1^2 + (n_2-1)s_2^2}{n_1 + n_2 - 2} = \dfrac{5(1.447)^2 + 5(1.894)^2}{6+6-2} = 2.844$

c. $t^2 = \dfrac{n_1(\bar{x}_1 - \bar{x})^2 + n_2(\bar{x}_2 - \bar{x})^2}{s_p^2} = \dfrac{6(10.125 - 10.75)^2 + 6(11.375 - 10.75)^2}{2.844} = 1.648$

d. Since $t = \sqrt{1.648} = 1.284$, $df = 10$, the p-value is 0.23. As such, we do not reject the null hypothesis that there is no difference in the mean age at first steps for these two treatments.

P2. a. Plot I has more within-sample variability.

b. Between-sample variability is the same.

c. Using the definition of t^2, we see that Plot II has a larger value of t^2 because it has smaller within-sample variability and has the same between-sample variability as I.

P3. a.

$$SS(between) = n_1(\bar{x}_1 - \bar{x})^2 + n_2(\bar{x}_2 - \bar{x})^2$$
$$= 6(10.125 - 10.75)^2 + 6(11.375 - 10.75)^2 = 4.688$$

$$SS(within) = (n_1 - 1)s_1^2 + (n_2 - 1)s_2^2 = 5(1.447)^2 + 5(1.896)^2 = 28.443$$

b. No. There probably will be insufficient evidence to declare that the true means differ because we divide SS(between) by SS(within) to compute t^2.

P4. a. No, since all classes seem to have scores heavy to the right of 80, but the seniors have fewer small values.

b. No, since the spread for each data set is comparable, the difference between the sample means and overall mean will be small, thereby causing SS(between) to be small. So, t^2 will be small in turn. There probably will be insufficient evidence to declare that the true means differ.

P5. a. Observe that

$$\overline{x} = \frac{n_1\overline{x}_1 + n_2\overline{x}_2 + n_3\overline{x}_3 + n_4\overline{x}_4}{n_1 + n_2 + n_3 + n_4} = 3.078$$

$$s_p^2 = \frac{(n_1 - 1)s_1^2 + \cdots + (n_4 - 1)s_4^2}{n_1 + n_2 + n_3 + n_4 - 4} = 0.016512$$

and so, $s_p = 0.128$.

b. SS(between) = 1.2664 and SS(within) = 0.4293

c. df for MS(between) = 3 and df for MS(within) = 26

d. $F = \dfrac{MS(between)}{MS(within)} = \dfrac{\frac{1.2664}{3}}{\frac{0.4293}{26}} = 25.57$

P6. a. Assuming that the null hypothesis is true, this probability is essentially zero because the 100 simulated F values are all at or below 3.5, while we got a value of 25.57.

b. There is strong evidence of differences among the four treatment means.

c. (1/100) = 0.01; There is sufficient evidence to say that there are differences among the four treatment means.

d. (74/100) = 0.74; There is not sufficient evidence to say that the treatment means differ.

P7. Using the formulae listed in the summary chart in Section 14.2, we see that

Total df = 34
Total SS = 0.985 + 5.907 = 6.892
MS(treatment) = 0.4925
MS(error) = 0.1846
F = 2.67

P8. We must identify the main quantities in the given table. Indeed, we have:

Number of treatments: k = 3
Number of observations per treatment: $n_1 = n_2 = n_3 = 3$
Total number of observations: 9
Treatment means: $\overline{x}_1 = 2$, $\overline{x}_2 = 6$, $\overline{x}_3 = 4$
Overall mean: $\overline{x} = 4$

Now, simply substitute these values into the appropriate formula to obtain:

a. SS(between) = 24, SS(within) = 12, SS(total) = 36

b. df for SS(between) = 2, df for SS(within) = 6, df for SS(total) = 8

c. MS(between) = 12, MS(within) = 2, MS(total) = 1

d. $F = 6.0$

Here is the summary table:

Source	SS	df	MS	F	p-value
Treatment	24	2	12	6.0	0.0370
Error	12	6	2		
Total	36	8			

P9. Depth is reflected in MS(between) and random error is reflected in MS(within).

P10. a. Scanning the data, we would expect MS(between) and MS(within) to be relatively close in value. So, expect the F statistic to be close to 1.

b. SS(between) = 59.11, SS(within) = 243.33, SS(total) = 302.44

c. Here is the summary table:

Source	SS	df	MS	F	p-value
Treatment	59.11	2	29.555	1.82	0.1957
Error	243.33	15	16.22		
Total	302.44	17			

P11. a. Team 6 has the most within-team variation, while Team 4 has the least. Yes, the longer distances tend to vary more.

b. The most between-team variation occurs between Teams 4 and 6, and the least occurs between Teams 1 and 4.

c. H_o: The mean launch distances for all teams are the same.
H_a: There are differences among the mean launch distances for the teams.

Using technology, we see that $F \approx 10.17$.

d. There is sufficient evidence to conclude that there are differences among the true mean launch distances for the teams.

e. Yes, there are concerns about pooling to get a common estimate of s_p because Team 6 has a very large standard deviation, compared to the others.

P12. a. The conditions that must be met in order to run an ANOVA are:

321

- One of the treatments is randomly assigned to each subject
- Samples come from populations that are approximately normal.
- Sample standard deviations are approximately equal.

These conditions are met in this scenario.

b. Here is the summary table:

Source	SS	df	MS	F	p-value
Treatment	0.1196	2	0.0598	4.61	
Error	0.5578	43	0.0130		
Total	0.6774	45			

c. There is sufficient evidence to conclude that there are differences among the three treatment means.

P13. a. Since $1 - \frac{0.05}{4} = 0.9875$, we have 98.75%.

b. Since $1 - \frac{0.02}{3} = 0.9933$, we have 99.33%.

P14. Construct 3 intervals, each at 98.33% confidence. With 43 *df* for error, t^* is about 2.198.

$$(eye\text{-}black) - (petroleum): \ 0.09 \pm 0.088$$
$$(eye\text{-}black) - (stickers): \ 0.12 \pm 0.092$$
$$(petroleum) - (stickers): \ 0.03 \pm 0.092$$

The mean for eye-black grease differs significantly from both the mean for petroleum jelly and the mean for antiglare stickers; the means for petroleum jelly and antiglare stickers do not differ significantly.

Exercise Solutions

E1. a. H_0: The mean heart rates are the same for men and women.
H_a: The mean heart rates are different for men and women.

b. The distributions are very similar, so it is unlikely that the null hypothesis will be rejected. Entering the data into Minitab, we find that

$$\bar{x}_{men} = 75.69, \ s_{men} = 7.11$$
$$\bar{x}_{women} = 73.08, \ s_{women} = 10.53.$$
$$\bar{x} = 74.39$$

c. Using the sample statistics from (b), along with $n_{men} = n_{women} = 13$, we see that

$$t^2 = \frac{n_{men}\left(x_{men} - \overline{x}\right)^2 + n_{women}\left(x_{women} - \overline{x}\right)^2}{s_p^2} = 0.551.$$

So, t = 0.742 with p-value of 0.47. So, there is insufficient evidence to reject the null hypothesis and so, we conclude that the mean heart rates are the same for men and women.

E3. **a.** Los Angeles has the most within-team variation, while Colorado has the least (but Florida is close)

b. Atlanta versus Chicago, or Los Angeles versus Arizona are about the same.

c. No, the distributions illustrated by the boxplots do not seem to be vastly different regarding spread.

d. No; all are in the same ball-park, so to speak.

E5. **a.** Short days because of the value 18.275.

b. Using technology with the formulae in the text, we see that
$$SS(total) = 84.72;$$
$$SS(between) = 49.63;$$
$$SS(within) = 35.09;$$
$$SS(total) = SS(between) + SS(within)$$

c. The test statistic value is 8.485 with a P-value of about 0.027. As such, we can conclude that the means are significantly different.

d. No, though the s.d. for the data sets are certainly different, they aren't drastically different.

E7. Assume for two samples that
$$\overline{x} = \overline{x}_1 = \overline{x}_2, \quad s_1^2 = s_2^2 = s^2, \quad \text{and} \quad n_1 = n_2 = n.$$
a. Observe that
$$s_p^2 = \frac{(n-1)s^2 + (n-1)s^2}{n+n-2} = \frac{2(n-1)s^2}{2(n-1)} = s^2.$$

b. Increasing either s_1 or s_2 will increase s_p. But, since $x_1 - \overline{x} = x_2 - \overline{x} = 0$, we see that t^2 remains at 0.

c. Decreasing either s_1 or s_2 will decrease s_p. But, since $x_1 - \overline{x} = x_2 - \overline{x} = 0$, we see that t^2 remains at 0.

d. Assume that we add 5 to all values in the first sample. Then,

$$\left(\overline{x}_1\right)_{new} = 5 + \overline{x}, \quad \left(s_1^2\right)_{new} = s_1^2 = s^2.$$

So, s_p remains unchanged, but one of the 2 terms in the numerator for t^2 will increase, thereby increasing the value of t^2.

E9. a. df are 2, 21 and 23, down the column; $SS(within) = 458.20$; $MS(between) = 372.680$; $MS(within) = 21.819$. The value of the test-statistic is $F = 17.08$ with a P-value near 0.

b. Yes, since the F statistic is large.

E11. For the alternative that there are differences among the mean ages at first steps, the F value is 2.07 and the P-value is 0.138. Evidence is insufficient to conclude that there are differences among the mean ages at first steps for these four treatments.

E13. a. Unroll each roll of towels and choose 5 towels at random from each roll. Then, randomize the order of the 20 towels to be used in the measurement process.

b.
```
Oneway Anova
     Source    DF    Sum of Squares    Mean Square    F Ratio      Prob>F
     Model      4      1267000.0         316750        271.5000     <.0001
     Error     15        17500.0           1167
     Total     19      1284500.0         67605
```

Since the p-value is <0.0001, there is strong evidence of significant differences among the mean strengths of the towels.

E15. a. Here are the histograms and descriptive statistics for each of the four sports:

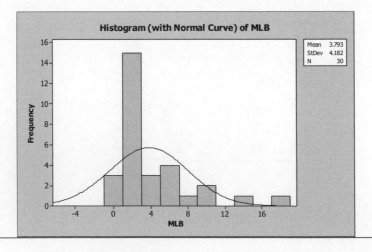

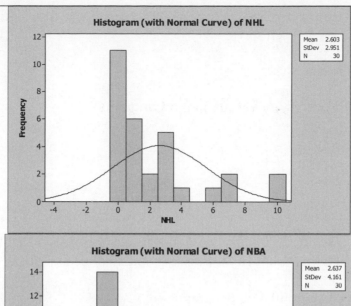

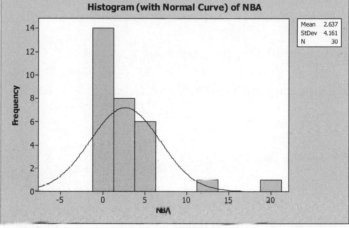

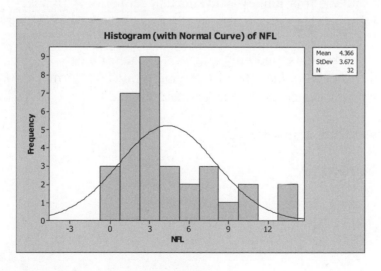

Descriptive Statistics: NFL, NBA, NHL, MLB

Variable	N	Mean	StDev	Q1	Median	Q3
NFL	32	4.366	3.672	1.900	2.900	6.650
NBA	30	2.637	4.161	0.300	1.300	3.925
NHL	30	2.603	2.951	0.400	1.400	3.400
MLB	30	3.793	4.182	1.000	2.000	5.250

Now, we see that the answer to the question is NO, because the distributions are skewed with at least one outlier in each sport.

b. Here is the ANOVA:

One-way ANOVA: Sport Type versus Mean Longevity

```
Source        DF      SS     MS     F      P
Team CODED     3    75.8   25.3   1.77   0.158
Error        115  1645.9   14.3
Total        118  1721.8

S = 3.783   R-Sq = 4.40%   R-Sq(adj) = 1.91%
```

Observe that $F \approx 1.77$ with a P-value ≈ 0.158. Hence, we conclude that there are no differences among the four longevity means that could not be explained by chance.

c. Here is the ANOVA with the outliers removed:

One-way ANOVA: Longevity with outliers removed versus Team

```
Source   DF     SS      MS     F      P
Team      3   77.21   25.74  4.72   0.004
Error   107  583.03    5.45
Total   110  660.24

S = 2.334   R-Sq = 11.69%   R-Sq(adj) = 9.22%
```

Observe that $F \approx 4.72$ with a P-value ≈ 0.004. Hence, we conclude that there is evidence of differences among the four longevity means that cannot be explained by chance alone.

E17. **a.** The conditions that must be met in order to run an ANOVA are:

- One of the treatments is randomly assigned to each subject
- Samples come from populations that are approximately normal.
- Sample standard deviations are approximately equal.

These conditions are met in this scenario.

b & c. Here is the ANOVA:

```
        ANALYSIS OF VARIANCE ON Growth
        SOURCE      DF        SS        MS        F        p
        Treat        3      771.2     257.1     5.24     0.004
        ERROR       36     1767.3      49.1
        TOTAL       39     2538.5

        99% confidence intervals on differences of means

        Mean 1 - Mean 2:  (-6.11, 10.91)
        Mean 1 - Mean 3:  (-13.56, 3.46)
        Mean 1 - Mean 4:  (-17.41, -0.39)*
        Mean 2 - Mean 3:  (-15.96, 1.06)
        Mean 2 - Mean 4:  (-19.81, -2.79)*
        Mean 3 - Mean 4:  (-12.36, 4.66)
```

Observe that $F \approx 5.24$ with P-value of 0.004. Hence, we conclude that that there are statistically significant differences among the treatment means.

Using 99% confidence intervals, the statistically significant differences are between means 1 and 4 and means 2 and 4.

d. Treatment 2 gives the smallest mean, although treatment 1 is close.

E19. a. To perform a simulation to approximate the distribution of the F-statistic, follow these steps:

- Mix all data values from the four treatments and randomly reassign the values to the four treatments.
- Calculate $SS(between)$, $SS(within)$ and the F-statistic;
- Record the F value.
- Repeat the process many times and plot the F values.

b. With $F \approx 2.0$, the P-value is about $(19/200) = 0.145$, close to the calculated value in E11.

c. We conclude that there is no significant difference among the treatments.

E21. a. Here is the regression analysis:

Regression Analysis: number of taps versus milligrams of caffeine

```
The regression equation is
number of taps = 245 + 0.0175 milligrams of caffeine

Predictor               Coef    SE Coef      T       P
Constant             244.750      0.632  387.20  0.0001
milligrams of caffeine  0.017500  0.004896    3.57  0.0013

S = 2.18967   R-Sq = 31.3%   R-Sq(adj) = 28.9%

Analysis of Variance

Source          DF       SS       MS      F      P
Regression       1   61.250   61.250  12.77  0.001
Residual Error  28  134.250    4.795
Total           29  195.500
```

Observe that the P-value ≈ 0.0013 for testing that the slope is 0 versus not 0. Hence, we conclude that there is strong evidence of a positive slope.

b. P-value ≈ 0.0062 for testing equality of means versus some differences among the means. Regression analysis tests for only one pattern in the treatment means, a linear one. That pattern happens to be present here, so the evidence for it is strong. ANOVA tests for any type of inequality among the treatment means, so is less sensitive to the particular linear pattern.

c. Regression tests for linear trends among the treatment means, which implies that the treatments must have a numerical order (as they do in the caffeine experiment). ANOVA tests for any type of inequality among the treatment means, and the treatments need not be ordered.

E23. a. Completely randomized, if measurements are done on specimens in a random order.

b. Here is a boxplot of the individual metals, followed by the ANOVA:

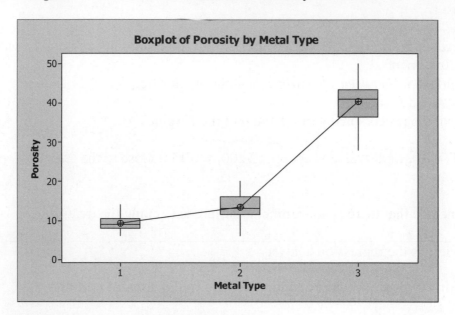

Boxplot of Porosity by Metal Type

One-way ANOVA: Porosity versus Metal Type

```
Source        DF       SS       MS       F       P
Metal Type     2   14132.3   7066.2   474.84   0.000
Error         72    1071.4     14.9
Total         74   15203.8

S = 3.858    R-Sq = 92.95%    R-Sq(adj) = 92.76%

                         Individual 95% CIs For Mean Based on
                         Pooled StDev
Level   N    Mean   StDev   --+---------+---------+---------+-------
1       25   9.360  2.018   (*-)
2       25  13.480  3.002       (*-)
3       25  40.320  5.618                                   (*-)
                             --+---------+---------+---------+-------
                               10        20        30        40

Pooled StDev = 3.858
```

Observe that $F \approx 475$ with a p-value near zero. Hence, the differences among mean counts are highly significant, mainly because electrolytic copper has much higher counts than the others.

c. Using Minitab, we have the following t-test, from which we will deduce the 95% confidence interval on difference in means for Linde and antimony.

Two-Sample T-Test and CI: Antimony, Linde

```
Two-sample T for Antimony vs Linde

             N    Mean   StDev   SE Mean
Antimony    25    9.36    2.02      0.40
Linde       25   13.48    3.00      0.60


Difference = mu (Antimony) - mu (Linde)
Estimate for difference:  -4.12000
95% CI for difference:  (-5.57992, -2.66008)
T-Test of difference = 0 (vs not =): T-Value = -5.70  P-Value = 0.000  DF = 42
```

Note that the 95% confidence interval is about 4.12 ± 1.45. We conclude that the means differ because the interval doesn't contain 0.

E25. a. Two completely randomized designs, one for each launch angle, with three teams each.

b. One-way ANOVA on teams ignoring angle followed by a one-way analysis on angle ignoring teams.

c. Here are the graphs to verify that the ANOVA is ok to run:

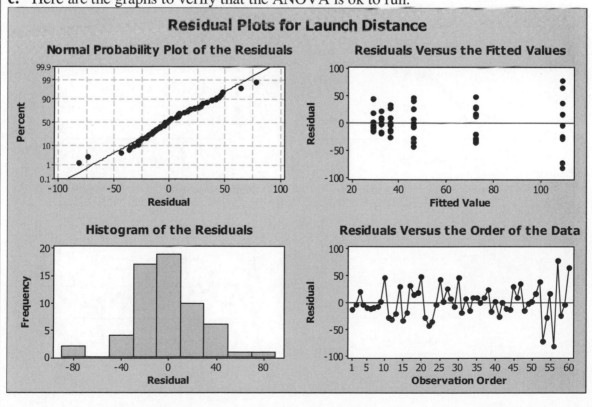

Now, here is the ANOVA:

One-way ANOVA: Launch Distance versus Gummy Bear Team

```
Source              DF     SS     MS      F      P
Gummy Bear Team      5  48099   9620  10.17  0.000
Error               54  51073    946
Total               59  99172


S = 30.75   R-Sq = 48.50%   R-Sq(adj) = 43.73%
                            Individual 95% CIs For Mean Based on
                            Pooled StDev
Level   N    Mean   StDev  -------+---------+---------+---------+--
1      10   29.30   18.36  (------*-----)
2      10   72.10   30.45                    (-----*-------)
3      10   46.00   31.11       (-----*------)
4      10   32.40   13.99  (------*-----)
5      10   36.20   19.23    (-----*------)
6      10  109.00   53.63                          (-----*------)
                           -------+---------+---------+---------+--
                               30        60        90       120

Pooled StDev = 30.75
```

Since F=10.17 with a p-value near zero, we conclude that the team means differ significantly.

d. Here are the plots showing that an ANOVA is ok to run, following by the actual ANOVA:

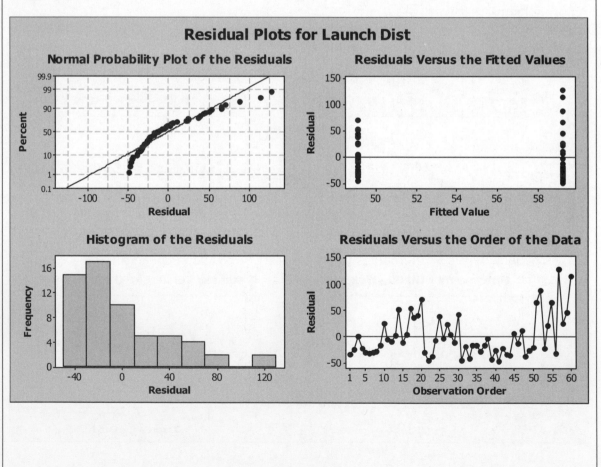

One-way ANOVA: Launch Dist versus Gummy Bear Team- BASED ON ANGLE

```
Source              DF    SS     MS     F     P
Gummy Bear Team-    1    1520   1520   0.90  0.346
Error               58   97652  1684
Total               59   99172

S = 41.03   R-Sq = 1.53%   R-Sq(adj) = 0.00%


                           Individual 95% CIs For Mean Based on
                           Pooled StDev
Level  N   Mean   StDev   --+---------+---------+---------+-------
1      30  49.13  31.84   (------------*-----------)
2      30  59.20  48.51              (-----------*------------)
                          --+---------+---------+---------+-------
                            36        48        60        72

Pooled StDev = 41.03
```

Since F=0.90 with p-value 0.346, we conclude that launch angle means do not appear to differ in light of the large variation within and between teams.

E27. a. The following ANOVA table shows the researchers are correct.

	Sum of Squares	df	Mean Square	F	Significance (p)
Between Groups:	1.512	2	0.756	0.054	0.947
Within Groups:	1,699.908	122	13.934		
Total:	1,701.420	124			

b. Yes, treatments were randomly assigned, the sample sizes are reasonably large so the normality condition is reasonable, and the standard deviations, which are found by multiplying the *SE* by $\sqrt{n}$, are approximately equal.

c. No, $F \approx 0.29$ and the *P*-value is about 0.75. Conclude that this experiment provides no statistically significant evidence of a Mozart effect.

E29. a. 6.65

b. They conducted 133 different tests of significance. To keep the overall probability of at least one Type I error at 0.05, each individual test must be run at a lower level of significance than 0.05. For 133 tests, the level of significance should be 0.05/133, or about 0.00038.

c. No, this is not significant because 0.004 is larger than $\alpha = 0.00038$.

Chapter 15

Practice Problem Solutions

P1. For these, simply determine which way the plane slants as you move in the positive direction of both variables. We have: A IV, B III, C II, D I

P2. a. Here is a table showing the information needed for the computations:

magnitude	latitude	depth	magnitude^	residual
3.9	-7	422	3.896	0.004
4.4	-34	35	5.210	-0.810
4.1	28	6	4.028	0.072
5.3	-4	35	4.610	0.690
3.1	3	24	4.492	-1.392
3.1	60	44	3.312	-0.212
5.7	-23	38	4.984	0.716
3.8	53	0	3.540	0.26

So, using the formulae provided in the text, we have
$$SSE = 3.700, \quad SST = 6.175, \quad s = 0.860, \quad R^2 = 0.401.$$

b. About 40% of the variation in earthquake magnitudes can be explained by latitude and depth.

P3. No. R^2 increases a little by adding depth and s decreases a little, but not by enough to make *depth* significant. Indeed, the p-value for depth is 0.299, which suggests that it is not significant.

P4.

1st equation: Add the first two entries in the second column of the ANOVA table to obtain the 3rd entry in that column.

2nd equation: Observe that substituting in the values into the equation
$$F = \frac{\frac{SSR}{k}}{\frac{SSE}{n-k-1}} = \frac{MSR}{MSE}$$

yields

$$419.61 = \frac{\frac{58.955}{2}}{\frac{0.084}{12}} = \frac{29.4775}{0.07025}.$$

3rd equation: Add the first two entries in the first column of the ANOVA table to get the 3rd entry in that table.

P5. No. PCS adds nothing significant when HGM is already in the model because its p-value is 0.620. PCS and HGM are correlated with each other.

P6.

a. H$_o$: The regression coefficient for *latitude* is zero when *depth* is also in the model for predicting *magnitude*.
H$_a$: The regression coefficient for *latitude* differs from zero when *depth* is also in the model for predicting *magnitude*.

H$_o$: The regression coefficient for *depth* is zero when *latitude* is also in the model for predicting *magnitude*.
H$_a$: The regression coefficient for *depth* differs from zero when *latitude* is also in the model for predicting *magnitude*.

b. In the first test, reject H$_o$; *latitude* plays a significant role since its p-value is 0.031. In the second test, do not reject H$_o$; *depth* does not play a significant role since its p-value is 0.299.

P7. a. This is an observational study, so there is no randomization.

b. There is some curvature in the relationship between *time of day* and *magnitude*.

c. The regression is not statistically significant at the 0.01 level. Using the t-tests, we see that neither *time of day* not *latitude* make a contribution to the prediction of *magnitude* that is greater than would be expected by chance alone, when the other is in the model.

P8. a. Yes because $F = 45.7$, $R^2 = 0.67$.

b. The variables *weaning, liter size, birth weight* are all statistically significant (since p-values are near zero).

c. 67%, which is given by $R^2 = 0.67$.

d. Relatively large since gestation periods are typically between 9 and 12 months.

e. Conditional on the other explanatory variables in the model, as liter size increases gestation period tends to decrease.

P9. First, we run the regression analysis using all of the explanatory variables:

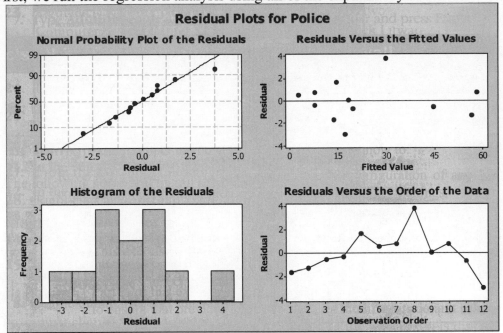

Regression Analysis: Police versus Expenditures, Population, ...

```
The regression equation is
Police = - 0.89 + 0.00896 Expenditures + 1.48 Population
         + 0.00433 Violent Crime Rate


Predictor              Coef    SE Coef       T      P
Constant             -0.889      1.527   -0.58  0.577
Expenditures       0.008963   0.001749    5.12  0.001
Population            1.4807     0.3071    4.82  0.001
Violent Crime Rate  0.004329   0.003765    1.15  0.283


S = 2.03957   R-Sq = 99.2%   R-Sq(adj) = 98.9%


Analysis of Variance

Source           DF       SS       MS       F      P
Regression        3   3969.4   1323.1  318.07  0.000
Residual Error    8     33.3      4.2
Total            11   4002.7


Source             DF  Seq SS
Expenditures        1  3871.3
Population          1    92.6
Violent Crime Rate  1     5.5

Unusual Observations

Obs  Expenditures  Police     Fit  SE Fit  Residual  St Resid
  8          1829  33.400  29.604   1.066     3.796     2.18R

R denotes an observation with a large standardized residual.
```

Note that VIOLENT CRIME is not significant. As such, we remove it from the analysis and re-run the regression to get:

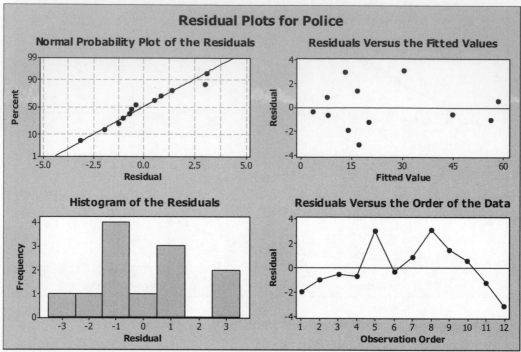

Regression Analysis: Police versus Expenditures, Population

```
The regression equation is
Police = 0.44 + 0.00972 Expenditures + 1.44 Population


Predictor          Coef     SE Coef     T        P
Constant          0.440       1.016   0.43    0.675
Expenditures   0.009715    0.001651   5.88    0.000
Population        1.4389      0.3104   4.64    0.001


S = 2.07572    R-Sq = 99.0%    R-Sq(adj) = 98.8%


Analysis of Variance

Source           DF       SS        MS        F        P
Regression        2    3963.9    1982.0   460.00    0.000
Residual Error    9      38.8       4.3
Total            11    4002.7


Source         DF   Seq SS
Expenditures    1   3871.3
Population      1     92.6


Unusual Observations
Obs  Expenditures  Police     Fit   SE Fit  Residual  St Resid
 10          2866  58.900  58.357    1.986     0.543      0.90 X

X denotes an observation whose X value gives it large influence.
```

Note that the regression equation from the second model is

$$\widehat{police} = 0.44 + 0.001 expenditures + 1.439 population.$$

Also, the test statistic F = 460 with a p-value near zero, and $R^2 = 0.99$, which indicates the model is very strong, and the *t*-tests for both coefficients show significance.

P10. a. Yes

b. Conditions are met fairly well. Indeed, there is a linear relationship of the response variable to each explanatory variable, the residual plots are patternless, and the univariate plot of the residuals is somewhat right skewed, but still reasonably symmetric.

c. *Flowers* by *starting time* and *residuals* by *starting time* are not necessary because there are only two starting times and the linear trends are the same for each, as can be seen in the original scatterplot.

P11.
a. H_o: The regression coefficient for *intensity* is zero when *start time* is also in the model for predicting *flowers*.
H_a: The regression coefficient for *intensity* differs from zero when *start time* is also in the model for predicting *flowers*.

H_o: The regression coefficient for *start time* is zero when *intensity* is also in the model for predicting *flowers*.
H_a: The regression coefficient for *start time* differs from zero when *intensity* is also in the model for predicting *flowers*.

b. Each of the explanatory variables is highly significant in a model to predict *flowers*.

P12. a. The categorical variables are as follows (with code in parentheses):
gender (1 for female, 0 for male);
smoking status (1 for smoker);
alcohol status (1 for heavy drinker);
use of hypnotics (1 for regular use)

b. gender since it is the only one with p-value <0.05.

c. The regression coefficient for gender is significantly different from zero.
<u>For females:</u> $\widehat{sleep} = (bo + 0.064) - 0.007cups + 0.042smoking - 0.026\ alcohol + 0.020hypnotics - 0.054age$

<u>For males:</u> $\widehat{sleep} = bo - 0.007cups + 0.042smoking - 0.026\ alcohol + 0.020hypnotics - 0.054age$

Note that only the intercept changes.

d. The regression coefficient for alcohol use is not significantly different from zero.

e. Sleep time tends to decrease with increasing cups of coffee, increasing alcohol use and increasing age. Only age is nearly significant (with p = 0.06).

f. Agree, because in the presence of the other variables, "number of cups of coffee per day" is not significant.

P13. a. We sketch the following lines:

Females: $\hat{y} = -0.007x + 0.064 + b_0$ and Males: $\hat{y} = -0.007x + b_0$

(Note that since the two lines have the same slope and the y-intercept for females is slightly larger than for males, the lines are parallel, with the one with greater y-values corresponds to the female curve. They appear, essentially, as horizontal lines since the slopes are so small.)

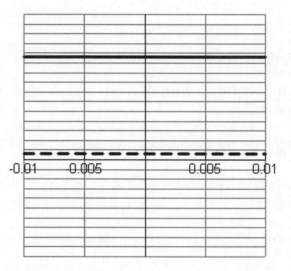

b. Now, we expect the two lines to intersect given that the slopes are different.

Exercise Solutions

E1. a. This is an observational study. Inference to a larger population is not possible because only, and all, Latin countries are sampled.

b. The country with the lowest life expectancy (Bolivia) is an outlier on the residual scale and might unduly influence the analysis. The normality assumption seems to be satisfied. The individual plots are not very linear, though the trend for the first is to decrease left to right and for the second, the opposite.

c. The regression analysis is significant (p = 0.021) and both variables are significant at the 5% level.

d. Both *forest proportion* and *military expenditure* make contributions to the prediction of *life expectancy* that are greater than would be expected by chance alone. The effect of the former is negative and the latter positive.

E3. a. There are $\binom{4}{2} = 6$ such models. There are too many to include here, but below are some of the 6 models using 2 explanatory variables that can be created:

Explanatory Variables are: RURAL CHANGE and INFANT MORTALITY

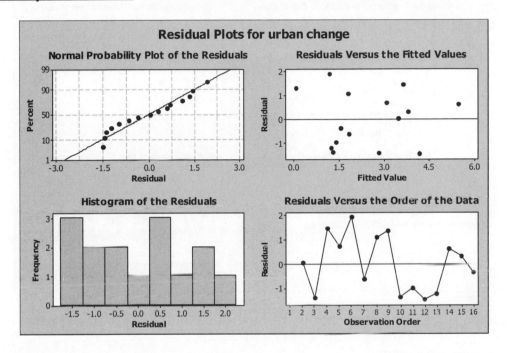

Regression Analysis: urban change versus rural change, infant mortality

```
The regression equation is
urban change = 1.24 + 0.689 rural change + 0.0241 infant mortality

15 cases used, 1 cases contain missing values

Predictor            Coef     SE Coef      T      P
Constant           1.2445      0.4507   2.76  0.017
rural change       0.6893      0.2736   2.52  0.027
infant mortality   0.024062  0.009153   2.63  0.022

S = 1.25505   R-Sq = 61.2%   R-Sq(adj) = 54.7%
Analysis of Variance

Source           DF      SS      MS     F      P
Regression        2  29.815  14.908  9.46  0.003
Residual Error   12  18.902   1.575
Total            14  48.717

Source           DF  Seq SS
rural change      1  18.930
infant mortality  1  10.885

Unusual Observations
```

```
        rural    urban
Obs     change   change   Fit    SE Fit   Residual   St Resid
 12     -0.40    2.700    4.169   1.047    -1.469     -2.12RX

R denotes an observation with a large standardized residual.
X denotes an observation whose X value gives it large influence.
```

Explanatory Variables are: RURAL CHANGE and MEN LIFE EXPECTANCY

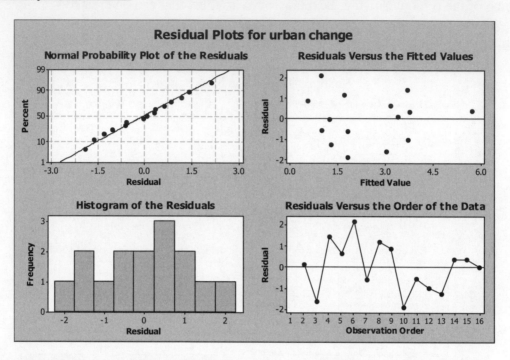

Regression Analysis: urban change versus rural change, men life expecta

```
The regression equation is
urban change = 8.13 + 0.740 rural change - 0.0915 men life expectancy


15 cases used, 1 cases contain missing values


Predictor                 Coef   SE Coef       T      P
Constant                 8.130     2.387    3.41   0.005
rural change            0.7397    0.2696    2.74   0.018
men life expectancy    -0.09149   0.03517  -2.60   0.023


S = 1.25985   R-Sq = 60.9%   R-Sq(adj) = 54.4%
Analysis of Variance

Source            DF      SS       MS      F      P
Regression         2   29.671   14.835   9.35   0.004
Residual Error    12   19.047    1.587
Total             14   48.717


Source            DF   Seq SS
rural change       1   18.930
men life expectancy 1  10.740
```

Explanatory Variables are: RURAL CHANGE and WOMEN LIFE EXPECTANCY

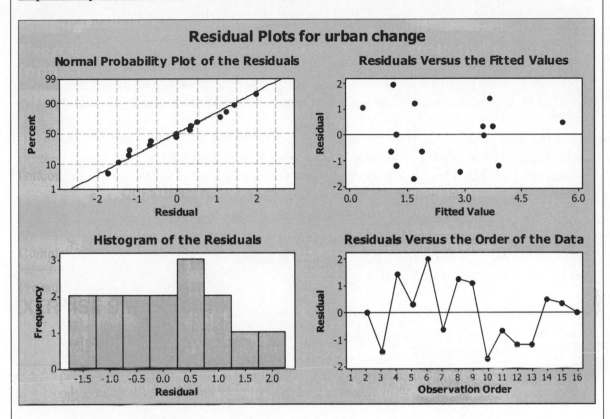

Regression Analysis: urban change versus rural change, women life expec

```
The regression equation is
urban change = 8.28 + 0.665 rural change - 0.0877 women life expectancy


15 cases used, 1 cases contain missing values


Predictor                    Coef    SE Coef       T       P
Constant                    8.279      2.284    3.63   0.003
rural change               0.6646     0.2695    2.47   0.030
women life expectancy     -0.08768   0.03147   -2.79   0.016


S = 1.22772   R-Sq = 62.9%   R-Sq(adj) = 56.7%


Analysis of Variance
Source            DF       SS       MS       F       P
Regression         2   30.630   15.315   10.16   0.003
Residual Error    12   18.088    1.507
Total             14   48.717
Source            DF   Seq SS
rural change       1   18.930
women life expectancy  1   11.699
```

The other 3 tests in the same manner. In the end, we find that *Urban Population Change* is best predicted by *Rural Population Change* and *Women's Life Expectancy*.

b. Referring to the output above, we see that $R^2 = 0.629$; about 63% of the variation in *Urban Population Change* can be explained by *Rural Population Change* and *Women's Life Expectancy*.

341

c. $F = 10.16$ with a *P*-value near zero; at least one of the two explanatory variables makes a contribution to the prediction of *Urban Population Change* greater than would be expected by chance alone.

d. The *P*-values for the t-tests, 0.03 and 0.016, respectively, show that each of the two explanatory variables make a contribution to the prediction greater than would be expected by chance alone, RPC in a positive direction and WLE in a negative direction.

e. Of the four possible explanatory variables for predicting *Urban Population Change*, the best model uses *Rural Population Change* and *Women's Life Expectancy* as predictors. Both yield highly significant regression coefficients, the first in a positive direction and the second in a negative direction. The regression model is

$$\widehat{UPC} = 8.279 + 0.665RPC - 0.088WLE$$

E5. a. The two regression models, with scatterplots and residual plots, are:
The two linear regression models are:

TOTAL PLAYER RATING versus RUNS SCORED

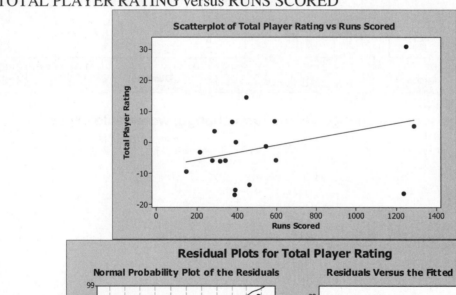

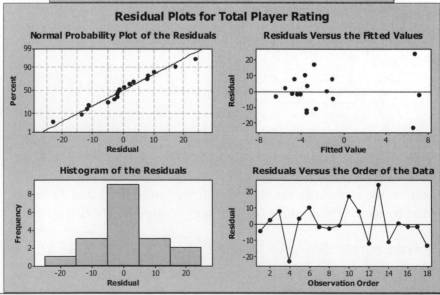

Regression Analysis: Total Player Rating versus Runs Scored

```
The regression equation is
Total Player Rating = - 8.10 + 0.0119 Runs Scored

Predictor         Coef    SE Coef      T      P
Constant        -8.098      5.004  -1.62  0.125
Runs Scored   0.011884   0.007917   1.50  0.153

S = 11.5965   R-Sq = 12.3%   R-Sq(adj) = 6.9%
Analysis of Variance

Source           DF       SS      MS      F      P
Regression        1    303.0   303.0   2.25  0.153
Residual Error   16   2151.7   134.5
Total            17   2454.6

Unusual Observations

                Total
        Runs   Player
Obs   Scored   Rating   Fit   SE Fit   Residual   St Resid
  4     1236   -16.50  6.59     6.23     -23.09      -2.36R
 13     1249    30.90  6.75     6.32      24.15       2.48R

R denotes an observation with a large standardized residual.
```

RATING versus TIMES CAUGHT STEALING

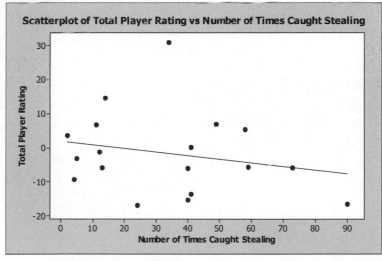

343

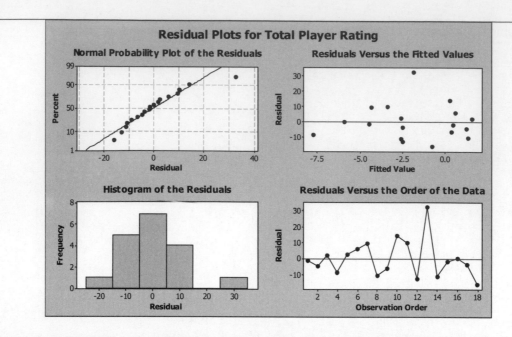

Regression Analysis: Total Player Rat versus Number of Times

```
The regression equation is
Total Player Rating = 1.77 - 0.106 Number of Times Caught Stealing

Predictor                              Coef   SE Coef      T      P
Constant                              1.771     4.833   0.37  0.719
Number of Times Caught Stealing     -0.1055    0.1153  -0.92  0.374

S = 12.0739   R-Sq = 5.0%   R-Sq(adj) = 0.0%

Analysis of Variance

Source           DF      SS      MS      F      P
Regression        1   122.2   122.2   0.84  0.374
Residual Error   16  2332.5   145.8
Total            17  2454.6

Unusual Observations

        Number
      of Times    Total
        Caught   Player
Obs   Stealing   Rating     Fit   SE Fit  Residual  St Resid
  4       90.0   -16.50   -7.73     7.07     -8.77    -0.90 X
 13       34.0    30.90   -1.82     2.85     32.72     2.79R

R denotes an observation with a large standardized residual.
X denotes an observation whose X value gives it large influence.
```

Note that neither model shows a significant relationship; neither scatterplot shows a pronounced linear trend.

b. Here is the multiple regression analysis:

Regression Analysis: Total Player versus Number of Ti, Runs Scored

```
The regression equation is
Total Player Rating = - 4.47 - 0.306 Number of Times Caught Stealing
                        + 0.0246 Runs Scored


Predictor                            Coef    SE Coef      T      P
Constant                           -4.473      4.492  -1.00  0.335
Number of Times Caught Stealing   -0.3060     0.1163  -2.63  0.019
Runs Scored                      0.024625   0.008320   2.96  0.010


S = 9.90801   R-Sq = 40.0%   R-Sq(adj) = 32.0%

Analysis of Variance

Source           DF        SS       MS      F      P
Regression        2    982.12   491.06   5.00  0.022
Residual Error   15   1472.53    98.17
Total            17   2454.65


Source                           DF   Seq SS
Number of Times Caught Stealing   1   122.18
Runs Scored                       1   859.94
```

Note that the p-values suggest that each of the two slopes is significant in the presence of the other.

c. Consider the following residual plots:

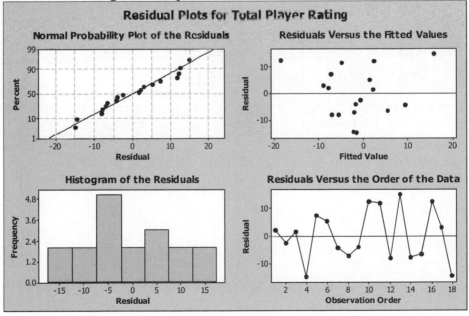

Variation is not homogeneous in either bivariate plot.

d. Neither predictor is significant by itself, but are both significant when each is used conditional on the other being in the model. For example, conditional on the number of times caught stealing being held below 90, the relationship between *rating* and *runs scored* is highly significant.

E7. a. Here is a table of all the data:

Distance	Heights	Heights-SQUARED
573	1000	1000000
534	800	640000
495	600	360000
451	450	202500
395	300	90000
337	200	40000
253	100	10000

Now, the regression analysis is:

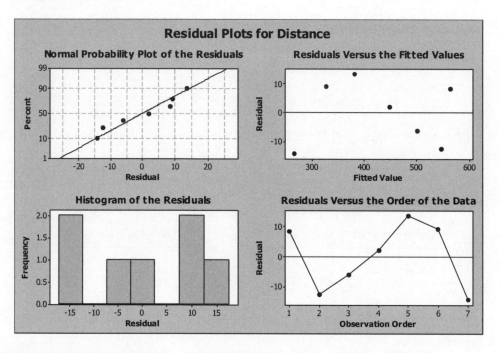

Regression Analysis: Distance versus Heights, Heights-SQUARED

```
The regression equation is
Distance = 200 + 0.708 Heights - 0.000344 Heights-SQUARED


Predictor                 Coef        SE Coef       T       P
Constant                 199.91         16.76    11.93   0.000
Heights                 0.70832        0.07482    9.47   0.001
Heights-SQUARED      -0.00034369    0.00006678   -5.15   0.007


S = 13.6389   R-Sq = 99.0%   R-Sq(adj) = 98.6%
Analysis of Variance

Source           DF      SS      MS        F       P
Regression        2   76278   38139   205.03   0.000
Residual Error    4     744     186
Total             6   77022
```

346

```
Source          DF  Seq SS
Heights          1   71351
Heights-SQUARED  1    4927
```

Note that the regression model is
$$\widehat{distance} = 200 + 0.708 height - 0.0003(height)^2.$$

b. Consider the following scatterplot with the regression curve:

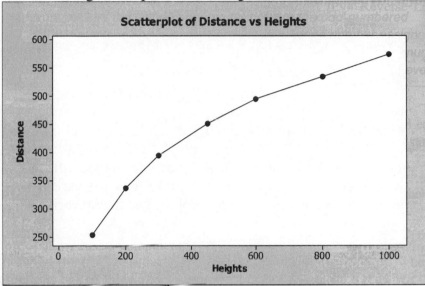

The curve fits the points well, with $R^2 = 0.99$.

c. This is consistent with the regression analysis.

E9. a. No, because the p-value is 0.299.

b. R^2 changed from 64.2% to 54.5%, while R^2 (adj) changed from 49.9% to 47.9%.
R^2-adjusted is a better indication of the importance of *depth* to the regression analysis.

c. <u>Regression 1</u>: Here, SSE = 2.2117, n = 8, k = 2, and SST = 6.1750.
Observe that
$$1 - \frac{\frac{SSE}{n-k-1}}{\frac{SST}{n-1}} = 1 - \frac{\frac{2.2117}{5}}{\frac{6.1750}{7}} \approx 0.49856.$$

<u>Regression 2</u>: Here, SSE = 2.8066, n = 8, k = 1, and SST = 6.1750.
Observe that
$$1 - \frac{\frac{SSE}{n-k-1}}{\frac{SST}{n-1}} = 1 - \frac{\frac{2.8066}{6}}{\frac{6.1750}{7}} \approx 0.46974.$$

E11. a. SST is the same in both regressions.
b. SST measures variation around the mean response and is free of any explanatory variables.

347

E13. a. 24 = Total DF + 1

b. At least one of the explanatory variables, *predator* or *maximum life*, is significant in predicting *speed*.

c. *Predator* is significant in predicting *speed* (p-value 0.037); *maximum life* (p-value 0.411) is not.

d. See part c.

$$\text{For predators: } \widehat{speed} = (35.9 + 12.5) - 0.148 MaxLife$$
$$\text{For non-predators: } \widehat{speed} = 35.9 - 0.148 MaxLife$$

Note that being a predator raises the intercept and, hence, the speeds.

E15. a. This is an observational study, and hence involves no randomization.

b. Bolivia is an outlier on the residual scale and may have undue influence in the analysis. Otherwise, there is a linear relationship of the response variable to each explanatory variable, the residual plots are patternless, and the univariate plot of the residuals is reasonably symmetric.

c. When *growth rate* is added to the model, none of the three explanatory variables shows up as significant. Without growth rate, both *forest proportion* and *military expenditure* were significant.

E17. a. Here are all of the regression models, first with all explanatory variables included, and then the models obtained by excluding one explanatory variable each time. All explanatory variables included:

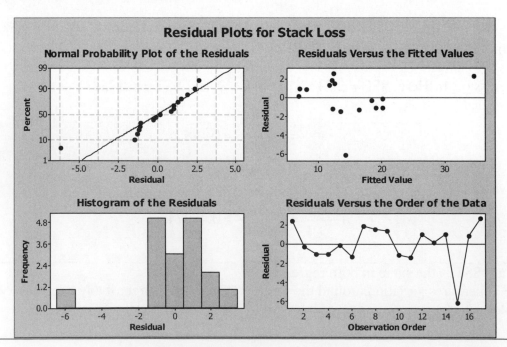

Regression Analysis: Stack Loss versus Air Flow, Temp, Acid

```
The regression equation is
Stack Loss = - 45.6 + 0.689 Air Flow + 0.778 Temp + 0.046 Acid


Predictor      Coef   SE Coef       T       P
Constant    -45.591     8.820   -5.17   0.000
Air Flow     0.6892    0.1217    5.66   0.000
Temp         0.7777    0.3012    2.58   0.023
Acid         0.0464    0.1100    0.42   0.680


S = 2.30737   R-Sq = 91.5%   R-Sq(adj) = 89.6%

Analysis of Variance

Source           DF       SS       MS       F       P
Regression        3   747.02   249.01   46.77   0.000
Residual Error   13    69.21     5.32
Total            16   816.24


Source     DF   Seq SS
Air Flow    1   707.90
Temp        1    38.18
Acid        1     0.95

Unusual Observations

              Stack
Obs  Air Flow  Loss     Fit  SE Fit  Residual  St Resid
 15      60.0  8.000  14.202   1.063    -6.202    -3.03R

R denotes an observation with a large standardized residual.
```

With AIR FLOW thrown out:

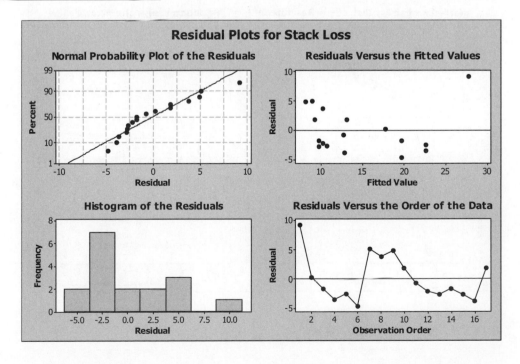

Regression Analysis: Stack Loss versus Temp, Acid

```
The regression equation is
Stack Loss = - 37.9 + 1.97 Temp + 0.143 Acid


Predictor      Coef   SE Coef       T       P
Constant     -37.93     15.64   -2.43   0.029
Temp         1.9667    0.3875    5.08   0.000
Acid         0.1432    0.1949    0.73   0.475


S = 4.14002   R-Sq = 70.6%   R-Sq(adj) = 66.4%

Analysis of Variance

Source            DF        SS       MS       F       P
Regression         2    576.28   288.14   16.81   0.000
Residual Error    14    239.96    17.14
Total             16    816.24


Source   DF   Seq SS
Temp      1   567.03
Acid      1     9.25


Unusual Observations
          Stack
Obs  Temp  Loss     Fit   SE Fit   Residual   St Resid
  1  27.0  37.00   27.77     2.62       9.23       2.88R

R denotes an observation with a large standardized residual.
```

With TEMP thrown out:

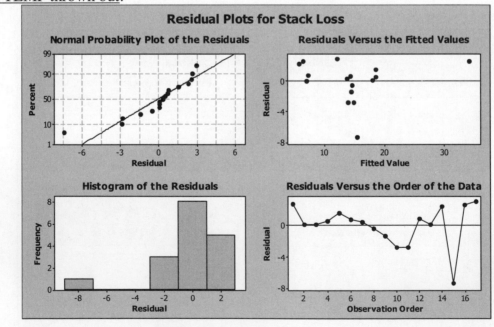

Regression Analysis: Stack Loss versus Air Flow, Acid

```
The regression equation is
Stack Loss = - 46.2 + 0.908 Air Flow + 0.090 Acid


Predictor    Coef   SE Coef     T      P
Constant    -46.21    10.45   -4.42  0.001
Air Flow    0.9083   0.1034    8.78  0.000
Acid        0.0898   0.1288    0.70  0.497


S = 2.73470   R-Sq = 87.2%   R-Sq(adj) = 85.3%

Analysis of Variance

Source          DF      SS       MS      F       P
Regression       2   711.53   355.77  47.57  0.000
Residual Error  14   104.70     7.48
Total           16   816.24


Source      DF   Seq SS
Air Flow     1   707.90
Acid         1     3.63

Unusual Observations

Obs  Air Flow  Stack Loss     Fit  SE Fit  Residual  St Resid
  1      80.0      37.000  34.348   2.244     2.652    1.70 X
 15      60.0       8.000  15.375   1.138    -7.375   -2.97R

R denotes an observation with a large standardized residual.
X denotes an observation whose X value gives it large influence.
```

With ACID thrown out:

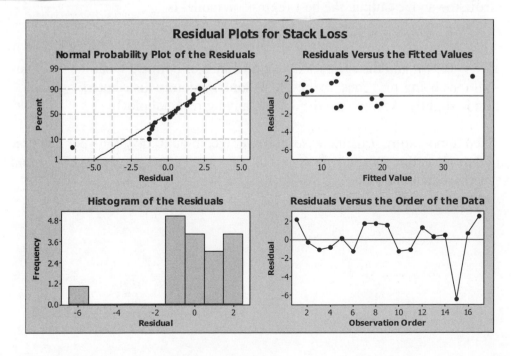

Regression Analysis: Stack Loss versus Air Flow, Temp

```
The regression equation is
Stack Loss = - 42.5 + 0.697 Air Flow + 0.797 Temp

Predictor      Coef   SE Coef       T      P
Constant    -42.484     4.703   -9.03  0.000
Air Flow     0.6972    0.1166    5.98  0.000
Temp         0.7971    0.2888    2.76  0.015

S = 2.23859   R-Sq = 91.4%   R-Sq(adj) = 90.2%

Analysis of Variance

Source           DF       SS       MS       F      P
Regression        2   746.08   373.04   74.44  0.000
Residual Error   14    70.16     5.01
Total            16   816.24

Source      DF   Seq SS
Air Flow     1   707.90
Temp         1    38.18

Unusual Observations

Obs  Air Flow  Stack Loss      Fit  SE Fit  Residual  St Resid
  1      80.0      37.000   34.814   1.788     2.186      1.62 X
 15      60.0       8.000   14.494   0.783    -6.494     -3.10R

R denotes an observation with a large standardized residual.
X denotes an observation whose X value gives it large influence.
```

Judging from the above output, the best regression model is

$$\widehat{loss} = -42.5 + 0.697 AirFlow + 0.797 Temperature.$$

Note that $R^2 = 0.914$, and the t-tests show significance of both explanatory variables. As far as conditions being met goes, the residuals look randomly distributed, but their distribution is slightly skewed (see the accompanying residual plot).

b. For fixed temperature, *loss* increases, on average, about 0.697 units per one unit increase *in air flow*. For fixed *air flow*, *loss* increases, on average, about 0.797 units per one unit increase in t*emperature*.

E19. a. The equations are:

$$\widehat{\%\ Correct} = 107.9 - 0.03 Time$$

$$\widehat{\%\ Correct} = 94 - 0.03 Time$$

Note that the slopes are equal, but the intercept changes by 13.9 units.

b. No, older chimp data should have steeper slope.

c. Here is a table of the new data:

Percent Correct	Age	Time	Age*Time
81	0	885	0
83	0	672	0
83	0	929	0
84	1	635	635
67	1	659	659
62	1	944	944

Running the new regression analysis yields

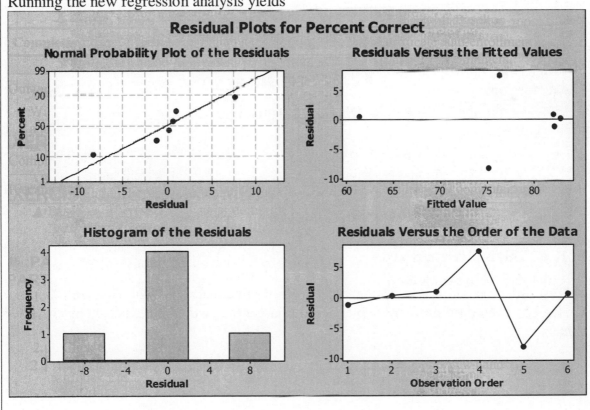

353

```
Regression Analysis: Percent Correct versus Age, Time, Age*Time

The regression equation is
Percent Correct = 84.8 + 22.5 Age - 0.0030 Time - 0.0457 Age*Time

Predictor        Coef  SE Coef       T      P
Constant        84.80    34.45    2.46  0.133
Age             22.51    42.57    0.53  0.650
Time         -0.00298  0.04120   -0.07  0.949
Age*Time     -0.04570  0.05275   -0.87  0.478

S = 8.00823   R-Sq = 72.2%   R-Sq(adj) = 30.5%

Analysis of Variance

Source            DF       SS      MS      F      P
Regression         3   333.07  111.02   1.73  0.387
Residual Error     2   128.26   64.13
Total              5   461.33

Source     DF  Seq SS
Age         1  192.67
Time        1   92.26
Age*Time    1   48.14

Unusual Observations
          Percent
Obs  Age  Correct    Fit  SE Fit  Residual  St Resid
  6 1.00    62.00  61.36    8.00      0.64      1.40 X

X denotes an observation whose X value gives it large influence.
```

The regression equation is

$$\widehat{\%\ correct} = 84.8 + 22.51\text{Age} - 0.003\text{Time} - 0.046\text{Age*Time}$$

d. Young: $\widehat{\%\ correct} = 84.8 - 0.003time$

Old: $\widehat{\%\ correct} = 107.31 - 0.049time$

Both the intercept and slope change, and the new models fit the respective data sets.

E21. a. This is observational data.

b. Test difference of means. Using technology, we find that $t = 2.674$, P-value $= 0.011$. So, the observed difference in sample means is larger than would be expected by chance alone.

c. Consider the following regression analyses, with and without the interaction:

Regression analysis without the interaction:

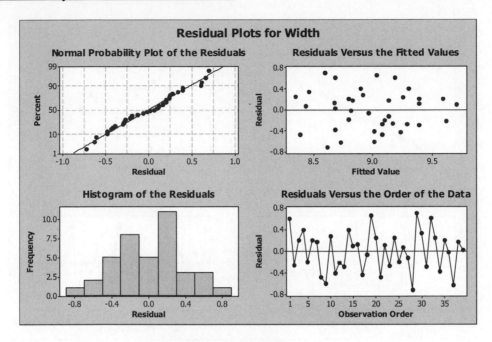

Regression Analysis: Width versus Sex, Length

The regression equation is
Width = 3.99 - 0.238 Sex + 0.224 Length

```
Predictor      Coef   SE Coef       T      P
Constant      3.994     1.134    3.52  0.001
Sex         -0.2379    0.1315   -1.81  0.079
Length      0.22398   0.04918    4.55  0.000


S = 0.384689   R-Sq = 46.0%   R-Sq(adj) = 43.0%

Analysis of Variance

Source           DF       SS       MS       F      P
Regression        2   4.5402   2.2701   15.34  0.000
Residual Error   36   5.3275   0.1480
Total            38   9.8677


Source   DF  Seq SS
Sex       1  1.4708
Length    1  3.0694
```

Regression analysis with interaction:

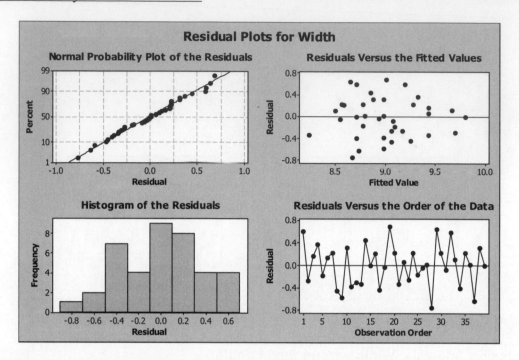

Regression Analysis: Width versus Sex, Length, Sex*Length

```
The regression equation is
Width = 3.05 + 2.62 Sex + 0.265 Length - 0.127 Sex*Length

Predictor        Coef    SE Coef       T       P
Constant        3.055      1.367    2.23   0.032
Sex             2.615      2.357    1.11   0.275
Length        0.26483    0.05936    4.46   0.000
Sex*Length    -0.1267     0.1045   -1.21   0.234

S = 0.382206   R-Sq = 48.2%   R-Sq(adj) = 43.7%

Analysis of Variance

Source             DF       SS       MS       F       P
Regression          3   4.7548   1.5849   10.85   0.000
Residual Error     35   5.1128   0.1461
Total              38   9.8677

Source          DF   Seq SS
Sex              1   1.4708
Length           1   3.0694
Sex*Length       1   0.2146

Unusual Observations

Obs   Sex   Width      Fit   SE Fit   Residual   St Resid
 21  0.00  7.9000   8.2454   0.2164    -0.3454      -1.10 X
 28  1.00  7.9000   8.6666   0.1106    -0.7666      -2.10R
 31  1.00  9.0000   9.0809   0.2305    -0.0809      -0.27 X

R denotes an observation with a large standardized residual.
X denotes an observation whose X value gives it large influence.
```

Observe that the mean lengths of feet do not differ significantly between boys and girls. There is no other measure of size to compare.

d. A good model is given by the following regression model:
$$\widehat{width} = 3.994 + 0.224 length - 0.238 sex$$

Sex is significant at the 10% level and length at the 1% level and there is no interaction between *sex* and *length*. The $R^2 = 0.46$, which is rather weak.

E23. a. Consider the following regression analysis:

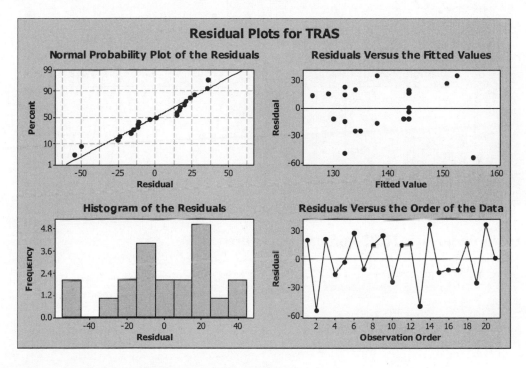

Regression Analysis: TRAS versus Banking

```
The regression equation is
TRAS = 120 + 0.986 Banking

Predictor    Coef   SE Coef     T       P
Constant    120.21    14.54   8.27   0.000
Banking     0.9860   0.7098   1.39   0.181

S = 26.3994    R-Sq = 9.2%    R-Sq(adj) = 4.4%

Analysis of Variance

Source           DF       SS       MS     F      P
Regression        1   1344.9   1344.9  1.93  0.181
Residual Error   19  13241.7    696.9
Total            20  14586.6

Unusual Observations
Obs   Banking     TRAS     Fit   SE Fit   Residual   St Resid
  2      36.0   101.07  155.71    13.49     -54.63     -2.41R
R denotes an observation with a large standardized residual.
```

357

No, it is not significant because t = 1.39 with p-value 0.181.

b. Consider the following regression analysis:

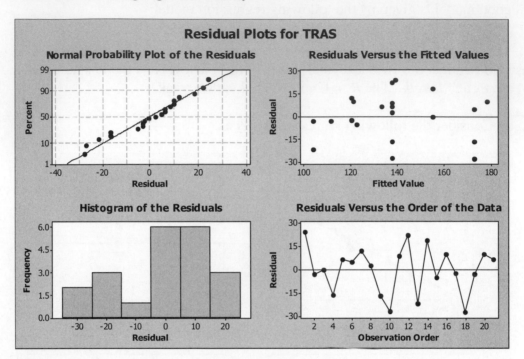

Regression Analysis: TRAS versus Circumference

```
The regression equation is
TRAS = 85.7 + 34.9 Circumference

Predictor        Coef   SE Coef      T       P
Constant       85.703     8.924   9.60   0.000
Circumference  34.859     5.423   6.43   0.000

S = 15.5515    R-Sq = 68.5%    R-Sq(adj) = 66.8%

Analysis of Variance

Source          DF        SS       MS       F       P
Regression       1    9991.4   9991.4   41.31   0.000
Residual Error  19    4595.1    241.8
Total           20   14586.6
```

Yes, it is significant because t = 6.43 with p-value < 0.0001.

c. Consider the following regression analysis:

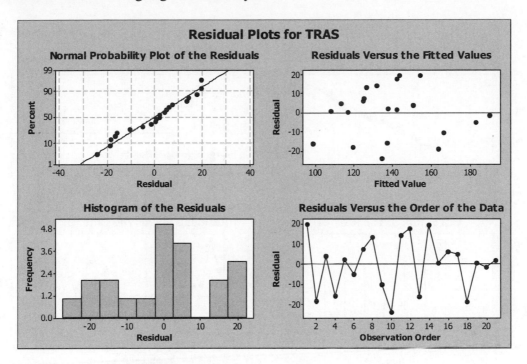

Regression Analysis: TRAS versus Circumference, Banking

```
The regression equation is
TRAS = 70.3 + 34.3 Circumference + 0.860 Banking

Predictor          Coef   SE Coef      T      P
Constant          70.34     10.55   6.67  0.000
Circumference    34.327     4.920   6.98  0.000
Banking          0.8598    0.3794   2.27  0.036

S = 14.0928    R-Sq = 75.5%    R-Sq(adj) = 72.8%

Analysis of Variance

Source          DF        SS      MS      F      P
Regression       2   11011.6  5505.8  27.72  0.000
Residual Error  18    3574.9   198.6
Total           20   14586.6

Source          DF   Seq SS
Circumference    1   9991.4
Banking          1   1020.2
```

Circumference is significant, and so is banking but to a lesser degree.

d. Conditional on the length of the tracks, *banking* plays a significant role. There is a strong linear relationship between *banking* and *TRAS* on long tracks, but not on short tracks.

E25. a. Consider the following regression analysis:

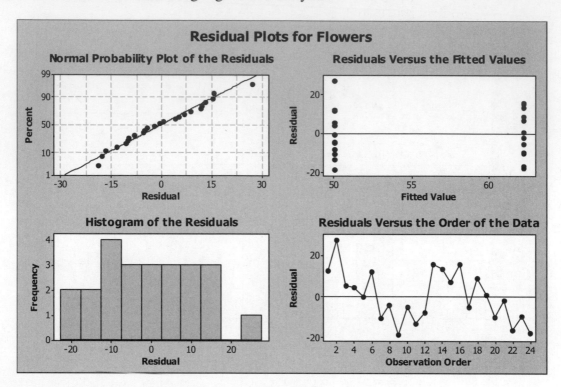

Regression Analysis: Flowers versus Starting Time

```
The regression equation is
Flowers = 50.1 + 12.2 Starting Time

Predictor          Coef   SE Coef       T       P
Constant         50.058     3.614   13.85   0.000
Starting Time    12.167     5.110    2.38   0.026

S = 12.5180    R-Sq = 20.5%    R-Sq(adj) = 16.9%

Analysis of Variance
Source          DF      SS      MS      F       P
Regression       1   888.2   888.2   5.67   0.027
Residual Error  22  3447.4   156.7
Total           23  4335.6

Unusual Observations
       Starting
Obs       Time  Flowers     Fit   SE Fit   Residual   St Resid
  2       0.00    77.40   50.06     3.61      27.34       2.28R

R denotes an observation with a large standardized residual.
```

The regression model is significant with F=5.67 and p-value = 0.027. In general, as start time moves from late to early the average flowering increases.

b. Test on the difference of two means. Using technology, we see that *P*-value = 0.027. We conclude that the mean flowering for early and late start times differ significantly.

c. The two tests are exactly the same.

E27. a. Yes, because the test statistic is $F = 47.17$, with a P-value near 0. Note that $R^2 = 0.977$; At least one explanatory variable is significant.

b. Forearm (0.06), waist (0.00), height (0.03)

c. 97.7%

d. Relatively small

e. *Thigh* is now significant, when controlling for *forearm, waist* and *height*. *Thigh* is correlated with some of the other explanatory variables and did not show up as important when all of those variables were part of the model.

f. No; R^2 and s remained about the same.

Chapter 16

Exercise Solutions

E1. a. The number of terminated employees under age 40 is less than 5.

b. Using $\hat{p}_1 = \frac{14}{27}$, $\hat{p}_2 = \frac{4}{9}$, $\hat{p} = \frac{14+4}{27+9} = 0.5$, we see that the test statistic is

$$z = \frac{(\hat{p}_1 - \hat{p}_2) - 0}{\sqrt{\hat{p}(1-\hat{p})\left(\frac{1}{n_1} + \frac{1}{n_2}\right)}} \approx 0.3849.$$

c. No, because it is nowhere close to being <0.05.

E3. a. Fill in the values of the table to "stack the deck:"

	Terminated	Retained	Total
Under 40	0	9	9
≥ 40	18	9	27
Total	18	18	36

b. You would expect the upper-right and lower-left, which would indicate younger employees were retained and older ones terminated.

c. Based on (b), use p=0.50.

d. No, since p-value is not < 0.05.

e. Quite different, by about 0.12, because conditions aren't met for the test used in E1.

E5. a. Use a one-sided test with the following null and alternative hypotheses:

H_0: The difference in the proportion of workers laid off who were 50 or older and the proportion of workers laid off who were less than 50 **is no larger** than you would expect if Westvaco were picking 28 people at random for layoff.

H_a: The difference in the proportion of workers laid off who were 50 or older and the proportion of workers laid off who were less than 50 **is larger** than you would expect if Westvaco were picking 28 people at random for layoff.

Using $\hat{p}_1 = \frac{19}{28} = 0.679$, $\hat{p}_2 = \frac{9}{22} = 0.409$, $\hat{p} = \frac{19+9}{28+22} = 0.56$, we see that the test statistic is

$$z = \frac{(\hat{p}_1 - \hat{p}_2) - 0}{\sqrt{\hat{p}(1-\hat{p})\left(\frac{1}{n_1} + \frac{1}{n_2}\right)}} \approx 1.9092.$$

So, the p-value is P(Z>1.9055) = 0.0284. Hence, we reject the null hypothesis, meaning that a difference in proportions of 0.270 or larger would occur by chance less than 3 times out of 100.

b. At age 50. The test using 50 as the cutoff age reveals a clear-cut pattern that is not detected by the test using age 40.

E7. a. No, although the test in E5 comes very close with a one-sided *P*-value of 0.028.
b. I
c. Type I error, if the null hypothesis is true.
d. A Type-II error would occur when deciding that a company does not discriminate on the basis of age when, in fact, it does.

E9. a. The observed frequencies are:

	Terminated	Retained	Total
Under 40	7	7	14
40 - 54	7	9	16
≥ 55	14	6	20
Total	28	22	50

b. Chi-square test of independence

c. Sample is large enough as each expected count is at least 5, but this isn't a random sample from a larger population.

d. We need the expected frequencies:

	Terminated	Retained	Total
Under 40	$\frac{14(28)}{50} = 7.84$	6.16	14
40 - 54	8.96	7.04	16
≥ 55	11.2	8.8	20
Total	28	22	50

Now, the test statistic is

$$\chi^2 = \sum \frac{(O-E)^2}{E} = 2.770, \ \text{df} = 3\text{-}1\text{=}2.$$

So, the p-value is 0.2503. As such, we cannot reject the Ho that age and termination/retention are independent.

E11. a. Yes, based on the drastic differences in percentages terminated in the older age group.

b. A two-sample *z*-test of the difference of two proportions is appropriate. Using $\hat{p}_1 = \frac{16}{28} = 0.571$, $\hat{p}_2 = \frac{4}{22} = 0.182$, $\hat{p} = \frac{16+4}{50} = 0.4$, we see that the test statistic is

$$z = \frac{(\hat{p}_1 - \hat{p}_2) - 0}{\sqrt{\hat{p}(1-\hat{p})\left(\frac{1}{n_1} + \frac{1}{n_2}\right)}} \approx 2.79 \ .$$

The p-value for the two-sided test is $2P(Z>2.79) = 0.0053$. As such, we conclude that the difference in the proportions terminated cannot reasonably be attributed to chance alone.

c. Yes, especially the large drop in the number of employees age 50 or older in the first two rounds.

E13. a. You need to refer to Display 16.1 to compute the sample means and SDs for salaried employees terminated and salaried employees retained. These two groups correspond to the data in the following manner:

Group 1: Salaried and Retained -- (Rows 15 – 32 in the display)
Group 2: Salaried and Terminated – (Rows 33 – 50 in the display)

Using technology, we have the following descriptive statistics for these two groups:

Descriptive Statistics: salaried and retained, salaried and terminated

```
Variable             N   N*    Mean  SE Mean   StDev  Minimum      Q1  Median      Q3
salaried and ret    18   0    47.17    2.50   10.60    29.00   36.25   48.00   55.50
salaried and ter    18   0    50.56    3.21   13.63    23.00   39.50   53.50   61.50

Variable          Maximum
salaried and ret    61.00
salaried and ter    69.00
```

So,

$$\text{Group 1: } \bar{x}_1 = 47.17, \ s_1 = 10.60, \ n_1 = 18$$
$$\text{Group 2: } \bar{x}_2 = 50.56, \ s_2 = 13.63, \ n_2 = 18$$

Now, we wish to test:
Ho: The difference in the mean ages of the terminated and retained employees is attributed to chance alone.
Ha: The difference in the mean ages of the terminated and retained employees is not attributed to chance alone.

Running a two-sample t-test, we see that the test statistic is given by

$$t = \frac{\bar{x}_1 - \bar{x}_2}{\sqrt{\frac{s_1^2}{n_1} + \frac{s_2^2}{n_2}}} \approx -0.83 \ .$$

Using technology, we find that the p-value is 0.4111. Hence, we do not reject the null hypothesis that the difference in the mean ages of the terminated and retained employees can reasonably be attributed to chance alone.

b. The *P*-value is larger here. The *t*-test uses more information, the exact ages, rather than just whether the employee is over or under age 40.

E15. a. For this analysis, we recode Retained as "1" and Terminated as "2"; these two factors, together, are labeled as "Group" in the analysis.

The output of the analysis is as follows:

One-way ANOVA: age versus group

```
Source  DF    SS   MS     F      P
group    1   103  103  0.69  0.411
Error   34  5067  149
Total   35  5170

S = 12.21   R-Sq = 2.00%   R-Sq(adj) = 0.00%


                               Individual 95% CIs For Mean Based on
                               Pooled StDev
Level   N   Mean  StDev  -------+---------+---------+---------+--
1      18  47.17  10.60  (--------------*--------------)
2      18  50.56  13.63        (-------------*--------------)
                       -------+---------+---------+---------+--
                          44.0      48.0      52.0      56.0

Pooled StDev = 12.21
```

Observe that the result is not significant since $F \approx 0.694$ and a (two-sided) *P*-value of about 0.4108.

b. The *P*-value should be double that in E13 because ANOVA, while otherwise equivalent to the two-sample *t*-test, is two-sided.

E17. a. For each additional round of layoffs, the mean age of the terminated employees tended to decrease by 1.57 years.

b. Not at the 5% level, but it is significant at the 10% level since p=0.057.

c. This was the youngest employee, yet got laid off in the first round, so has a large residual.

E19. a. Just look at the y-value on the curve corresponding to age = 40 and age = 60 to obtain the percentages terminated are about 45% and about 57%, respectively.

b. No, because the p-value is greater than 0.10 (although not by very much).

E21. a. 50.56 – 47.17, or about 3.39 years

b. Between about 233 to 258 out of 1000, or about 0.233 to 0.258, had mean difference of at least the one found in (a).

c. If 18 of the 36 salaried employees had been selected at random for termination, the probability is about 0.233 to 0.258 that *mean age of those terminated – mean age of those retained* would be as large or larger than 3.39.

d. Getting a difference in mean ages of 3.39 or larger can reasonably be attributed to chance alone.

E23. The *P*-value in E21 of 0.233 to 0.258 from the simulation is close to that in E13 of about 0.2056. The conclusions will be the same.

E25. a. 0.074
b. Answers will vary.
c. Westvaco randomly selecting 18 salaried employees for termination.
d. This would represent a little more than 0.5.
e. The p-values are almost equal.

E27. Answers will vary here and will, to some extent, depend on your opinion as to which test seems to most accurately present a balanced argument illustrating the cases when discrimination would not be present and the present situation (using the actual data).

NOTES

NOTES

NOTES

NOTES

NOTES

NOTES

NOTES

NOTES